UNDER THE EDITORSHIP OF

BENTLEY GLASS

*State University of New York
at Stony Brook*

A. M. Winchester
Colorado State College

GENETICS

A SURVEY OF THE PRINCIPLES
OF HEREDITY □ THIRD EDITION

Houghton Mifflin Company

Boston / New York / Atlanta
Geneva, Ill. / Dallas / Palo Alto

EDITOR'S INTRODUCTION

THE INTEREST OF THE MAJORITY OF COLLEGE STUDENTS IN GENETICS centers on questions of human heredity. In this new edition of *Genetics,* the author takes account of many new aspects of human heredity which have developed in the nine years since the second edition. The author gives attention to the newer knowledge of the chemical basis of heredity and the genetic code, the human chromosomes, the biochemical patterns of gene action, microbial genetics, and the genetic hazards of radiation.

At the same time, Professor Winchester has successfully avoided the technicalities which would have impaired the level of treatment which was so successfully adopted for the non-specializing student. The textbook for the beginner ought to answer many fundamental questions, yet be no mere catalogue of hereditary traits, even if such could be made interesting to the reader. One needs a knowledge of principles and of the physical basis of heredity far more than an index of anomalies. For this purpose a study of the breeding experiments with suitable plants and animals and of the cytogenetic investigations of chromosome behavior is far more fruitful than the study of human pedigrees.

It is the excellent balance between these aims and emphases that makes this book so suitable an introduction to the study of heredity. Clear, straightforward, and readable, it carries one from example to general principle, with worthwhile attention to the bearing of genetics on controversial social issues. It makes clear the vital and thought-provoking relevance of the science of heredity to human well-being, not only in the improvement of domestic animals and cultivated plants but also in the potential solution of problems of individual and social maladjustment and disease. The discussion has, moreover, been enriched by an original and striking collection of illustrations, and some fine new ones have been introduced in this edition. Professor Winchester has thus provided a revision that will continue to be of engrossing interest to students and the general public alike.

BENTLEY GLASS
State University of New York
at Stony Brook

PREFACE

IN THE PREPARATION OF THIS THIRD EDITION the author has adhered to the tenets which prompted him to write the book in the first place, namely the production of a textbook covering the basic principles of genetics in a manner which is both informative and stimulating to those being first introduced to a course in this important field of biological learning. Since genetics has become an integral part of freshman college and even high school biology courses, it is now possible to present the subject on a somewhat more advanced level than previously. Nevertheless, this continues to be a book characterized by thoroughness of explanation so that the beginning student will not be lost in a maze of new information and terminology.

The explosive rate of growth of knowledge in the various branches of genetics during the past few years has created problems in the selection of material to be included. In order to prevent the book from becoming too large, there have had to be choices on the elimination of material previously included. Microbial genetics has the spotlight in present day genetic research and, as a result, there is a temptation to overemphasize it in a textbook. Soon, however, the microbial genetics of today may be replaced in popularity by other exciting new discoveries about which we can only speculate. Hence, an attempt has been made to achieve a balance by interweaving the newer microbial genetics into the complex pattern of the entire spectrum of genetic knowledge, old and new.

As was true of previous editions, human genetics is emphasized. This branch is intensely interesting to beginning students and is one which can now be used to illustrate many of the important genetic principles that apply to all forms of life.

Since illustrations are an important part of any science textbook, they have received considerable attention. Many photographs are used, most of which have been made by the author. Those which have been borrowed are acknowledged in the captions. These photographs stimulate interest by giving the student a vicarious contact with the materials which are discussed in the text. Careful attention has also been given to the diagrams and charts which can be quite helpful in clarifying difficult topics.

Although the geneticists who have furnished suggestions for this revision are too numerous to list individually, the author hereby acknowledges their contributions with great appreciation. However, he particularly wishes to express his gratitude for the generous help received from Bentley Glass, whose counsel has greatly benefited the text. He also wishes to thank the editors and artists of the Houghton Mifflin Company for their helpful cooperation in the production of this book.

A. M. WINCHESTER
Colorado State College

CONTENTS

1

Heredity —
Fact and Fancy

ONE OF THE MOST IMPORTANT SUBDIVISIONS of the science of biology is genetics, the study of heredity. The word "genetics" is derived from the Greek root *gen,* which means "to become" or "to grow into." Thus, the derivation of the word indicates that genetics deals not only with the way in which characteristics are transmitted from one generation to the next, but also with the actions of the units of heredity as they bring about the characteristics which they control.

As human beings, it is quite natural that we should be particularly interested in the method of inheritance in man. Fond parents take pride in seeing their own characteristics develop in their children and more distant descendants. Man is not the best form of life for studies of genetics for many obvious reasons, yet our knowledge of human genetics has increased greatly in recent years as new techniques for study have been developed. We shall use many examples of human genetics as illustrations for genetic principles throughout this book, but at the same time we shall discuss many of the discoveries in other forms of life which have pointed the way to discoveries in human genetics.

To the untrained person, genetics may appear to be a vague, indefinite subject. A red-haired child may be born to parents with red hair on some occasions, and to parents with dark brown hair on other occasions. Such results are not vague, nor indefinite; they are in accordance with the laws of heredity which are very definite and precise. The appearance of characteristics in a child which were not present in either parent is to be expected as a result of the normal operation of these laws. There would be something wrong if we did not get such cases in a certain proportion of the children.

1

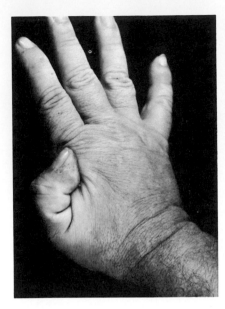 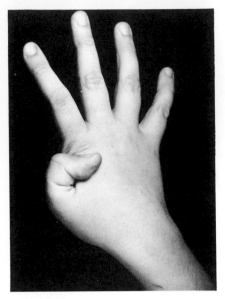

FIG. 1–1 *Like father, like daughter. The extreme flexibility of the thumb as shown here in a father and his daughter is an inherited trait which has come down through many generations.*

Everyone recognizes the fact that heredity plays a very important role in the development of physical characteristics — common observation shows this to be true beyond any question. Yet many do not realize that heredity is equally important in the development of the less tangible qualities such as intelligence, aptitudes, temperament, and even the minor body mannerisms which contribute to one's total personality. One would hardly think that the way a person folds his hands would be influenced by heredity; yet the placement of the fingers in this simple gesture can be traced through heredity just as the shape of the nose or the texture of the hair can be traced. A group of students with the same basic background in mathematics may show considerable variation in ability to handle mathematical problems as a result of differences in aptitude due to heredity.

As we progress with our study, you will find that the genetics of other forms of life can also be quite interesting. At first thought, it may appear that there could be little of interest in a study of the genetics of a mold, yet some of the most fascinating discoveries on the method of functioning of the units of heredity and the inheritance of physiological traits have come about as a result of such studies. In fact, two men have received the Nobel prize in recent years for their discoveries on the genetics of a pink mold, known as *Neurospora*. Bacteria and even viruses have come in for a great share of attention in genetic research because it is possible to study their physiological reactions and the transmission of hereditary traits so precisely. In addition, breeding experiments with larger forms of plants and animals can be quite interesting and help round out the total picture of genetics.

PRACTICAL APPLICATIONS OF GENETICS

In addition to the theoretical interest in genetics, as a means of extending our knowledge of the unknown, genetics has many practical applications which are of great value to man. To the animal breeder who is concerned with the application of genetic principles to the improvement of domestic animals, a knowledge of the principles of heredity is of utmost importance. The fat, beef-producing cattle that roam our western plains today are a far cry from the scrawny animals that grazed in this region a short time ago. Almost all of our domestic animals have been greatly transformed by practical applications of genetic principles. The

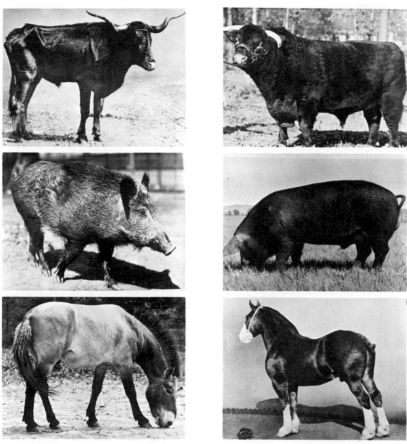

FIG. 1–2 *Breeds of domestic animals can be improved through the application of genetic principles. At the left are the original types from which the breeds at the right have been derived. Top: A longhorn steer and a modern shorthorn bull bred for quantity and quality of beef. Middle: A European wild boar and a Black Hampshire boar bred for high meat production. Bottom: A Prjevalsky wild horse and a modern draft horse.* WILD HORSE AND BOAR, NEW YORK ZOOLOGICAL SOCIETY; OTHERS, U.S. DEPARTMENT OF AGRICULTURE.

FIG. 1–3 *Many highly improved varieties of fruits have been produced through the application of genetic principles. The sweet, juicy, thin-skinned, almost seedless orange at left is far superior, so far as man is concerned, to the sour, heavily seeded wild orange at right from which it was derived.*

same can be said of the plant breeders who have been so successful in producing superior varieties of many of our food crops that we have a surplus of these crops at the time of this writing. In other regions of the world where such a genetic program has not been carried out, the yields are low and many shortages exist. Not only has quantity been increased, but the quality of the food has been improved. The sweet juicy oranges with few or no seeds which we enjoy today differ from the oranges with sparse, sour juice and many seeds found growing on wild orange trees.

The application of our knowledge of genetics to the field of medicine has revealed the great importance of heredity in the case of many of the diseases of mankind. We know that tuberculosis is a germ-caused disease, yet it has been very definitely established that heredity plays a most important part in determining whether or not one will contract this disease. Such serious human afflictions as cancer, heart trouble, diabetes, and cataract of the eye all are influenced by heredity. For many diseases, an accurate diagnosis is more quickly and accurately made through a study of one's family history than through elaborate and expensive laboratory tests. Also, it is possible to avoid many serious mistakes in diagnosis through application of our knowledge of genetics. Macklin reports a case of a child who was taken to a doctor for treatment. This child had a very dry skin, sparse hair, poorly developed teeth, and a tendency to become feverish upon the slightest exertion. Hypothyroidism (under-secretion of the thyroid gland) results in such symptoms, and the child's condition was thus diagnosed and he was put on thyroid medication. But the treatment seemed to aggravate the condition, and another doctor was called in. Fortunately, this man had had training in genetics, and made a study of the child's family history. He found that one of the parents and a number of the relatives showed similar, but less severe, symptoms, and he diagnosed the condition as an inherited disease, *ectodermal dysplasia*. Among the symptoms of this disease is an absence of normal sweat glands. In a normal person the body is cooled through the evaporation of perspiration from the skin, but in persons without normal perspiration the body is easily over-heated. Hence, the administration of thyroxine was the worst sort of treatment, for thyroxine speeds body metabolism and increases the heat output. This case shows us how a disease may sometimes be diagnosed by genetic means when the symptoms might indicate other conditions than those which actually exist.

Another case reported by Snyder also illustrates the importance of heredity in the diagnosis of the disease. A young man had developed a serious affliction which involved spasmodic facial contortions which were not only embarrassing, but would be an actual detriment to him in any profession which required contact with the public. A doctor diagnosed this condition as *tic douloureux,* a type of neuralgia which can be corrected through a delicate nerve operation performed by a competent specialist in this type of surgery. Before arranging for the operation, however, the doctor consulted another physician, who recalled that there were a number of cases of diabetes in the family of the afflicted young man. Realizing that the contortions of the face might be an unusual manifestation of diabetic neuritis, the consulting physician examined the patient for diabetes. The tests were positive, the patient began taking insulin as a treatment for diabetes, and the *tic* soon disappeared. Thus a serious and expensive operation which would have been useless was avoided through the application of the principles of genetics.

Genetics also has its value in preventive medicine. In many cases it is possible to anticipate the development of a disease or other body abnormality and to take appropriate steps to prevent its occurrence. A person with a family history of tuberculosis might take precautions against infection which would not be necessary for one without such a background. Also, every effort might be made for early recognition of such infection and steps might be taken for its eradication before the disease developed fully.

An excellent example of the importance of this phase of genetics is illustrated by an actual case. There is a type of anemia known as *hemolytic icterus,* which is inherited. Early symptoms of this disease involve abnormal shape and fragility of the red blood corpuscles. Later there may be a greatly enlarged spleen which removes from the blood abnormally large numbers of red blood cells, resulting in serious anemia and perhaps death. About one half of the children of persons with this condition will develop it unless preventive surgery is performed. It is wise, therefore, to examine the blood of all children of afflicted parents to detect early, preclinical symptoms of this disease. Should the symptoms be found, the spleen should be removed as a preventive measure. Snyder reports the case of a man who died of the disease and who had two sons showing the preclinical symptoms of it. One was operated on and the spleen was removed as a precautionary measure. The other refused to have the operation since, as he put it, he "felt sound as a dollar." A number of years later, however, this man developed the disease and was operated on, but too late, and he died before his body could recover from the anemia which had developed.

To a person who would become a marriage and family relationship consultant, an extensive study of genetics is a vital part of training. Prospective parents who exhibit some abnormality naturally wish to know the chances of the abnormality appearing in their children. Likewise, they are equally interested in the possible transmission of desirable traits such as musical ability, mathematical aptitude, or pleasing facial features. To answer such questions intelligently one must be thoroughly familiar with the mechanism of heredity and the methods of distinction between the effects of heredity and environment. Also, one must be familiar with the reference sources which give accounts of the inheritance of similar traits in other family pedigrees, and he must know how properly to interpret these pedigrees in the light of the case which has come to his attention.

There are even legal applications of the principles of heredity. Court cases involving questions of disputed parentage may often be solved by an analysis of blood types or other inherited characteristics. Divorce, custody of children, estate inheritance, and support of illegitimate children are some of the legal problems for which the courts may turn to the science of genetics for their solution.

The science of genetics also includes a branch of study which is sometimes called **eugenics.** This is the application of the principles of heredity to the improvement of mankind. Through rigid application of our knowledge of inheritance we have been able to produce domestic animals with very superior qualities. It is impossible, of course, to apply the same techniques to human beings, but it is possible to apply some of our knowledge of heredity to man without violating any moral or ethical codes which govern human conduct. Under such limitations we cannot hope for the spectacular results which have characterized the improvement of domestic animals, but at least we may be able to prevent deterioration of the civilized races through the action of dysgenic forces.

Thus we see that the study of genetics promises to be not only very interesting but intensely practical and well worth while.

FALSE CONCEPTS OF HEREDITY

Mankind seems to be of such a disposition as to want an explanation for the various facts he observes. One of the commonest words in the vocabulary of a young child is "Why." We like to have some explanation for our observations. Unfortunately, however, we are prone to seize upon the first explanation that comes to hand without inquiring sufficiently into the possible truth or falsity of the explanation. Lacking any explanation of any nature, we are likely to invent one from our own imagination and accept it without further question. Because of this tendency to accept "hearings" without question, many superstitions and false beliefs have become widespread and are accepted with credulity by large numbers of people. Since there is a natural human interest in heredity, it is not surprising that there are many false beliefs and superstitions concerning the methods of inheritance. Hence, before we begin our study of the scientific facts behind inheritance, it will be worth while to point out some of the more common false beliefs and to indicate the actual facts involved.

Blood as a Heredity Force

One of the oldest concepts of inheritance holds that the blood is the hereditary determiner. We still hear the terms "blood will tell," "blue blood," "blood line," "blood relative," and "bad blood" used in connection with heredity, and even though these phrases may now be largely metaphorical, literal belief in what they imply is not uncommon. Although science has long since proved that blood has nothing to do with heredity many people still feel that blood has some mysterious effect. There are many who will refuse a transfusion of blood from a member of another race for fear that in some way they will acquire some of the characteristics of that race. As a matter of fact, this is far from being true. Indeed, there have been a number of cases where a person's entire blood supply has been replaced by blood from others without influencing his characteristics in any way. Blood is merely one of the tissues of the body, a fluid tissue, which has no more relation to heredity than the other body tissues such as bone, muscle, and nerve. The heredity determiners which are transmitted to future generations in higher animals are located only in a small mass of germinal tissue more or less isolated from other tissues.

Influence of Age of Parents

Another popular misconception about heredity concerns the supposed influence of the age of parents. Some believe, for instance, that children born during the earliest part of the parents' reproductive life will be consistently inferior in hereditary qualities compared with the children of more mature parents. It is sometimes thought that a person is unable to transmit qualities which he may not yet have developed in the normal course of maturity. It is also believed that children of older parents will not inherit the same vigor which characterizes the children of parents who have not yet shown any of the signs of senescence. So far as heredity is concerned, the time of a person's life during which he contributes a germ cell to the formation of a new human life has nothing at all to do with the inheritance of that child. Of course, we realize that the child of an immature mother may be somewhat handicapped by the smaller size of the mother's uterus, and that a normal birth may be more difficult because of the smaller size of the mother's pelvis. But these are environmental factors which may influence the child before or during birth and have nothing to do with heredity. Some time ago, however, a perfectly normal baby was delivered (by caesarean operation) from a mother who was only 5 years old. This baby has now grown into a man who is fully developed mentally and physically. It is also true that environmental effects may result in the production of a child with certain defects if the mother is approaching the end of the reproductive period, because degenerative changes occur in the reproductive organs at that time. In the father, degeneration of the sex organs has no effect on embryonic development.

In spite of this knowledge, however, many breeders of domestic animals still hold to the belief of the influence of age. The author has known of cattle breeders in Texas who will not use a bull for breeding purposes after he has passed a certain age for fear that the offspring will not be as hardy because of the comparatively old age of the bull. Many fine bulls have been discarded because of this belief and, as a result, many fine hereditary qualities have been lost to future generations of cattle.

Influence of Drugs

Various drugs are often considered as having a possible influence on the germ cells and hence on the children which may result from these cells. Alcohol, nicotine, morphine, and similar drugs are often mentioned in this connection. It is thought by some that excessive use of these substances may in some way weaken the germ cells and produce defective offspring. Occasionally one may hear of a misinformed temperance lecturer describing the terrible deformities which may result in the children of a man who becomes a chronic alcoholic. It is known that habitual, excessive use of this drug over a long period may cause a man to be sterile through destruction of his sperm cells; but if these cells remain viable and one fertilizes an egg cell, the resulting child will possess the same characteristics as if the man had been a model of sobriety. A drinking expectant mother can harm the embryo she is carrying since the alcohol can pass directly into the blood stream of the fetus and a child may be born who actually bears the symptoms of alcoholism. But this is an environmental influence.

Similarly, a child may be born who is a morphine addict if the mother is accustomed to taking this drug during her pregnancy, but morphine cannot influence the determiners of heredity in either a man or a woman.

Influence of Acquired Characteristics

One of the most widespread of the false beliefs on the subject of heredity is that characteristics acquired during the lifetime of the parents can influence the offspring. There was a time during the last century when the idea was considered plausible by some rather competent scientists, but, as the science of genetics has expanded, it has become increasingly evident that the hereditary factors are not influenced by acquired characteristics. The idea that acquired characters can be inherited means that those body parts which an organism develops to the greatest extent will tend to be more highly developed than normal in the descendants. For instance, if a blacksmith uses his arms in handling the heavy implements of his trade the muscles of his arms naturally will become well developed and his children would be supposed to have better developed arms as a result of this activity of the father. The history of the rise and decline of this belief is discussed more fully in the next chapter.

It is interesting to note the recent resurgence and decline of the belief in the inheritance of acquired characteristics in Russia. At the Russian Agricultural Institute there was an obscure plant breeder, Trofim D. Lysenko, who conducted some experiments on wheat and tomatoes and thought that he obtained evidence that they could be improved genetically by improved growing conditions. He had never had training in fundamental genetics, but sought to extend this viewpoint to inheritance in all forms of life. In his own words, as translated from the Russian, he sets forth the idea that "changes in the conditions of life bring about changes in the type of development of vegetable organisms. A changed type of development is thus the primary cause of changes in heredity. . . . Heredity is the effect of the concentration of the action of external conditions assimilated by the organism in a series of preceding generations." This is a far cry from the facts of genetics as accepted by the leading geneticists of the world today — it represents a return to the long-since disproved theories of the past century. In spite of this, however, Lysenko was clever enough to have his unorthodox viewpoint approved by the Central Committee of the Communist Party. When it was publicly announced, several years before the outbreak of the second World War, that the communist party had approved Lysenko's work, there was consternation among the geneticists of Russia. Several of the top-ranking geneticists were deprived of their positions because they refused to accept Lysenko's absurd theories and have not been heard from since. The author was personally acquainted with two such unfortunate individuals. When Lysenko's theories failed to work out, however, he was relegated to an obscure position and he no longer has any influence on the course of genetics in Russia today.

Maternal Impressions

The belief in maternal impressions or "prenatal influence" is a superstition which is not strictly related to heredity, but it is so widely held and so closely connected to the topics studied in genetics that it warrants some consideration here. Briefly, the belief is that an expectant mother may "mark" or in some other way influence the development of her child during the period of her pregnancy. For instance, a woman may crave strawberries while she is carrying her child, and when the child is born it may bear a strawberry-shaped birthmark. To some persons, such an occurrence is absolute proof of the validity of maternal impressions. To experienced geneticists, however, such reasoning is illogical. There is no known way by which such an effect can be brought about. True, the mother may have craved strawberries and the child been born with a strawberry birthmark, but such birthmarks also appear in regions of the world where the mothers have never seen strawberries.

All of you who read this book, no doubt, have heard of a number of stories handed down by old-timers to illustrate the influence of maternal impressions. It is a subject which appeals to human fancy, and the stories are told and retold with embellishments which often make it difficult to sift out truth from imagination. During the development of a child, it is certain that the mother will have countless thousands of impressions from many different sources. Whenever a child is born with some slight deviation from normal, it is very easy for her to pick out one of these many impressions to which she can assign the cause of the abnormality. As a test of this explanation for the belief in maternal impressions one physician made it a point to ask each of his expectant mothers if she had had any strong mental impressions which she thought might mark her child. A few listed some impressions, but most did not remember any. When the children were born there was found to be no correlation between the impressions which were listed and any effects which they might have had on the children. Many of the mothers who bore children with some slight defects, however, were quick to remember something they had failed to list which they thought might have been the cause of the abnormality.

The Belief in Heredity or Environment

One of the most common misbeliefs on the whole subject concerns the relative influence of heredity and environment on the production of the total individual. It is a common mistake to assume that characteristics are the result of heredity *or* environment, whereas in actuality most characteristics are the result of heredity *and* environment. A tall, husky man may produce tall, husky sons. The physical resemblance is very marked, so that one may conclude that body height is an inherited characteristic and may therefore rule out any consideration of environment. On the other hand, we know that had these same sons been fed on a diet which was deficient in the vitamins and minerals necessary for the formation of a normal skeleton, they would not have developed so well. Their bones would be under-developed and deformed, and the sons might actually be considered dwarfs by human standards. Hence, we should think that the sons inherited the ability to grow into tall, husky men under the influence of the proper environmental factors. In the absence of the proper environment the full potentialities of the heredity cannot be realized. On the other hand, no amount of adequate diet will make a tall, husky individual out of a person who does not inherit these potential qualities. Favorable environment will allow such a person to attain the maximum development which his heredity will permit.

As another example, suppose a woman living in Indiana develops hay fever about August 15 of each year. She marries and has two children. As they

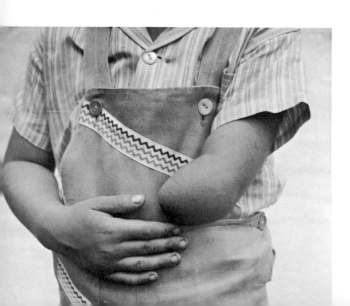

FIG. 1–4 *Maternal impression? This child was born with only one hand. The mother attributes this abnormality to the fact that she witnessed a serious automobile accident late in her pregnancy. Knowledge of genetics and embryology indicates that such impressions do not occur.*

mature one of the children begins to show the symptoms of hay fever at about the middle of each August. Now it so happens that on this approximate date the ragweed plants in this region of the country begin to release pollen in prodigious quantities; pollen, which is carried by the air currents, produces hay fever upon contact with the mucous membranes of a person allergic to it. In our hypothetical case, it is easy to jump to the conclusion that one of the children inherited an allergy for the ragweed pollen. On the other hand, it is known that allergies are acquired; one cannot be allergic to a substance which he has never had contact with. Had this family moved to Hawaii before the children were born, the child would not be allergic to ragweed pollen, for there is no ragweed in Hawaii. Even though he visited the United States during the ragweed season, he would show no symptoms of hay fever during the first visit. Hence, we see that what the child inherited was an ability to become easily sensitized to foreign proteins, and, since ragweed pollen is very abundant in Indiana, it is not surprising that he developed an allergy to the same substance which causes his mother's trouble. The elements of heredity cannot express themselves without a suitable environment.

One might compare heredity and environment to the exposure and development of a photographic negative. A properly exposed film (a desirable heredity) when properly developed (given a good environment) will result in a clear, sharp negative capable of printing good pictures. On the other hand, if the development is poor — if the developer is too warm, light rays strike the film during the developing process, and the film is not properly agitated during development — a good negative cannot result. Conversely, if the exposure (heredity) is poor, no amount of good development (good environment) can produce a desirable negative. A good development will enable one to obtain the very best negative possible under the circumstances, but the final results cannot compare with the quality resulting from a good exposure plus a good development. It will be well for all social workers to realize the scientific truth of this last point.

METHODS OF GENETIC STUDY

Now that we have disposed of some of the false beliefs concerning the subject of heredity we may be better prepared to understand the facts with which the study is concerned. Most of the principles of heredity were not discovered from human beings but from other forms of life. The reason for this is not hard to find. Most biological phenomena must be investigated by the experimental method. In the field of genetics this involves crossing individuals that differ in one or more characteristics and studying their descendants in an effort to determine the method by which these characteristics are inherited. Human beings, however, have the rather stubborn habit of wanting to choose their own mates and of deciding for themselves just how many children to have. As a result, we must be content with observation of the results of marriages which have already occurred and must attempt to draw conclusions from such observations. Also, man has an extremely long generation, or life cycle, and an extremely small number of offspring (in comparison with other forms of animal life) — both of which are factors which militate against the use of the human species in studies of heredity. Fortunately, however, the observations which have been made on heredity in man have shown us that the method of human inheritance shows no essential difference from that found in other forms of life. As a matter of fact, the first important discoveries in genetics were made in studies on plants (garden peas). Yet these discoveries have been found to hold true for most other forms of life, including man.

Experimental Breeding

Many discoveries in genetics today are being made through breeding experiments with various forms of plant and animal life. There are four important factors which must be considered when one chooses an organism for experiments in heredity. First, it should have a **short life cycle** — it would take centuries to learn much about heredity in elephants through experimental matings because of their extremely long life cycle. Mice, on the other hand, are ready for breeding within 6 weeks after birth. Hence, mice are widely used as laboratory animals for genetic experiments, but the author has never heard of anyone undertaking an experimental breeding program with elephants. Second, the organism chosen should have a **large number of offspring** in order to provide the numbers necessary for statistical analysis of the results of experimental crosses. This rules out many of the higher forms of animal life that ordinarily bear only one offspring at each birth. Third, the organism chosen should show considerable **variation** in its characteristics. We could learn nothing about the inheritance of eye color in man if all people had blue eyes. It is only because other eye colors exist that we can learn how blue eyes are inherited. The greater the number of variable characteristics, the more valuable is the organism for genetic studies. Fourth, the organism chosen should be of such a nature that it can be **raised conveniently and cheaply,** and also it should be one which does not require too much space for handling. Many plants, such as corn, are good from this standpoint when land for cultivation is available. In the use of animals, we are somewhat restricted to those smaller ones which require little food and attention.

Among mammals, the mouse is the most ideal animal for genetic experiments because of its prolific breeding, its ease of handling, and its rather large number of variable characteristics. In Figure 1–5 several of these characteristics are shown. It is therefore used extensively in genetic experiments. There is another animal, which is used more extensively and from which a major portion of our knowledge of genetics has been derived. This is the little fruit fly, *Drosophila melanogaster,* which may often be seen buzzing about bunches of bananas, vegetable bins, and garbage cans in large numbers like a swarm of gnats. This little insect, only about one-fourth the size of a house fly, is an excellent organism to study from the standpoint of genetics. It has a very short life cycle and a very large number of offspring — within 10 or 12 days after a pair of these flies have been placed in a vial with food they will have produced about 100 offspring, and

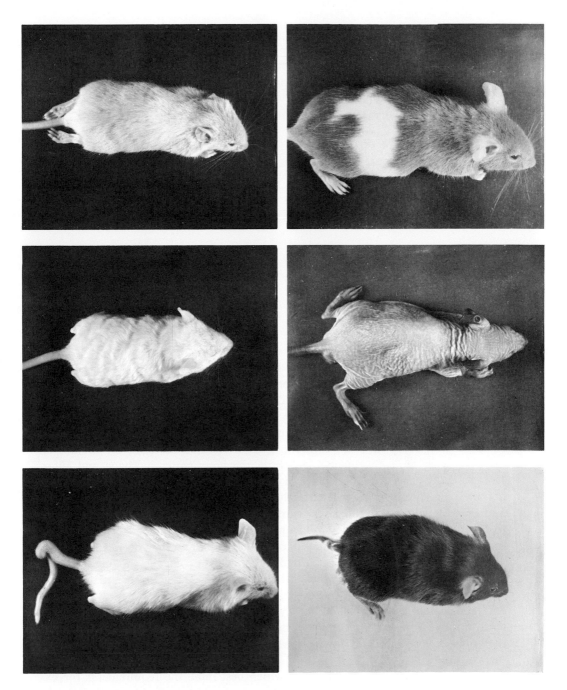

FIG. 1–5 *The mouse is an excellent mammal for genetic study because, among other things, it shows many inherited variations. The inherited traits shown here are: short ear, waved fur, kinky tail, belted body, hairless body, and short tail.*

FIG. 1–6 *The fruit fly, Drosophila melanogaster, is very widely used for genetic study. The male is on the left and the female is on the right. These are living flies which have been etherized for study.*

a second generation may be obtained from these within another 10 days or so. (See Figure 1–6.) Within a few months' time at this rate, if reproduction is allowed to continue unchecked, the descendants of a single pair of flies will actually number in the millions. Also, the many variations in the characteristics of these flies offer rich material for study. The eyes vary in color, texture, and size; the wings vary in size, curvature, and distribution of veins; the bristles on the body vary in arrangement and nature; the body varies in shape and color. In fact, every part of the body is subject to some form of variation which is inherited. Finally, *Drosophila* is certainly ideal from the standpoint of convenience and cost of upkeep. A few cents worth of materials will prepare enough food to raise thousands of fruit flies. They may be raised in small vials on the laboratory shelf, they do not give off any unpleasant odors, and they do not need daily attention. Of course, there is one disadvantage which comes to the minds of students of genetics when they first see *Drosophila:* "They're so little." True, they are small; but through the use of modern stereo microscopes they can be enlarged so that every bristle on the body can be clearly seen. Also, these flies tend to move around rather quickly, but through the use of an etherizing bottle they may be induced to lie passively while one examines them and decides upon their future.

Without question, the organisms which best fulfill the requirements for experimental genetics are the microorganisms, such as bacteria, yeasts, and molds. Some of the bacteria have a generation cycle of only about 20 minutes and billions of them can be raised in a small dish in the laboratory. For many years these small organisms were neglected as objects for genetic investigation because they seemed to show so few variable characteristics. When we developed methods of study of their physiological traits, however, we found that they showed a great number of inherited characteristics which could be detected rather easily. It was at this time that microbial genetics had its beginning and it has since become a major area of genetic research.

Statistical Analysis as a Tool for Genetic Study

Without doubt, a program of experimental breeding is the most rapid and the most satisfactory means of determining the method by which specific characteristics are inherited. We have already pointed out that it is impossible to conduct such a breeding program in some forms of life, however, and we must rely upon observations of the results of inheritance as they have already occurred and draw our conclusions from these. When such results are analyzed on a statistical basis, in the light of the knowledge which has been obtained through experimental breeding of related forms of life, we can obtain very accurate information. In the case of man, we now clearly understand the mechanism of inheritance for hundreds of characteristics which have been studied by this method.

In order to conduct such a study, however, it is necessary that a standard method be used to represent the family pedigrees in a concise, easily understood form. A case of a condition known as *polydactyly* gave rise to the form of pedigree illustrated in Figure 1–8, which has become standard usage among geneticists. This pedigree was discovered when a woman brought her young daughter to a doctor for examination. The girl had an extra finger on one hand and an extra toe on one foot. This relatively uncommon condition is known to be inherited. Upon further investigation it was found that the child's father showed

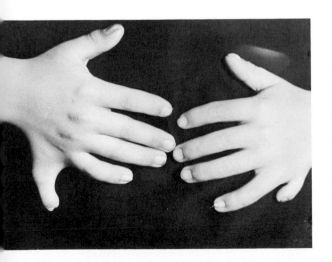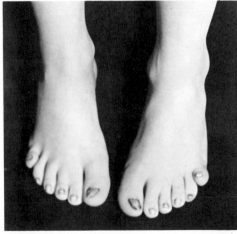

FIG. 1–7 *Polydactyly. Extra digits on the hands and feet sometimes appear because of heredity. Pedigree analysis helps us to understand the method of transmission of this trait.* COURTESY C. NASH HERNDON.

this character (though the extra finger had been removed surgically), and that her brother also showed the condition, whereas two other children in the family had the normal number of fingers and toes. One can visualize this entire family history by a glance at the chart in Figure 1–9. In charts of this kind men are customarily represented by squares and women by circles. Marriage is indicated by a connecting horizontal line and children are shown by attachment to a vertical line extending downward from the horizontal one. Those individuals which show the particular character which is being traced are represented by a solid square or circle while those not showing it are indicated in outline.

From such a chart it is easy to obtain a statistical ratio of the appearance of this trait in an attempt to analyze the method by which it is inherited, but the numbers are far too small to have much significance statistically. We may obtain larger numbers through a study of the ancestors of this family and can construct a more extensive family history chart for this purpose. We now begin to have sufficient numbers to have some statistical significance. As we survey the chart we note that no person shows the trait unless at least one parent shows it. This type of inheritance is typical of characters which are known as **dominant.** This gives us one clue to the method of inheritance. Among the children of marriages where one parent showed the trait there are eleven children who are normal and ten who are polydactyl. This is a ratio of about 1 : 1 (that is, ½ show it and ½ do not). In the group which shows the trait are four boys and six girls, so there seems to be no influence of sex on the development of the character. After other family histories showing this character have been studied, we may determine the method of inheritance through comparison with results obtained on lower animals through experimental breeding of various characters. It may then become pos-

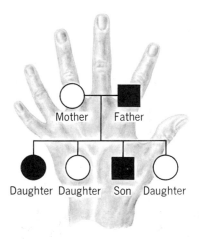

Mother Father

Daughter Daughter Son Daughter

FIG. 1–8 *Pedigree of a family showing polydactyly in a father and two of his children.*

● ■ POLYDACTYL

○ □ NORMAL

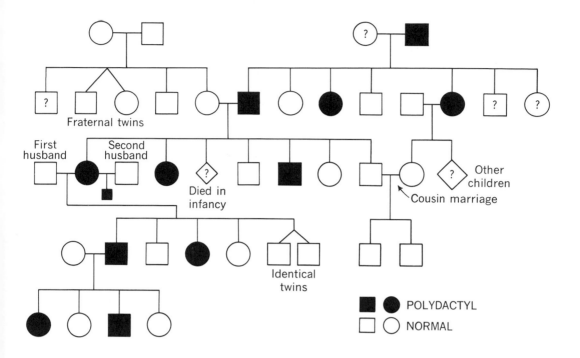

FIG. 1–9 *Pedigree showing more complete ancestry of the inheritance of polydactyly. This chart also indicates the method used in preparing such family histories. Squares represent males; circles, females.*

sible to predict the chances of polydactyly appearing in the future children of this family.

Two other family pedigrees are shown in Figure 1–10 for **attached ear lobes.** From the chart it appears that this character must be inherited in a different manner, for two parents with freé ear lobes produce two children with attached ear lobes in a family of five children. And in another family, both parents have attached ear lobes and so have all four of their children. This type of inheritance is typical of characters which are known as **recessive.**

Some persons have the power to roll the tongue, while others are not so gifted. A couple, both of whom are **tongue rollers,** have three children, two of whom can roll their tongues and one who cannot, as shown in the family pedigree, Figure 1–12. This seems to be a different method of inheritance than that for attached ear lobes.

Finally, the pedigree in Figure 1–13 shows the inheritance of **red-green color blindness,** a condition which causes a person to have difficulty in distinguishing red and green. An examination of this chart shows that only males are afflicted. This gene must be inherited in a still different manner, which is somehow connected with sex.

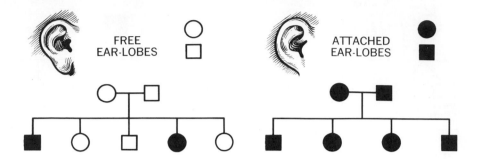

F I G. 1 – 1 0   *Pedigree of two families showing inheritance of attached ear lobes.*

These examples are given to show, in general terms, how one must go about the study of genetics by the use of statistics. There are many forms of statistics, of course, but in genetics we use the **statistics of probability** primarily. Such statistics involve a study of the laws of chance. This discussion of statistics should not be taken to imply that statistical analysis is used only for those forms of life which cannot be studied by the experimental method. Statistical studies are also of great value in analyzing the results of experimental breeding.

Cytology as a Tool for Genetic Study

The science of genetics is very closely related to cytology, a specialized study of the nature of cells as seen under the microscope. We know that the cells carry the units of heredity, for a tiny sperm cell from a man and an egg cell from a woman furnish the physical link between one human generation and the next. The units of heredity, whatever they are, must be contained within these cells. It stands to reason, therefore, that a detailed study of the structure and activities of cells will throw light on the method of inheritance. With the proper methods of study one can see within the cells small, thread-like bodies which undergo a series of marvelous transformations as an accompaniment to cell division. These bodies, known as **chromosomes,** play a vital part in the transmission

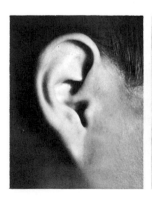
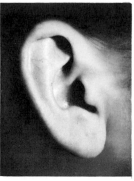

F I G. 1 – 1 1 *The ear lobes of some individuals are attached directly to the head (left), while in others the lobes hang free (right).*

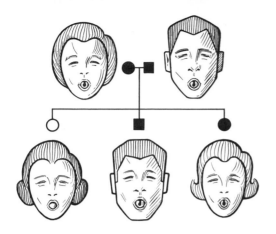

FIG. 1–12 *Pedigree showing inheritance of ability to roll the tongue.*

of hereditary factors. Through a careful study of the chromosomes and their activities we can unravel many of the puzzling mysteries which arise from the results of experimental breeding. In a similar manner, many of the puzzles of cytology become resolved after an experimental breeding program has been carried out. Thus, the two branches of biological science are closely integrated — in fact, so closely integrated that sometimes the word **cytogenetics** is used as a term relating to the whole study of heredity. Cytology will form an important part of our study in this book.

Physiological Genetics

Finally, physiology also is a tool for the study of genetics, and is a phase of the subject which has been receiving an increasing amount of attention during recent years. Older geneticists were more or less content to learn what genetic factors produce specific effects on an organism, but through physiology we are now beginning to learn just how these effects are produced. We know, for instance, that red hair is inherited, but just how do the hereditary factors exert their influence? Why are pigments formed to make the hair reddish rather than

FIG. 1–13 *Pedigree showing inheritance of color blindness. Note that no girls in this family history are color-blind.*

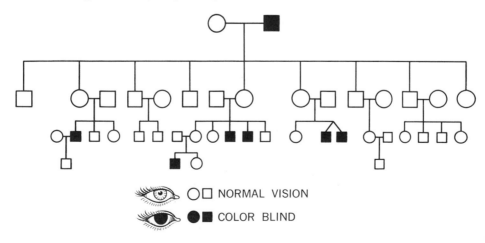

○ □ NORMAL VISION
● ■ COLOR BLIND

black, brown, or blonde? How do the hereditary determiners influence the chemistry of the hair-follicle cells to produce these variations in color? Do these determiners also influence other body parts, such as the glands or nerves, so that people with red hair are not only passionate lovers, but also hot-tempered individuals? As any plant or animal embryo develops, just how do the hereditary units within the cells cause the cells to become arranged in a form characteristic of the species to which it belongs? Why does a cat always have kittens and a dog always have puppies rather than the other way round from time to time? Physiological genetics attempts to answer these and many other questions relating to the action of hereditary factors.

PROBLEMS

1. Poliomyelitis (infantile paralysis) is known to be caused by an infective virus. Does this mean that heredity can play no part in the acquisition of this disease? Explain.

2. Animal breeders have produced superior commercial breeds through application of the principles of genetics. Why cannot we use these same methods to improve the human race?

3. The proportion of defective calves born from heifers (young cows) is somewhat greater than those born from mature cows. Does this mean that the age of the mother has an influence on the heredity of the calves? Explain.

4. Formulate an experiment which would give an objective test of the inheritance of acquired characters. Describe fully.

5. A certain woman who was very talented in art, wishing to transmit some of this talent to her child, spent much time painting and studying art masterpieces during her pregnancy. As her child matured he showed great artistic talent. Does this prove that the mother influenced her unborn child through her mental activities before birth? Explain.

6. The little fruit fly, *Drosophila,* normally has a gray body, but if a small quantity of silver nitrate is added to its food, its body will turn yellow. There are some strains of this fly, however, which develop a yellow body regardless of the food on which they are raised. Would you say that body color depends on heredity or environment? Explain.

7. A certain man cannot keep a dog in his home because he develops severe hay fever when he inhales the dander from the skin of a dog. This man has a son who develops a rash on his body every time he eats eggs. Can there be any hereditary connection between these two different allergies? If so, explain how.

8. Name three plants which you think would be good for experimental breeding purposes. Give the reasons for your choice in each instance.

9. Name three animals which you think would be good subjects for experimental breeding. Give reasons for your choice of each.

10. Choose two characters, which you think are inherited, that you possess and which are also found in some of your relatives. Make a survey of your ancestry for the presence of these characters as far back as you can go. After you have collected the material make a pedigree chart for each of these two characters. Later in the course you will be asked to make a tentative determination of the method of inheritance of each of these traits from your charts.

Backgrounds of Modern Genetics

MOST OF OUR KNOWLEDGE IN THE FIELD OF GENETICS has been developed since the beginning of the present century. As we search back into the history of scientific thought, however, we find that speculations on the nature of heredity are as ancient as the history of mankind. One old Babylonian tablet, dating back to 4,000 B.C., shows the pedigree of five generations of horses with an indication of the appearance of the head and mane, showing how these characteristics were transmitted through the generations. It is shown in Figure 2–1. Other tablets of this same period show the artificial cross-pollination of the date palm, indicating that as long as 6,000 years ago people understood the principle back of sexual reproduction in plants. Studies of the remains of ancient Chinese civilization dating back to 4,000 B.C. reveal that superior varieties of rice had been developed, no doubt by the application of some of the principles of genetics through hybridization and selection. Many of our present crops have never been known to exist in the wild state. Cotton, for instance, was probably developed by some of our ancient ancestors through an application of the principles of genetics and passed down to us through the centuries, while the wild plants from which it was developed have ceased to exist. The many varieties of domestic animals on the face of the earth give mute testimony to the effectiveness of the application of these same principles to animal breeding before the dawn of modern history.

Early Speculations

The first accounts of definite attempts to formulate an explanation for inheritance were largely speculation, with little, if any, support in fact. Pythagoras,

21

a famous Greek philosopher who died about 500 B.C., proposed the theory that a moist vapor descends from the brain, nerves, and other body parts of the male during coitus and that from this, similar parts of the embryo are formed in the uterus of the female. This would give us some explanation of the transmission of likenesses from the male parent to the offspring, but leaves it unclear how the female could transmit any qualities.

Another Greek philosopher of this period, Empedocles, made some amendments to these ideas. He recognized that both parents must contribute something to the formation of the embryo. To explain this, he proposed the theory that each parent produces a semen which arises directly from the various body parts which are formed by it in the embryo. He realized that both parents did not contribute all of their characteristics and sought to explain this by assuming that not all of the body parts contributed to the pool of semen which united during coitus. This could account for the fact that children show some characteristics from each parent.

Some 200 years later the great Aristotle suggested an interesting, although highly imaginative, theory. He proposed that the semen of the male was produced from his blood — that it was, in fact, highly purified blood and, in such a purified state, it possessed the ability to give form to a new life. The menstrual fluid of the woman was supposed to be semen also, but the woman supposedly did not have the power to achieve so high a degree of purification of blood as the man. This less-concocted semen was supposed to furnish the substance or material from which the embryo was produced, while the semen of the male added the form-giving substance. In other words, the female furnished the building material while the male furnished the life-giving power, "dynamis," which enabled this material to be formed into an embryo.

Harvey's Deer

For about 2,000 years Aristotle's theory was generally accepted. In some of the medical books of the seventeenth century there were illustrations showing the stages of the coagulation of the embryo within the uterus from a mixture of the semens of the parents. It was at this time, however, that William Harvey,

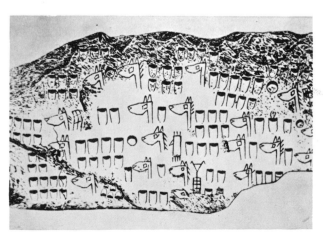

FIG. 2–1 *A 6000-year-old pedigree which shows transmission of characteristics through five generations of horses. Three types of mane are shown here: erect, pendant, and maneless; and three types of profile: convex, straight, and concave. This pedigree was found engraved in stone on a seal excavated in Elam, east of Ur in Chaldea.* FROM AMSCHLER, JOURNAL OF HEREDITY.

who is better known for his discoveries on the circulation of the blood, began to doubt the validity of the venerable theory of Aristotle and decided to undertake some experiments to test it.

He obtained permission to use twelve mature female deer from the private reserves of Charles I, who supported Harvey in this work. He mated these and killed six of them later at various stages of pregnancy. At no time did he find anything which resembled coagulating fluids in the uterus. When he first observed an embryo, it was in a deer killed several weeks after mating and the embryo was very small. Upon close examination he found that this embryo did not even look like a deer. As he examined embryos at later stages of development he found that they gradually took the shape of a deer as they increased in size. The six remaining deer had normal fawns about 8 months after mating.

As a substitute for Aristotle's theory he suggested that as iron, by friction with a magnet, becomes possessed of magnetic properties, so the uterus, through the friction of coitus, acquires some magnetic power to conceive an embryo and that the semen plays only some sort of vitalizing role in the process. It was not long before the development of the instruments of science made a better understanding possible.

The Discovery of Sperms and Eggs

A great step from the realm of speculation to the realm of reality was taken during the latter part of the seventeenth century when the early microscope maker, Anton van Leeuwenhoek, observed living sperms in the semen of various animals. He also noted the association of the sperms with the eggs of frogs and fishes and believed that the sperms furnished the life of the embryo while the egg of the female provided a place for the nourishment and development of the embryo. He even went so far as to devise a simple experimental cross of rabbits to prove his point. He mated a white female with a gray male and obtained all gray offspring. This, to him, was proof of his belief that the male parent contributed the life to the embryo. Had he mated a white male with a gray female he would have been forced to an opposite conclusion, or had he allowed these gray offspring to mate among themselves he would have had a hard time explaining the appearance of white rabbits in the second generation.

A short time later another Dutch scientist, Jan Swammerdam, developed a theory which captured the imagination of scientists and philosophers for 2 centuries. This was the **preformation theory,** which held that the development of the embryo was actually only the enlargement of parts that were already present in the sperm or egg. Some of the scientists of this time even imagined that they could see a miniature human being inside the head of a sperm, and drawings of this period show such a sperm structure. Of course, this showed stupendous powers of imagination, yet it represented an important biological advance at the time, since it offered a more mechanistic viewpoint of heredity than the vague concepts of the past. It also offered a basis for repudiating the old superstitions that women, under the spell of witchcraft, might deliver kittens or puppies rather than a human baby.

FIG. 2-2 *The concept of pre-formation. These are miniature human embryos supposedly seen in human sperms by scientists of the seventeenth century.* LEFT, AFTER HARTSOEKER, 1694, AND RIGHT, AFTER DALEMPATIUS, 1699.

Still another Dutch scientist, Regnier de Graaf, disagreed with this idea that the sperm was the sole agent of heredity and pointed out that children often show physical characteristics which are found in the mother and not in the father. He also described cases in animal breeding where the female transmitted characteristics to the offspring. In a search for the physical basis of this observation, he carefully studied the ovaries of mammals and noted the protuberances which he interpreted as eggs which were similar to the eggs of birds. The protuberances bear his name today (Graafian follicles). He believed, correctly, that the egg breaks from the ovary, is fertilized, and makes its way to the uterus for development. He found cases of extra-uterine gestation which he cited as proof that there is nothing within the uterus which is necessary for conception, but that the egg is the necessary factor. These were far reaching and sound observations, but unfortunately they were obscured during the eighteenth century because of the controversy between the ovists (those who believed that the egg contained a preformed embryo) and the animalculists (those who believed that the sperm contained a preformed embryo).

Theories and Experiments of Maupertuis

During the eighteenth century a Frenchman born in 1698, Pierre-Louis Moreau de Maupertuis, studied, experimented, and wrote of concepts of heredity which were far in advance of his time. When we view his beliefs in the light of the thought and knowledge of his day, we cannot but feel that he deserves to be classed as one of the great scientists of all time. Maupertuis thoroughly disagreed with the idea of preformation, for he felt that a consideration of the plain facts of biparental inheritance ruled out the idea that one parent formed a miniature preformed embryo and that heredity took the form of some vague, spiritual essence which could impress itself upon the preformed embryo. He felt that some more tangible mechanism was necessary for the transmission of characteristics from parents to offspring. He proposed the belief that both parents produced semen which united during coitus to form the embryo. The semen, he believed, was made up of tiny particles, each of which was destined to form some specific body part, and that two such particles, one from each parent, unite to form each body part. One of these might dominate the other, however, to make a child more like one parent in this one respect. All of this sounds remarkably like the conclusions drawn by Mendel over a century later. Maupertuis also supposed that these particles were formed in the various parts of the body and migrated to the reproductive organs to form the semen. Furthermore, these particles were supposed to retain some sort of "recollection" of their previous surroundings and to produce body parts in the embryo similar to those from which they came. This represents essentially the theory which was to be proposed by Charles Darwin as an explanation of the mechanism of heredity. Maupertuis also proposed an experiment to test his theory — if one should mutilate a portion of animals' bodies for a number of generations, perhaps one would note a diminution of the mutilated parts, since the hereditary particles could have no "recollection" of parts which the animal body did not possess. This very experiment was to be performed by Weismann nearly 150 years later.

In addition, Maupertuis deserves credit as one of the first persons to collect family pedigrees and to analyze the results by statistical means in an effort to learn of the specific method by which a trait is inherited. He collected a very accurate pedigree of polydactyly and drew conclusions about the inheritance of this trait which we know today are correct. He also studied albinism in Negroes and recognized the fact that this trait was inherited in a manner somewhat different from polydactyly.

Not content with this, he undertook actual breeding experiments with animals to test his theories, and thus became one of the first known persons to use

FIG. 2–3 *Pierre-Louis Moreau de Maupertuis, an eighteenth century Frenchman who developed concepts of heredity far in advance of his time.* FROM H. BENTLEY GLASS IN THE QUARTERLY REVIEW OF BIOLOGY.

experimental breeding as a genetic tool. We gain some idea of the extent of these experiments from an account given by Samuel Formey, secretary of the Berlin Academy. "The house of M. de Maupertuis was a veritable menagerie, filled with animals of every species who failed to maintain the properties. In the living rooms, troops of dogs and cats, parrots and parakeets, etc. In the forecourt, all sorts of strange birds. It was sometimes dangerous to pass by the run of these animals, by whom some had been attacked. I was especially afraid of the Iceland dogs. M. de Maupertuis amused himself above all by creating new species by mating different races together; and he showed with complaisance the products of these matings, who partook of the qualities of the males and of the females who had engendered them."

Truly, we must conclude that Maupertuis possessed a keen power of observation and insight plus a sense of scientific reasoning which placed him far ahead of the scientific thought of his time.

Encasement Theory of Bonnet

Charles Bonnet, born in Geneva in 1720, differed with Maupertuis on the idea of the seminal particles uniting by pairs to form the body parts. He thought it ridiculous to argue that the organic entity which we call an animal could arise from an amorphous semen, but rather that there must be some organization to begin with. In this reasoning he was right, of course, but in seeking an explanation for some organization he adhered to the preformation concepts, and he believed, further, that each preformed embryo contained the still smaller preformed embryos of the generations to come! Other men of his time also held this doctrine, maintaining that every female contains within her body miniature prototypes of all the creatures which will ever descend from her, one generation within the other somewhat like a series of Chinese boxes. This was known as the **encasement theory.**

Bonnet, however, seemed to realize that the embryos within one another were not perfect replicas of the animals which they were to form. He expressed the view that, instead of miniature individuals, there were particles which had been preorganized in such a manner that the individual could develop from them. This is much like the view of Maupertuis, except that Bonnet thought of the particles being within a reproductive cell of one parent, while Maupertuis thought of them as being present at large in the semen from both parents. The truth, as we understand it today, actually lies in a synthesis of the two ideas: the hereditary particles come from both parents and are contained within the reproductive cells.

Epigenesis

During the latter part of the eighteenth century a German investigator, Caspar Friedrich Wolff, did extensive work on the development of the chick embryo and showed that there is no preformed individual in the egg — that the organs of the body develop in a logical sequence from undifferentiated material. He proposed, instead of the preformation theory, that the germ cells contain certain definite but undifferentiated substances which, after fertilization, become organized into the various complex body organs which go to make up the adult. This idea was referred to as **epigenesis.** This theory is similar to our present day gene concept of heredity.

Lamarck

Having thus achieved a rather accurate knowledge of the nature of the conception and development of the embryo, biological scientists turned their attention toward a more serious consideration of the causes of heredity and variation, and (as a natural outgrowth of such studies), to evolution. Jean Baptiste Lamarck, born in France in 1744, proposed theories which had a tremendous impact on scientific thinking during the nineteenth century. He is best known for his proposal of the significance of "use and disuse" and the **inheritance of acquired characters.** We can summarize his conclusions in his own words: "It is not the organs — that is to say, the form and character of the animal's bodily parts — that have given rise to its habits and peculiar properties, but, on the contrary, it is the habits and manner of life and the conditions in which its ancestors lived that have in the course of time fashioned its bodily form, its organs, and its qualities." He held, for instance, that moles have lost their sight as a result of living underground, that the antbear has lost its teeth simply because it began swallowing its food whole and ceased using its teeth. Conversely, he believed that the wading birds have acquired their long necks through stretching them in their search for food, and that the webbed feet of frogs and swimming birds have developed because these animals have stretched their toes, and the skin between them, in their swimming activities. Lamarck even went so far as to say that if a number of children were deprived of the left eye at birth and these bore children of one another who in turn had their left eyes removed, and so on, eventually we would have a race of one-eyed people. He also believed that climate, geographical conditions, and abundance of food have an influence on animals, causing new organs to appear and old ones to disappear or become modified. In this way evolution has occurred. Lamarck recognized a supreme deity as the creator of life, but he believed that his greatness was evidenced by the fact that he had created life in such a way that it had within itself the power to develop into its present profuse multiplicity without divine control or interference.

To summarize, we may say that Lamarck's theory proposed that an animal's needs determine its desires; its desires determine the use and disuse of its body parts; use or disuse brings about modifications of those parts; and these modifications are in due course inherited. While today most scientists realize the falsity of this line of reasoning, the idea is adhered to by many who have not had genetic training.

Darwin's Contribution

In the year 1809 a baby was born in England who was to have the most profound influence on the subject of heredity and evolution of anyone in the nine-

FIG. 2–4 *Charles Darwin, famous Englishman best known for his advocation of the importance of natural selection in evolution.* SCIENCE SERVICE.

teenth century. This was Charles Darwin, who was trained as a clergyman and early accepted the traditional dogmas on the origin of the species. On his famous voyage around the world, however, Darwin made observations which conflicted with the idea that each species was created in exactly the form in which it exists today. He thoroughly disagreed with Lamarck's theory of the modification of the species — "Heaven forfend me from Lamarck's nonsense of a tendency to progression." Still, his observations indicated that there must be some method of change of species, and he set out to find the explanation for it. Pigeon breeders showed him how, through selection of the parents, they could achieve modification in the progeny at will. But how could modification be obtained in nature? At this stage of his speculations, Darwin ran across Malthus's book on the overproduction of offspring and the necessity that a large proportion of these offspring should perish. This was it — there is variation within any natural population, overproduction causes a struggle for existence, and only the most fit survive. Thus there arises a natural selection which tends to propagate those characteristics which make the organisms more fit to survive in their environment. With this as a coordinating thought, Darwin wrote his famous book, *The Origin of Species by Means of Natural Selection or the Preservation of Favored Races in the Struggle for Life,* using his many observations as illustrations to support his theory.

In an effort to find an explanation for heredity and variation which would fit into this theory, Darwin proposed the provisional hypothesis of **pangenesis.** According to this, every cell, tissue, and organ produces minute pangenes (gemmules); these are scattered throughout the body by the currents of blood or other fluids and conjoin to recreate those cells, tissues, and organs from whence they came. Reproductive cells also would contain these pangenes, and thus a child represents a blending of the qualities of its two parents. Through this mechanism, acquired characters could be inherited. If a blacksmith used the muscles of his shoulders and arms extensively in the practice of his trade, then pangenes from these body parts would become a part of his germ cells, so Darwin reasoned, and his children would have well-developed shoulders and arms as a result. All of this seems very similar to Lamarck's theory of "use and disuse," yet Darwin refused to acknowledge this fact. In an attempt to apply the theory of pangenesis to plants, Darwin proposed that the pangenes migrated not only to the pollen grains and ovules, but also to the buds of woody plants. This would account for the known fact that a new limb or an entire tree can be produced from a bud which has been completely removed from a tree. To explain the regeneration of body parts by some lower animals, he supposed that an aggregation of pangenes gathered at the region of injury and stimulated the regrowth of the missing member. To explain why, in some cases, a child will show characteristics which appeared in grandparents but not in parents, Darwin assumed that all of the pangenes were not used to produce the embryo and that those left over could remain to enter the germ cells and thus transfer the characteristics to the next generation. However wrong the pangene theory seems today, it must not be forgotten that Darwin was a keen observer and possessed a great power for organizing facts. Many of his conclusions were sound and his work as a whole is of great value to modern science. His hypothesis of pangenesis, however, makes clear the fact that he was not a natural scientist but rather a speculative philosopher who never attempted to devise practical experiments to prove or disprove his speculations. In spite of this, however, his views were widely championed by the leading scientists of the time.

Weismann's Germ Plasm Theory

Toward the latter part of the nineteenth century a German biologist, named August Weismann, began to question certain of Darwin's contentions. He studied one-celled animals and noted the potential immortality of their protoplasm. As these animals divide by fission, a part of their protoplasm is passed to each of the two individuals which are produced. This in turn enlarges through the process of growth and in time is divided and passed to the next generation. Thus there is an indefinite continuity of protoplasm. Why could not something of this nature be true in the multicellular animals? With this as a beginning, Weismann formulated the **germ plasm theory** of heredity, according to which all multicellular organisms possess a special germ plasm which preserves itself by repeated divisions as do the one-celled animals. A child is produced by a mingling of the germ plasm from the two parents. Through repeated divisions, this germ plasm produces the various organs of the child's body, but a portion of the germ plasm remains isolated and undifferentiated and is carried by the child, to be passed on to future generations. This germ plasm is in no way influenced by the activities of the child or by variations in the surrounding environment. According to this interpretation, the body consists of two distinct parts, the **somatoplasm,** which makes up all of the body organs with the exception of the reproductive cells, and the **germ plasm,** which is set apart from time of early embryonic development solely for the purpose of reproduction. The cells of the somatoplasm become highly differentiated during the formation of the complex organs of the body and lose their capacity for reproduction, but the cells of the germ plasm remain undifferentiated and retain their power to generate new life. The germ plasm thus goes on in a continuous stream from generation to generation, an immortal flow of protoplasm, existing somewhat in the manner of a parasite on the succession of somatoplasmic bodies which it creates. Figure 2–5 gives a graphic portrayal of the germ plasm theory as contrasted with the theory of pangenesis.

In an effort to prove his theory, Weismann conducted many experiments, primarily consisting of mutilations of experimental animals to test the possibility that such acquired characteristics might be inherited. In one of his best-known experiments he cut the tails off a group of mice for twenty-two generations, yet, when the tails were allowed to grow out in the twenty-third generation, they were fully as long and well developed as the tails of mice in control groups whose tails had never been cropped in any generation. This seemed to disprove the theory of pangenesis. For how could pangenes for tails migrate to the germ plasm of mice when their ancestors had had no tails for many generations past? Weismann agreed with Darwin, however, on the importance of natural selection and believed that this factor alone was sufficient to account for the variations among living things without the necessity of any method for the alteration of the germ plasm itself. By this time, the chromosomes of cells had been discovered, and Weismann believed that the hereditary-determiners were located in the chromosomes. He deserves great credit for dispelling beliefs in the theory of pangenesis and the inheritance of acquired characters, and for substituting for these beliefs the

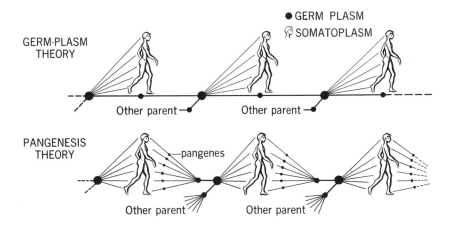

FIG. 2–5 *Weismann's concept of the continuity of the germ plasm compared with Darwin's theory of pangenesis. Weismann believed that the germ plasm continued in a stream from generation to generation, producing somatoplasm, but was in no way influenced by it. Darwin's theory, on the other hand, held that the germ plasm produced the somatoplasm which in turn produced the germ plasm from which the next generation was formed.*

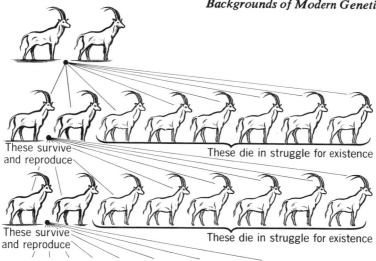

FIG. 2–6 *Darwin's concept of natural selection. A pair of antelopes living in Africa will produce about eight offspring, on the average, during their lives. Of these, only two, on the average, will survive, the others dying as prey to carnivorous animals, through disease, through competition for food and water, or for other reasons. The surviving pair will produce about eight offspring, and the same elimination of the less fit will be repeated.*

idea of the continuity of the germ plasm which carries the hereditary units. His views on the isolation of the germ plasm, however, are contrary to many now-known biological facts and have failed to stand the tests of time. In an effort to make his views universally applicable, he went off into flights of fancy in an attempt to correlate such facts as regeneration in the lower animals and the rooting of cuttings in plants with the isolation of the germ plasm.

De Vries' Mutation Theory

Shortly following the work of Weismann, a Dutchman educated in Germany, Hugo De Vries, performed some significant experiments on the evening primrose, *Oenothera lamarckiana*. While teaching at Amsterdam, De Vries noticed a number of unusual forms of evening primroses growing among the more common types in the outlying meadows. Among these were dwarf plants with unusual leaf types, plants with double petals, and other variants. These interested him greatly, and he transplanted the rosettes from a number of them into his own garden. When he planted seed from these variant forms he found that they bred true, and he therefore reasoned that these must be new species which had arisen from the old. He thought that some of the old species had suddenly exploded and given rise to these supposedly new ones. In 1901, De Vries proposed his **mutation theory** — that living organisms occasionally but regularly produce new types of offspring through sudden changes (mutations) in the hereditary mechanism. These mutations, he believed, were a much greater factor in evolution than the small, gradual changes which Darwin thought to be so important. In other words, it was evolution through a series of sudden jumps rather than a gradual but continuous series of small changes. Darwin had observed the appearance of such mutants in some species of animals and referred to them as "sports," but he failed to attach any great significance to them. Further experiments have shown that the variants which De Vries obtained in the evening primrose were not true mutations as we use the word today. Neither were they new species. But the mutation theory was sound. We know now that sudden changes in the units of heredity do take place and produce new types of organisms and that these changes are very important in providing new characteristics for selection. Quite aside from his own great contributions to the science of genetics, De Vries deserves our gratitude for his part in the rediscovery of the highly significant work of an obscure Moravian monk, Gregor Mendel.

Mendel, "The Father of Modern Genetics"

Mendel was one of those rare geniuses who would not be denied by the unfavorable circumstances of the life which confronted him. Born in 1822 of peasant parents in Moravia (now a part of Czechoslovakia), he displayed a great love of learning during his early school years. He later entered the Institute at Olmutz, but poor health and financial conditions forced him to drop out. It was at this point that he decided to join the Augustinian monks in the monastery at Brunn, for he felt that this would give him an opportunity to pursue his studies without the press of financial worries. After a brief assignment as a parish priest, an occupation for which he was obviously unfitted, he was sent to Vienna to study to become a schoolmaster. At the University of Vienna he spent 3 years studying mathematics and natural history, a happy combination of subjects which contributed to the precision of his later experiments and conclusions. He then returned to Brunn and began a very successful career as a teacher of science in the high school. At the same time he began his famous experiments with garden peas on a small plot of ground beside the monastery building. In one of his papers Mendel states that he was led to undertake these experiments because of his observations on the results of hybridization of ornamental plants to produce new varieties. He noted the striking regularity of the results obtained in the hybrids, and this led him to wish to find out what types of offspring would be obtained if these hybrids were crossed to yield further generations. He chose garden peas as a subject for his experiments only after a careful consideration of a large number of possible plants. There were several reasons for his choice. Garden peas were obtainable in quite a number of pure-breeding varieties; the reproductive organs were inclosed within the petals, hence they were normally always self-pollinated and there was no chance for accidental hybridization; and, finally, the hybrids which resulted from a cross between two varieties were completely fertile. Having made this decision, Mendel obtained seeds of thirty-four different varieties from a number of different sources. From these, he selected twenty-two varieties which he used during the course of his experiments. These extended over a period of 8 years.

FIG. 2–7 *Gregor Mendel, Moravian monk who discovered the basic principles of genic inheritance, which form the foundation of modern genetics.* COURTESY ROYAL HORTICULTURAL SOCIETY, LONDON.

Mendel's great discoveries were made possible by the meticulous care with which he planned his experiments, his careful records of all phases of his work, and the mathematical precision with which he analyzed his results. Also, he continued his experiments long enough to accumulate data in large enough quantities to have statistical significance. For instance, in one of his crosses of plants having yellow, round seed with plants having green, wrinkled seed he obtained a total of 556 seeds for analysis in the second generation. This was a sufficient number to provide a significant ratio. Other people had performed hybridizations of plants long before Mendel's time, but none had discovered the principles of heredity which Mendel uncovered. Camerarius had conducted experiments with plants in 1694 and had found that reproduction of plants was sexual in nature, involving the union of pollen and ovules. In 1760 the German botanist, Koelreuter, had crossed two species of tobacco plants by artificial transfer of the pollen from one to the other. He also performed other plant hybridizations and noted the consistency among the first generation offspring. Knight, early in the nineteenth century, had done artificial plant hybridization and had noted the fact that some characters from a parent always showed in the first generation (were dominant), while other characters did not show (were recessive). It remained for Mendel, however, to work out the details of the process by which hereditary factors were transmitted through the generations.

Details of some of Mendel's results are given in Chapters 5 and 6. Suffice to say at this point that he discovered the fundamental basis for the inheritance of characteristics which holds true for all forms of life. Mendel presented the results of his eight years' work before the Society for the Study of Natural Science in Brunn in 1865. In spite of the great significance of his findings, the world took little note of his paper. Darwin's *Origin of Species by Means of Natural Selection* had appeared in 1859, and the scientific world was immersed in argument over this monumental work. As a result, even those most keenly interested failed to realize that the answers to many of the unexplained problems which Darwin had raised were to be found in this work on garden peas. Mendel conducted other experiments on plants and also on bees; but, unfortunately for the scientific world, he was elevated to a position as head of the monastery. He spent his last years embroiled in a conflict of church and state and died in 1884 without ever knowing the great veneration which was to be his as "the father of modern genetics."

Modern Geneticists

It was not until the beginning of the present century that the results of Mendel's work and their great significance were to become known to the scientific world. At this time, several investigators, including De Vries, began checking the literature in an effort to explain some of their own results from hybridization, and they soon recognized the monumental importance of the work Mendel had done. Most of our present-day knowledge on genetics dates from the time of this rediscovery.

Within the scope of this brief discussion we cannot begin to give credit to all of the many outstanding men who have built our genetic knowledge during the present century. We will, however, have occasion to refer to the work of many of them in later chapters. Suffice it here to mention some of those who are outstanding for their pioneer work at or near the beginning of this century. The Danish botanist, Wilhelm Johannsen, is well known for his pure-line concept, developed as a result of his work on beans and indicating that selection for the size of beans was effective for only one generation. After that, the lines bred true regardless of selection. There was variation within each pure line, of course, but this variation was constant no matter whether large or small beans from the pure line were planted. Johannsen also coined the word **"gene,"** from which has been derived the word "genetics." Briefly, a gene is a single unit of heredity, similar to the factors of Mendel, and we know that the genes are located on the chromosomes within the cells. We will learn considerably more about the nature of a gene in the chapters to come. Also, we will have more to say about Johannsen's experiments with beans.

Other advances were made in other European countries. In Sweden H. Nilsson-Ehle achieved great success in the practical application of the newly discovered principles of genetics in the improvement of plants. He made important discoveries about the nature of heredity in his work on wheat. Later we will learn more of this work also. In Germany Erwin Baur and Carl Correns are credited with the discovery that all hereditary factors (genes) do not segregate independently as described by Mendel, but that some tend to hang together in subsequent generations. In England W. Bateson and R. C. Punnett did work on sweet peas and worked out the principle of linkage between genes which was first discovered by Correns and Baur.

In the United States, the early investigations in heredity were centered at Columbia University. E. B. Wilson was the pioneer in cytogenetics, the study of the cell in heredity, and made many important discoveries concerning the relation of the chromosomes of the cells to heredity. Thomas Hunt Morgan, also a professor at Columbia, was the first to begin laboratory breeding experiments with the little fruit fly so well known to geneticists of today, *Drosophila melanogaster*. To attempt to go further and to name more recent geneticists and their contributions would surely do injustice to the many great men whom we have not space to mention. Throughout the rest of this book, however, we shall refer from time to time to the work of some of the more recent workers as a tribute to their discoveries. Like any field of scientific study, genetics has grown by accretion, and each new worker builds on what has gone before.

PROBLEMS

1. Why did all the theories of heredity before the seventeenth century ignore the possible part played by individual sperms as carriers of heredity particles?
2. What was the basic error in the theory of Pythagoras and how did the theory of Empedocles seek to correct this error?
3. How did the experiments of Harvey disprove the long accepted theory of Aristotle?
4. How does the theory of preformation differ from epigenesis and how was the latter viewpoint proved to be correct?
5. Maupertuis proposed a number of theories concerning heredity. List those which were essentially correct and those which have been disproved as our knowledge of genetics has expanded.
6. If the skin of a white person is exposed to the sun over a period of time he will develop a tan. Races of people who live in regions of the world where there is much bright sunlight are dark-skinned through their heredity. How would Lamarck have explained this?
7. How would Darwin have explained the above observations?
8. How would De Vries have explained the above observations?
9. From a geranium plant we can take a portion of the roots, a portion of the stem, and even a portion of a leaf and from each of these we can grow an entire geranium plant. How does this fact agree with the theory of the isolation of the germ-plasm? Explain.
10. How do garden peas stand up to the requirements for good experimental organisms for genetic breeding as discussed in Chapter 1?

3

Two Cells from One

OF ALL THE FASCINATING PROPERTIES OF LIFE there is none more remarkable than the ability of living matter to reproduce itself. Let us consider an amoeba, moving slowly along on a rock at the bottom of a pond. It throws out pseudopodia, engulfs food, digests the food, and throws off the undigested residue. Yet, a part of the food remains behind to become an integral part of the protoplasm of the organism. This is growth, but growth cannot continue indefinitely. Eventually a size will be reached beyond which the amoeba cannot grow if it is to maintain the efficiency of its life processes. Then it happens — the amoeba ceases its normal movements and certain vital internal changes begin to occur. The contents of the nucleus change in appearance; certain bodies not previously visible now become so; the nucleus begins to elongate; finally the nucleus separates into two parts, and this is followed by a splitting of the entire body into two equal parts, each of which contains one of the new nuclei. We say that the amoeba has divided, but it has not divided in the sense that an apple is divided when it is cut into two equal halves. Rather it has duplicated itself. True, each of the resulting organisms is only about one-half the size of the original, but each of them is a complete organism bearing all the potentialities of the one original amoeba. Through growth again, each of these organisms can attain the size of the original, and each has all of the units of heredity possessed by the original one. Each has the power to duplicate itself in the manner just described.

Before we can understand heredity, we must gain some understanding of this process of cell duplication, which is similar in all forms of life. Just what takes place in the cell before and during division which produces two cells from one?

36

The Gene — The Unit of Heredity

To begin with, let us consider those submicroscopic units of heredity which we call **genes.** These are contained within the nucleus of the cell in rather large numbers. While we have no way of obtaining an exact count, there are genetic methods of obtaining reasonable estimates of their number. In the fruit fly, for instance, which has been investigated by geneticists more extensively than has any other form of life, there is good evidence that the number of genes in each normal body cell is around 10,000. In man there may be four times as many in each cell. You are what you are, to a large extent, because of the nature of these genes which are present in the cells of your body. If your eyes are brown, it is because you carry genes in your cells which are responsible for the production of brown pigment in the iris of the eye. If your hair is naturally curly you may thank your ancestors, for through them you have received genes which cause the hair to grow out in such a shape that it is curly. If you have artistic talent which makes it easy for you to learn to draw and paint, this again is due to the fact that you have inherited a gene combination which imparts this ability. We are the product of our genes, with due allowance for the effect of environment on the attainment of the potentialities contained within the genes.

Gene Duplication

When an amoeba divides (duplicates itself), it is necessary that each of its genes also duplicate itself, if the two amoebas which result are to possess all the properties of the original animal. Gene duplication, therefore, must precede cell division. The one original cell from which you have descended contained, let us say, 40,000 genes, each with some specific function which it carries out in the formation of a human body. Each of your body cells today contains some 40,000 genes which are of the same types as those present in the original cell from which your body developed. (There is an exception in the case of mature red blood cells which have no nucleus, but even these each have a nucleus and genes while immature.) This condition could only be possible if there were an exact

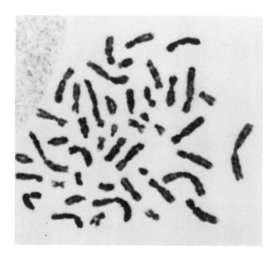

FIG. 3–1 *Human chromosomes as they appear in a cell which has been crushed flat. This cell was in early prophase, but each chromosome is already double and held together by a centromere. This photograph is of particular historical interest because it was one of the first to be made which demonstrated clearly that the human chromosome number is 46.* COURTESY JOE HIN TJIO.

duplication of the genes preceding each cell division, and provided that one of each pair of duplicate genes was incorporated within the nucleus of each new cell produced. Here we have a major problem of cellular mechanics on our hands — how to achieve duplication and accurate distribution to the daughter cells of each of the 40,000 tiny particles which form only a small part of the microscopic nucleus.

A solution to this problem is found in the arrangement of the genes on **chromosomes.** A chromosome is a thread-like body containing many genes as well as various protein components. It is much easier for a cell to handle these chromosomes during cell division than it would be to handle the many thousands of individual genes. Each species of plant or animal bears its genes on a specific number of chromosomes, which remain constant in number from generation to generation. Table 3–1 shows the chromosome number of certain well-known plants and animals.

TABLE 3–1

The Diploid Chromosome Number of Some Well-Known Forms of Plant and Animal Life

Organism	Diploid chromosome number
Ascaris univalens (parasitic worm)	2
Drosophila melanogaster (fruit fly)	8
Garden pea	14
Onion	16
Corn	20
Opossum	22
Bullfrog	26
Ambystoma (tiger salamander)	28
Honey bee	32
Domestic swine (hog)	38
Mouse	40
Man	46
Monkey (cebus)	54
Crayfish	200

Genes are made of long double strands of a complex substance known as **deoxyribonucleic acid, DNA.** When the time comes for gene duplication, these two strands separate from one another and each strand attracts to itself the parts which were present in its former partner strand. These spare parts are present in the cell as individual small units. In this manner two complete genes are formed, each an exact duplicate of the other. Details of this process are given in Chapter 17. The amount of DNA in a cell can be measured by delicate microtechniques and the time of actual gene duplication can thus be determined. Such studies by Daniel Mazia indicate that, in human cells growing in tissue culture, this doubling of the quantity of DNA occurs about 5 hours before there is any visible sign of a beginning of mitosis. Thus, we know that the genes duplicate during the period which we have called interphase and before we can see the chromosomes condensing in preparation for cell division. Each chromosome, when first seen in the early prophase of mitosis, is actually a double chromosome with two complete sets of genes. Later in the process of mitosis, each chromosome splits in two, forming two chromosomes, each containing a single set of the genes. One of these chromosomes then moves to one edge of the cell and the other moves to the opposite edge. The same is true of the other chromosomes, so two complete sets of chromosomes are formed into two groups. Finally, the cell divides and each of the newly formed cells includes one of the groups of chromosomes. Thus, two daughter cells have been produced, each containing the same number and kind of genes and chromosomes as were found in the cell from which they were formed.

Since an understanding of the mechanism of heredity is dependent upon an understanding of cell division, we shall study mitosis in some detail. It is customary to study mitosis as a series of stages or phases, and we shall do so here, but it is well to keep in mind the fact that the process of cell division is continuous and not a series of jumps from one stage to another. You might think of it in terms of a moving picture, and of the stages which are used for illustrations as merely a series of individual still frames which have been chosen at various points from the thousands of frames that make up the complete picture.

MITOSIS

We shall start our study of mitosis with a cell which is not in any stage of the process. Sometimes this is called the "resting" stage, but the cell may actually be at the height of its metabolic activity. It is perhaps better, therefore, to call this stage in the life cycle of a cell the **interphase** — the stage between mitoses. In typical cells at this stage the chromosomes are very long and are often so thin that they are not easily visible; instead they appear as a diffuse **chromatin network.** Even at this stage, however, the chromosomes are not extended to their full potential length. Rather, they are coiled somewhat, and, since genes are arranged in a linear order on the chromosomes, this coiled condition provides room for a much larger number of genes than if the chromosomes were the same length, but not coiled. It is estimated, for instance, that the completely uncoiled human chromosomes within a single cell would be about 5 feet long if arranged end to end. You can realize how thin these would be, however, when you consider the fact that all of these lie within the nucleus of a single cell which cannot even be seen with the naked eye because of its small size. Even in the interphase coiled state the chromosomes are very thin and are known as **chromonemata.** Each chromonema bears a specialized body known as the **centromere.** This centromere is located at some specific point on the chromonemata, a point which remains constant and is a characteristic of each chromosome. The centromere plays an important role in the separation of chromosomes during mitosis as we soon shall learn.

If the cell is from one of the higher animals or from one of the lower plants, there will also be a body outside the nucleus which deserves mention. This is the

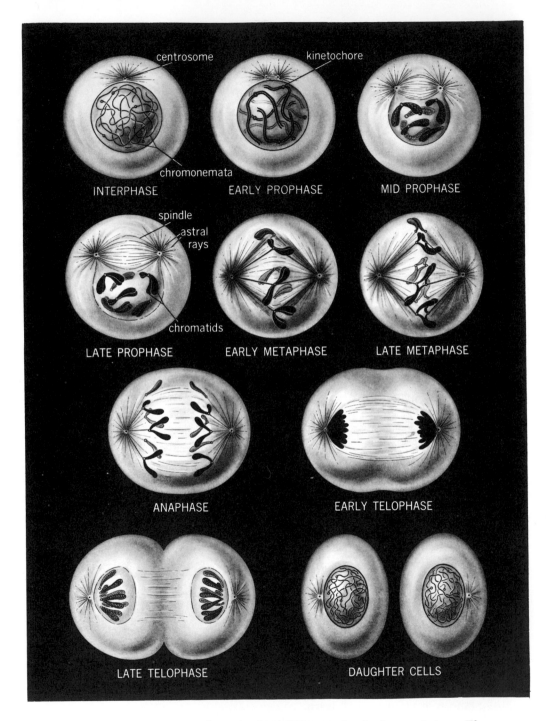

FIG. 3–2 *A number of selected stages of mitosis are shown. The minor coils of the chromonemata are not shown.*

centrosome. It consists of a central particle, the **centriole** (in some cells there are two of these), in a sphere from which rays radiate out into the cytoplasm, the **aster.** We shall see that this centrosome plays an important part in the events which take place during mitosis.

Prophase

The coils of the chromonemata which were mentioned during the interphase are known as **minor coils;** these persist throughout all stages. During the early prophase additional coils in the chromonemata appear, known as **major coils.** Their presence has a tendency to make the entire chromosome shorter and thicker. At the same time a **matrix** begins to be deposited around the double-coiled chromonemata, eventually giving the chromosome much the shape and appearance of a thick sausage. With the use of the ordinary staining reagents the entire chromosome at this stage looks like a uniform, darkly-stained body. By special techniques, however, it is possible to reveal the major coils of the chromonemata within the chromosomes. Also by such refined techniques and by a study of living cells under the phase microscope, it is possible to see not only the major, but also the minor coils in some cells. During the latter part of the prophase the dual nature of the chromosomes becomes more apparent, especially at the ends, where there may be a slight separation of the two halves. We call these two half-chromosomes **chromatids.**

While these changes are taking place in the chromosomes there are other changes occurring outside the nucleus. The centriole divides, if it is not already double, and the two centrioles begin to move apart. At this time the nuclear membrane begins to disintegrate, and between the centrioles a bundle of lines becomes visible which forms the **spindle figure.** These lines seem to be formed by a rearrangement of materials, probably protein chains, into a longitudinal sequence to form fiber-like bodies which make up the spindle. The precise origin of the spindle is difficult to determine, but it seems to be formed primarily from the nucleoplasm, probably with the cooperation of some cytoplasmic components. Also, the **astral rays** which radiate out into the cytoplasm become much more prominent than during the interphase. This entire structure — centrioles, asters, and spindle figure — is known as the **achromatic figure.** The spindle figure moves down to envelop the chromosomes during the final part of the prophase. In cells which do not have centrosomes, the spindle makes its appearance during the latter part of the prophase. The spindle commonly begins development at two poles on either side of the nucleus, just outside of the nuclear membrane. As the nuclear membrane disappears the spindle is completely formed across the central portion of the cell.

Metaphase

This stage of mitosis begins when the chromosomes move to the central portion of the spindle and become arranged in an **equatorial plate** at the equator of the spindle, which is the region equidistant from the two **poles** at either end. If the chromosomes happen to be rather long they may extend out into the sur-

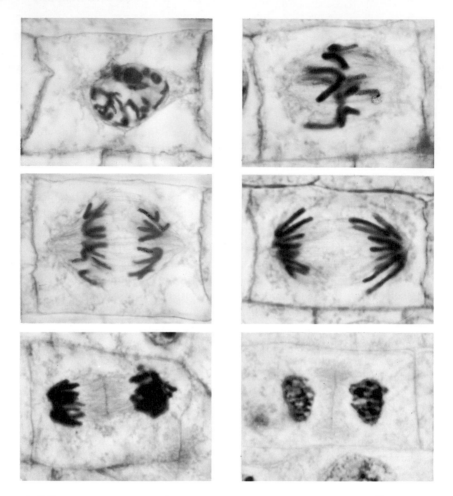

FIG. 3–3 *Photographs of various stages of mitosis of cells in the tip of the root of an onion. From left to right and top to bottom, these are: prophase, chromosomes show clearly, but nucleolus is still visible and spindle is forming; metaphase; anaphase; late anaphase, chromosomes have almost reached the poles; telophase, the cell plate is forming across the center of the spindle; late telophase, chromosomes becoming less distinct and cell plate has almost separated the cell into daughter cells.*

rounding cytoplasm, but the centromeres will always be anchored at the equator of the spindle. Within the spindle, one can see what appear to be fibers, and they are called **spindle fibers,** but delicate microdissection studies indicate that they are not true fibers. It has been suggested that they represent a lengthwise orientation of molecules, indicating lines of force in the spindle. The centromeres now separate and make their way toward the poles at opposite ends of the spindle, so that the two chromatids are pulled apart to form two new chromosomes from each original chromosome.

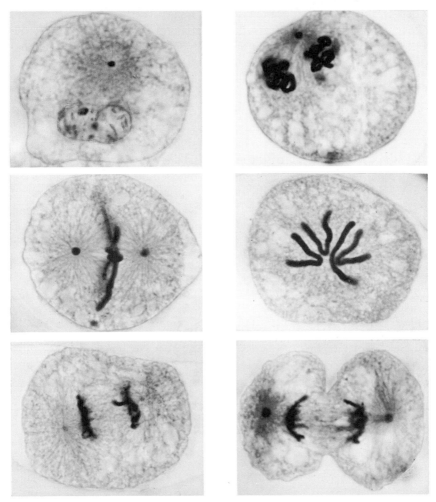

FIG. 3 – 4 *Mitosis in a simple animal as seen in the division of the zygote of a roundworm, Ascaris. From left to right and top to bottom, these are: prophase, the chromosomes of the sperm nucleus and egg nucleus are becoming visible; late prophase, chromosomes are much thicker; metaphase, side view showing spindle figure and astral rays; metaphase, end view showing the four chromosomes in one plane; anaphase; telophase.*

Anaphase

This new phase begins when the chromosomes separate into two distinct groups, and ends when the centromeres reach the poles. Always, the centromere leads the way and the balance of the chromosome is dragged along behind. If the centromere is attached at the end, the chromosome will have a rod-like appearance during this stage; if attached at the center, the chromosome will be V-shaped; if attached between the center and the end, the chromosome is likely to be shaped something like a J. As the centromeres move toward the poles they are seen to follow one of the fibers of the spindle, a phenomenon which supports the idea that the fibers represent lines of force in the protoplasm.

Telophase

During this stage the two groups of chromosomes become organized to form the two new nuclei. In many respects it is an exact reverse of the prophase — the chromosomes become longer and thinner as the chromonemata within them lose their major coils; the matrix around the chromosomes disappears gradually and the chromosomes become more difficult to see as individual units; the spindle figure disappears and the nuclear membrane reappears. While these changes are taking place, the actual division of the cell is occurring. In plants a **cell plate** is formed between the two daughter nuclei, eventually forming cell walls for the two cells. In animal cells there is a **cleavage furrow** which

FIG. 3–5 *Mitosis in a somewhat complex animal as seen in the cells of the early embryo of the whitefish. From left to right and top to bottom, these are: interphase; prophase; metaphase; anaphase; late anaphase; telophase.*

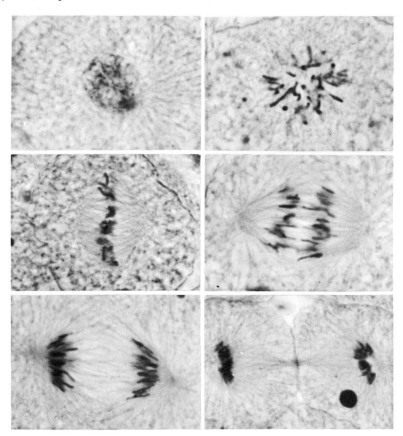

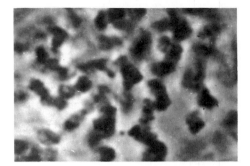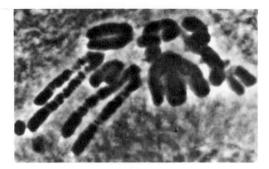

FIG. 3–6 *The coiling systems of chromosomes. The photograph at left shows a highly magnified view of early prophase chromosomes of Triturus, a salamander, in which the coiling is clearly visible. The photograph at right shows a group of chromosomes near the pole of an anaphase in a cell from a grasshopper. As the slide was made a slight sidewise movement on the cover glass caused a slight uncoiling of the otherwise highly contracted chromosomes. These are phase photographs of fresh, unstained chromosomes.*

develops on the outside of the plasma membrane, which constricts to pinch the cell into two parts.

MITOSIS AND HEREDITY

Daughter Cells

The process of mitosis is now complete. We are again at the interphase, but this time with two complete cells in place of the one original cell. The entire process in most cases consumes less than an hour. In so short a time there can be little opportunity for cell growth, so that each of the two cells which have been formed is approximately one-half the size of the original one. Yet — we cannot emphasize it too often — each of these cells has the same number and kinds of genes as were present in the original, thanks to the precision of the process of mitosis which we have just studied. This fact may be demonstrated in a rather dramatic way. If the original cell happened to be a fertilized egg from which a salamander will develop, then each of the daughter cells will, normally, go on to produce one-half of the body of the salamander. If the two daughter cells are separated from one another experimentally, however, each cell will form an entire salamander which will be no different from a salamander which was formed from an unseparated zygote (fertilized egg). Furthermore, these two salamanders will bear identical genes and will, therefore, show identical hereditary characteristics.

Twins and Multiple Offspring

The same principle can be demonstrated in human beings. Normally, a single zygote produces a single individual, but frequently some cells of the early embryo will break away and form a second embryo to yield identical twins. Each of these is a complete individual, lacking none of the characteristics which are present in people who develop from a zygote whose daughter cells have not been separated in this fashion. Each has the same genic make-up which would have been present in the single individual which would have been formed had the embryo remained all in one piece. Since identical twins have identical genic make-up, they make an excellent subject for the study of the effects of heredity and environment. With the same genes, it stands to reason that any differences which are shown will be differences due to environment.

Those rather rare cases where twins have been separated at birth and brought up in different environments are especially valuable for such studies. H. H. Newman has carefully studied quite a number of such cases, and has concluded that heredity is more extensive in its effect on one's total personality than is commonly realized. One should take care in such studies, however, to distinguish between the twins which are identical twins (**monozygotic,** those which arise from a single zygote) and fraternal twins (**dizygotic,** those which arise from two separate zygotes). Identical twins are always of the same sex and show identical inherited characteristics, while fraternal twins may be of opposite sexes and show variations in inherited traits. Fraternal twins are formed when two separate ova break from a woman's ovaries during one month and the two are fertilized by two different sperms. Thus twin children develop-

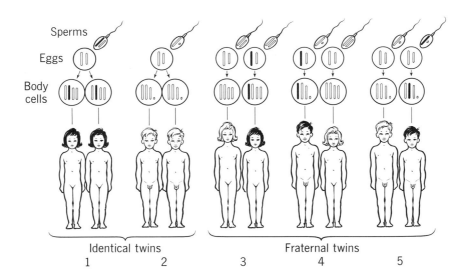

FIG. 3–7 *The difference in the formation of identical and fraternal twins. Identical twins are formed from a single fertilized egg and are of the same sex and the same heredity. Fraternal twins, on the other hand, are formed from two fertilized eggs, may be of opposite sexes, and will show differences in hereditary traits.* REDRAWN FROM CLYDE E. KEELER.

FIG. 3-8 *Identical twins. These two young ladies started life as a single individual, but became separated into two parts early in the embryonic stage. They have identical genes; any differences which they show are induced by their environment.*

ing from these ova are no more alike than brothers and sisters born at different times as far as their genes are concerned. The method of formation of the two types of twins is shown in Figure 3–7.

It is also possible for more than two embryos to arise from a single zygote. Additional groups of cells may break away to yield triplets, quadruplets, with the present record standing at quintuplets. Further, it is possible to have combinations of identical and fraternal offspring at the same time. For instance, two eggs could be fertilized and one young embryo could form two. This would yield triplets — a pair of identical twins and an additional fraternal sibling. The armadillo, which is very widespread in the southern states, has an interesting embryology which results in identical quadruplets. Only one egg is fertilized during one reproductive cycle of the female, but the developing cells typically break into four separate units to give the identical quadruplets.

Embryonic Power of Cells

In the higher animals the cells become so specialized in their functions in later development that beyond the very early stages they do not retain this power to produce entire new organisms, even though they do contain the full complement of genes and chromosomes. In some of the lower animals, however, such as planaria, an entire new animal can be produced from a small portion of the body. We say that such cells retain their embryonic power. In the plants we find that this power is very common, for in plants there is usually some portion of tissue which retains this embryonic power as long as the plant lives. In a tree, for instance, there is a cambium layer in the stems which produces all types of cells. If we cut a twig from a tree and place it in the ground, this cambium can produce new roots and an entire tree can grow from the twig.

These examples are given to emphasize the fact that mitosis is a true duplication of the hereditary elements within a cell. No matter how many times the cell has divided, these elements may exert their influence to produce all the parts of a new organism, provided that the cells have not become so highly specialized that their division can no longer produce other types of cells. For instance, it is unthinkable that a bone cell of a man could be removed and that it would grow into an entire human being. Since the cells must be highly and diversely specialized to produce so complex an organism as a human being, provision for reproduction is made by the isolation of **germ plasm** early in the life of the embryo. The cells in this germ plasm do not become highly specialized to perform other functions, and therefore they retain their power to produce new life. Details concerning the development of these cells are presented in the next chapter.

PROBLEMS

1. A scientific expedition to Africa finds some very large frogs which look similar to our bullfrogs except that they are much larger. A study of the chromosomes of this African frog shows the diploid number to be 30. The American bullfrog has a diploid of 26. On the basis of these results alone do you think it is possible that these are the same species? Explain.

2. How is the process of cell division simplified by having the genes on chromosomes rather than as individual units?

3. Explain how the greatly extended chromosome of the interphase becomes the much shorter and thicker chromosome of the late prophase.

4. In what respects is the telophase similar to the prophase in reverse, and how is it different?

5. Explain how it can be demonstrated by experiment that mitosis achieves a duplication rather than a halving of the cell when it divides.

6. A childless couple has adopted three children, identical twin boys and an unrelated baby boy of about the same age. The children are now in their teens. How would you study this family in an effort to determine the relative effects of heredity and environment?

7. In armadillos, the pattern of bands around the body shows considerable variation, but it is always the same among the offspring which arise at any one birth. Explain.

8. Within the lymph of man there are certain amoeboid cells which accumulate at the site of a wound and become transformed into the various kinds of tissue which are required for the healing process. It has been suggested that if we removed one of these cells to a nutrient medium under proper conditions, it might grow into an entire human being. Why would such a cell be more likely to do this than some of the other cells of the body, such as those found in the bone, muscle, etc.?

4

Bridge of Heredity

THE CONTINUATION OF LIFE FROM ONE GENERATION to another over the slender bridge of protoplasm contained in the reproductive cells is one of the great wonders of the world. Each of us — male or female, short or tall, handsome or homely, brilliant or dull — began life as a tiny pinpoint of protoplasm, the fertilized egg. Contained within this cell were all of the hereditary potentialities which caused the development of a mature human being. The extreme smallness of this bridge of heredity can perhaps be better visualized if we consider all of the human beings living on the earth today. There are over 3 billion of them at the time of this writing, yet the delicate thread of genetic material which ties all of these people to the generations of the past could be contained in a mass no larger than two aspirin tablets. All of the fertilized eggs from which the present inhabitants of our globe were produced could be placed in 12 ordinary drinking glasses. These fertilized eggs contain much cytoplasm in addition to the genes. This, we must admit, is a very slender and delicate bridge over which all of the qualities characteristic of human beings have been transmitted. Yet this is the only hereditary link — the genes in these fertilized eggs must carry all of the qualities that produce organisms which not only are human beings, but human beings which have characteristics of the races from which they have descended and the characteristics of their immediate ancestors within the race. This chapter will take up the method of production of the components of the bridge of heredity.

ONE-CELLED ORGANISMS

One-celled organisms have no division of the body into reproductive and somatic tissues; hence their bridge of heredity is somewhat simpler than that in the multicellular forms of life. Cell division and reproduction are one and the same thing in one-celled organisms. Each time a cell divides, two new individuals are created from a single parent. The genes which are transmitted to each of the two new cells represent the bridge of heredity. Even in the forms which have no definite nucleus, such as the bacteria and the blue-green algae, there is evidence of some internal division, a duplication and separation of genic material, preceding the division of the cytoplasm.

Such a method of reproduction, however, tends to yield a monotonous dearth of variety. It is somewhat as if people had an indefinite succession of identical children. But this monotony would not in itself be of any great significance were it not for the fact that evolution and the continued existence of the species are dependent upon variety. There can be no natural selection of any significance

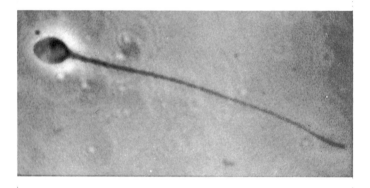

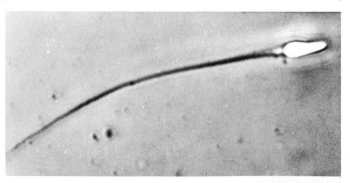

FIG. 4-1 *The male bridge of heredity. These are living human sperms, as seen under the phase microscope at a magnification of about 2,500X. In the lower photograph the head of the sperm is turned so that it is seen in a side view. All of the inherited characteristics which a child receives from his father are passed by way of genes contained in the tiny head of the sperm.*

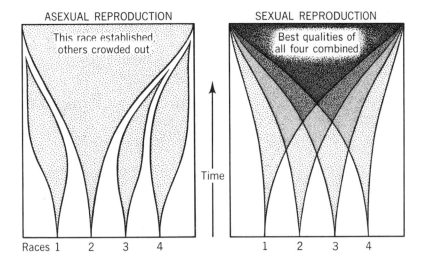

ASEXUAL REPRODUCTION

This race established; others crowded out

SEXUAL REPRODUCTION

Best qualities of all four combined

Time

Races 1 2 3 4

1 2 3 4

FIG. 4–2 *The advantages of sexual over asexual reproduction in natural selection. In asexual reproduction, bad qualities survive along with good in whichever race has the overall advantage in a given environment. In sexual reproduction, there is an opportunity for the mingling of characteristics and for the most advantageous qualities of the various races to be selected and yield one superior to all.*

if all the individuals to be selected from are identical. Outside of the comparatively rare occurrence of mutations, there can be no variety through indefinite reproduction by fission. It is therefore interesting to discover that, even though fission is a thoroughly efficient method of reproduction, many of the one-celled forms of life also have some method of sexual union in which there is a mixing of genic material and a resultant production of new gene combinations which, in turn, cause variety in the offspring. This can hardly be called sexual reproduction because the sexual union does not increase the number of individuals, but it achieves the same variety in genic combination as sexual reproduction.

Recent research indicates that such sexual exchange of genic material may be much more common in the one-celled and even simpler forms of life than was previously recognized. For many years it was taken for granted that there was no mechanism of sexual union among the bacteria. Yet a number of years ago the work of Lederberg and Tatum showed that such exchanges not only can but do occur in the case of at least one strain of a bacterium, *Escherichia coli.* Like other forms of life, this organism must have certain essential food elements in order to survive. We will designate four of these as A, B, C, and D. Normal strains of *E. coli* can synthesize all four of these elements from other components in their food media. There is one mutant strain, however, which has lost the power to synthesize A and B, and which must therefore be grown on media which contain these two substances. Another strain has lost the power to synthesize C and D and hence cannot survive unless its medium contains these two elements. These two strains were mixed together and grown on media containing all four elements. Now when some of these mutants were transferred to media which did not contain any one of the four elements, most of the bacteria, as would be expected, soon died and failed to grow into colonies. There were a few, however, which lived and reproduced normal colonies. Hence we can conclude that there must have been some exchange of hereditary traits between our two strains of *E. coli,* for neither of the two could synthesize all four elements at the time they were placed together, and only by developing the power of synthesis could any survive at all. In the little one-celled animal, *Paramecium,* there is a form of sexual union known as conjugation, in which genes are exchanged. In *E. coli* a conjugation bridge forms, and a part of a chromosome from one bacterium is injected into the other.

SEXUAL REPRODUCTION IN ANIMALS

In all the multicellular forms of animal life the pattern of sexual reproduction is based upon the production of two different kinds of gametes, sperms and eggs, and the subsequent union of these to form zygotes, from which new organisms are produced through the process of cell division and growth. Since cell division occurs by mitosis, all of the body cells contain the same number and kinds of chromosomes as were present in the original zygote. Still, if the gametes which are produced contain the full number of chromosomes and the next generation is formed by a union of gametes, then we would be faced with a great difficulty — there would be a doubling of the number of chromosomes in each generation. This, of course, could not go on for long. There would soon be so many chromosomes in the cells that mitosis could not take place normally. It

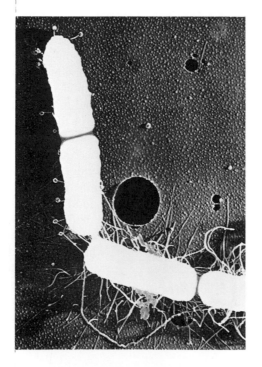

FIG. 4–3 *Sexual union, conjugation, in bacteria, Escherichia coli. This is an electron photomicrograph at a magnification of about 20,000X. To distinguish between the two strains, one was first placed in some mucoid threads which adhered to the outside of the cell. The other strain was infected with a mild virus, which can be seen as tadpole-shaped virus particles clinging to the sides of the cells. A portion of a chromosome from one cell is passed across the protoplasmic bridge into the conjugating partner.* COURTESY T. F. ANDERSON, E. WOLLMAN, AND F. JACOBS.

is evident that there must be some mechanism by means of which the number of chromosomes is kept constant from generation to generation. If either the egg or the sperm were to furnish all the chromosomes, this would be accomplished; but we know this is not true, since there is equal inheritance from each parent. Instead, there is a reduction of the chromosome number during the formation of the gametes, so that each gamete has only one-half the number of chromosomes found in the body cells. This is accomplished by a special type of cell division called **meiosis.**

Meiosis and Gene Reduction

Before we can understand meiosis, however, it is necessary for us to go back to a consideration of the genes and chromosomes. We have learned that genes are tiny strands of DNA which are arranged on the chromonemata of the chromosomes in a linear order. Furthermore, each gene has a specific function to perform in the development and maintenance of a living organism. At this point in our study it is well to learn that there are regularly two of each kind of gene in each cell. These two genes are found on two different chromosomes within the cell, and the two chromosomes will always be similar to one another. The chromosomes within any cell may be of different lengths and shapes, but for every chromosome of a given length and shape there is, as a rule, another like it. These are called homologous chromosomes, and each bears the same type of genes in the same sequence. For instance, if you carry a gene near the center of a V-shaped chromosome which affects the deposition of pigmentation in your eyes, there will always be another V-shaped chromosome of the same kind which will carry a gene for eye pigmentation at a corresponding locus. This is true for all of your chromosomes (with an exception which will be studied in connection with sex determination). Thus the human chromosomes may be grouped into twenty-three pairs. These twenty-three pairs are found in all of the body cells except the reproductive cells; the mature human sperm and egg each contain only twenty-three single chromosomes, one of each of the different kinds. This condition is brought about through the process of meiosis, which reduces the chromosome number to one-half just before the gametes are produced. The production of sperms and eggs differs somewhat and we will study the two separately. First, we will take up sperm production, or spermatogenesis, and then compare this with egg production, or oogenesis.

Spermatogenesis

In man there are forty-six chromosomes in the body cells. This is called the **diploid number (2n)** (Gr. *diploos,* double) since there are two of each type of chromosome. The term **haploid (n)** (Gr. *haploos,* single) is used to refer to the chromosome number which represents only one of each type of chromosome. In man this would be twenty-three. The testes of a man contain tubules which bear germinal epithelium, a layer of tissue made up of cells which may form sperms. These are spermatogonial cells, or just **spermatogonia.** Like the cells of the somatoplasm, they contain forty-six chromosomes each. From

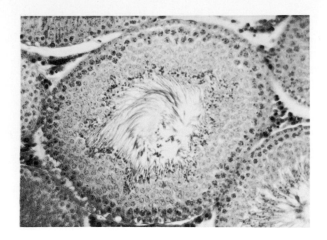

FIG. 4–4 *Cross-section of a human seminiferous tubule showing spermatogenesis. The germinal epithelium is around the outside of the tubule. Just inside this layer are the large primary spermatocytes, next are the secondary spermatocytes, then the dark spermatids, and, finally, the mature sperms can be seen coming off into the interior cavity of the tubule.*

time to time certain of these cells begin migration toward the center of the tubule and enlarge somewhat to become **primary spermatocytes.** These are the cells which are to divide by meiosis. As these cells go into the prophase an important difference from ordinary mitosis becomes apparent. The chromosomes are in pairs — each chromosome lies beside its partner, held by cross connections between them. Each chromosome has also become duplicated — the amount of DNA in the cell doubled in the interphase. The centromeres do not duplicate, so this gives a unit consisting of four chromatids with two centromeres, which is called a **tetrad.** These chromatids go into the metaphase plate and the centromeres (still undivided) begin their journey to the poles, pulling the paired chromosomes apart. Cell division follows, giving two cells, each with twenty-three chromosomes. Thus the chromosome number is reduced from the diploid to the haploid; but we must remember that each of these chromosomes consists of two chromatids which failed to separate during the division. There is a second division in meiosis which allows the chromatids to separate.

To survey the first division of meiosis, we find that the primary difference between this and ordinary mitosis lies in the fact that here the homologous chromosomes pair and separate, while in mitosis the chromosomes line up singly and divide, and a full diploid number goes to each daughter cell.

The two cells which are formed by the first division of the primary spermatocyte are called **secondary spermatocytes.** As these go into the prophase of the second division of meiosis, there is a division of the centromere without a division of the chromonemata — just the reverse of the events in the prophase of the first division. These chromosomes, now containing two centromeres as well as two chromatids, are called **dyads.** They arrange themselves on the equator of the spindle much as do the chromosomes in ordinary mitosis. The centromeres separate and pull the chromatids apart as they migrate to the poles at opposite ends of the spindle. Cell division follows to give two cells, which are called **spermatids,** each of which contains twenty-three single chromosomes.

Let us review the events of meiosis as illustrated by spermatogenesis. During the first division of meiosis there is a duplication of the genes and of the chromonemata which form the chromosomes. This gives cells which have only

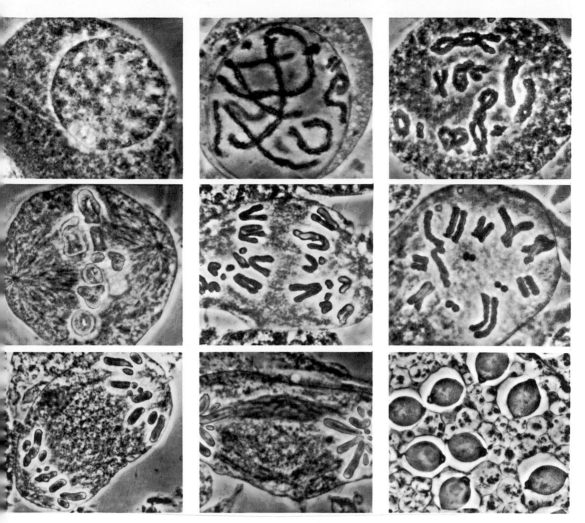

FIG. 4–5 *Spermatogenesis in the grasshopper. These photos are of living cells taken from the follicles of the testes without any staining. Top row: primary spermatocyte in interphase before meiosis begins; early prophase, the chromosomes show as long slender threads which are already paired; late prophase, the chromosome pairs have become shorter and thicker, the double nature of each chromosome being visible at some places. Middle row: metaphase, the chromosomes have reached their maximum shortening and are lined up in the center of the spindle figure ready for separation; anaphase, the chromosomes have pulled apart and are moving to the poles, the chromatids of each chromosome having become separated everywhere except at the centromere end, making the double nature of each chromosome evident; late prophase of the second division of meiosis, the double nature of each chromosome again clear. Bottom row: anaphase of second division, the two chromatids of each chromosome have pulled apart, making single chromosomes which are moving toward the poles; telophase, chromosomes have reached the poles; spermatids, the daughter cells produced by meiosis, are undergoing transformation into sperms.*

55

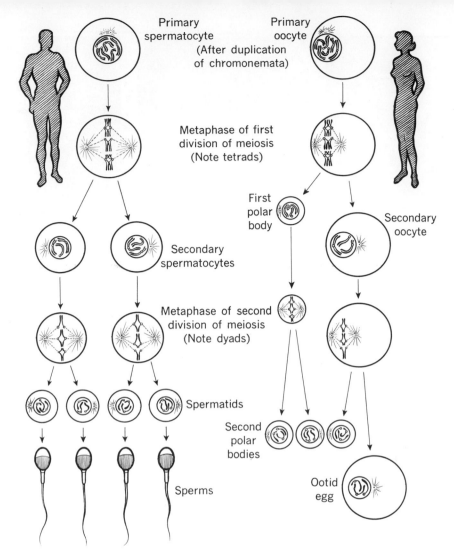

Primary spermatocyte (After duplication of chromonemata)

Primary oocyte

Metaphase of first division of meiosis (Note tetrads)

First polar body

Secondary oocyte

Secondary spermatocytes

Metaphase of second division of meiosis (Note dyads)

Spermatids

Second polar bodies

Sperms

Ootid egg

FIG. 4–6 *Spermatogenesis and oogenesis. Note the similarities and differences in the formation of sperms and eggs. This diagram shows 6 chromosomes as the diploid number. Actually, this number is 46 in man.*

one-half the original chromosome number, but which have the full diploid number of genes. During the second division of meiosis, there is a division of the chromosomes without a duplication of the genes. So we end with four spermatids, each of which contains only one of each kind of chromosome and one of each kind of gene. Each of the spermatids then becomes a **sperm** through a condensation of the nuclear material into a head, the development of a small middle-piece which carries the small amount of food necessary for life, and the formation of a tail for locomotion.

We have described the process using the chromosome number of man as an example, but the method is similar for all multicellular forms of animal life which have been studied. Figure 4–6 shows a graphic representation of this spermatogenesis in animals with six as the diploid number, and a comparison of the process with oogenesis which we shall discuss next.

At this stage of our study, the student may wonder why meiosis occurs in two stages rather than one. Why is there not one meiotic division without gene division or chromosome division so that reduction of both is accomplished at the same time? Of course, we have no reason to assume that every biological phenomenon has a definite function or that it accomplishes something worth while to the organism. Nevertheless, whenever any event is so widespread in its occurrence as meiosis, and is found always to require two cell divisions for completion, we can rest assured that it must have some biological advantage, else it would not have been maintained in so many forms of life throughout the ages of evolutionary development. We gain some clue to the possible evolutionary value of meiosis during the prophase of the first division of meiosis. Here the four chromatids become rather closely intertwined and may actually exchange segments. We must wait for our study of such crossing over in Chapter 14, however, before we learn the biological value of such an exchange.

Oogenesis

The primary difference between oogenesis and spermatogenesis lies not in the method of meiosis, but rather in the method of cytoplasmic division. The **oogonial cells** are found in the germinal epithelium of the ovaries. Oogenesis might be said to begin with the formation of a group of **follicle cells** around an immature oogonial cell. The oogonial cell then undergoes a period of great growth in which it builds up a large supply of yolk which, in the event that the egg is later fertilized, will serve as food for the developing embryo until it forms an attachment to the wall of the mother's uterus. Meanwhile the follicle cells around the oogonium increase in number, and follicular fluid develops between the follicle cells. This fluid contains an important female hormone which is related to the reproductive cycle. Eventually, the oogonium stops growing and is ready for the first division of meiosis. It is now called a **primary oocyte.**

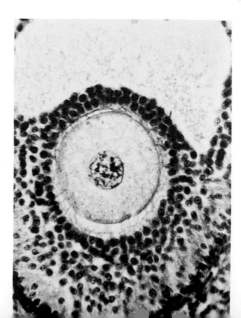

FIG. 4–7 *Portion of a human Graafian follicle. This photograph of a thin section of the ovary shows the egg surrounded by the follicle cells and the follicular fluid above.*

The entire body — follicular cells, follicular fluid, and primary oocyte — is now so large that it bulges out from the surface of the ovary and is known as a **Graafian follicle.** The nuclear changes within the primary oocyte parallel those which take place in the primary spermatocyte, but the spindle is formed near the edge of the cell and one pole is near the plasma membrane while the other points toward the center of the cell. Then, during the telophase, the cleavage furrow develops and pinches off a very small part of the cytoplasm along with one of the newly formed nuclei. This small cell is called a **polar body,** since it is pinched off at the pole of the spindle. The other nucleus remains with the greater part of the cytoplasm to form the **secondary oocyte.** This unequal division of the cytoplasm also occurs during the second division of meiosis to yield one **ootid** and a second polar body. The first polar body also divides in a second meiotic division, so that the process concludes with three polar bodies and one ootid. The ootid becomes an egg without further division. Some of the follicle cells adhere to the egg even after it breaks from the ovary and form a **corona.** The polar bodies have no function in reproduction and eventually disintegrate. It seems probable that their sole function is to receive the extra chromosomes in meiosis, leaving the ootid with the reduced number of genes and chromosomes. Since the egg must furnish nourishment for the developing embryo until it can form a connection with the mother, it is necessary that it be a rather large cell if it is to contain a sufficient amount of yolk. This type of division provides the largest possible cell from the materials present. Of course, only one-fourth as many eggs are produced as there would be by an equal division, such as we find in spermatogenesis, but judging by the success of the human species, biologically speaking, it appears that large numbers of eggs are not necessary. A woman will release no more than three to four hundred eggs from her ovaries during her entire lifetime, which is a very small number compared with the millions of sperms which are released from a man's body during one ejaculation.

We have used human oogenesis as an example of this important process in multicellular animal life. The mature eggs of different kinds of animals show considerable variation. Compare the tiny speck of protoplasm which makes up a human egg with the egg of a chicken, for example. As to the process of meiosis itself, however, and the formation of polar bodies, the process of oogenesis is remarkably uniform throughout the animal kingdom in those forms which have been investigated. The variations primarily involve the amount of yolk which is deposited and the total number of eggs which are produced — all of which are factors that have little significance from the standpoint of genetics.

The Shuffle of the Chromosomes

We might compare meiosis with the shuffling and dealing of cards. One cannot receive all of the chromosomes of either parent, and through meiosis the chromosomes are shuffled and dealt so that the hereditary "hand" which every child receives comes half from one parent and half from the other. We are so prone to emphasize the hereditary factors which we receive that we sometimes overlook those which we do not receive. We have irretrievably lost one-half of the factors which were carried by our parents when the eggs and sperms were formed from which we were descended. Neither we nor our descendants can ever show these characters, unless they are brought back into the family through marriage. A mother should not feel that the laws of heredity have failed because she has a daughter who lacks her good looks and musical ability. Perhaps genes for these traits disintegrated in some lost polar body while less desirable genes remained to be included in the egg.

We should consider those genes which are lost when we are tempted to take undue pride in our ancestry. Perhaps you have a very illustrious great-great-grandfather, but by the laws of probability you will have received only $\frac{1}{16}$ of your genes from this gentleman. And you should not forget your other ancestors of the same generation, some of whom you might like to prune from the family tree but who have nevertheless contributed a like amount of the genic material which makes you what you are.

It is this shuffling of the chromosomes that makes possible the great variety which may appear among the children of the same parents. Omitting the cases of identical twins for the moment, we can say that the chances of a couple having two children with the same assortment of chromosomes are so minute as to be considered impossible. Even though people lived for thousands of years and had children by the thousands, the possible number of combinations is still so great that we would not find two children with the same genes. Calculated by the laws of probability, the chance that you will have the same assortment of chromosomes possessed by your brother or sister is one in 281 trillion. Even this astronomical figure fails to allow for the exchange of parts of chromosomes, a process known as crossing over, which is known to occur. Taking this process into account would greatly increase the number of possible variations.

Fertilization

In Chapter 3 we learned how a complex, multicellular organism can be produced from a single cell, a zygote, through repeated mitotic divisions. In this chapter we have concentrated on the formation of germ cells with reduced numbers of chromosomes and genes in preparation for the formation of zygotes which will give rise to the next generation. To complete our cycle we will now study the process of fertilization — the union of egg and sperm to produce the zygote. Again, since we as human beings are quite naturally most interested in the process which produces the human zygote, we will use man as our primary example of fertilization.

The oogonial cells are produced near the outer edge of the ovary and then migrate in toward the center. Here the follicle cells develop around each oogonium to form a Graafian follicle. This continues to enlarge until it bulges out on the surface of the ovary. In a normal woman, during the reproductive period of her life, one of these follicles will burst about every 28 days, releasing the egg, together with its corona of follicle cells, into the body cavity. Normally, the egg drops into the ostium or entrance of the Fallopian tube and begins its migration down toward the uterus. Should semen be introduced into the vagina at this

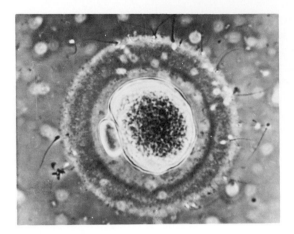

FIG. 4-8 *Human fertilization. A living human egg within its surrounding corona of follicle cells is shown as it is being attacked by living human sperms. Some of the sperms on the right have achieved a partial penetration of the barrier of follicle cells. A polar body can be seen on the left. This process normally occurs within the Fallopian tubes, but the egg was removed and mixed with sperms under the microscope.* COURTESY LANDRUM B. SHETTLES.

time, some of the sperms will begin migration up through the uterus and into the Fallopian tube where one will meet and fertilize the descending egg. In certain rare cases a human egg will drop free in the body cavity of a woman rather than into the ostium of the Fallopian tube. In such cases the sperms may swim all the way up the tube and out into the body cavity, where they fertilize the egg. This results in extrauterine pregnancy which, of course, cannot terminate successfully. The embryo must be removed surgically in such rare cases.

Only one sperm unites with the egg, but sperms, in males of all species, are produced in prodigious numbers. In man, the average ejaculation of semen, which will be about 4 or 5 ml. in volume, will contain the almost unbelievable number of about 400,000,000 sperms — sufficient to fertilize almost every woman on this hemisphere if every sperm were used. In spite of the fact that only one sperm cell is needed for fertilization, a man is likely to be sterile if he produces a smaller quantity of semen, say 1 ml. instead of 4 or 5, or if the number of viable sperms per ml. is smaller. There may be fewer sperms per ml. than normal or there may be a large number which are non-motile and cannot take part in the race for the egg. This is true because the sperms have a long and arduous journey and can succeed only through their combined efforts. First of all they must make their entrance into the uterus. The **cervix** of this organ, through which the sperms must enter, is usually filled with a **mucus plug** which the sperms must penetrate. Sperm cells produce an enzyme which can digest cervical mucus, but one sperm produces only enough enzyme to digest a tiny bit of this substance. Through the combined action of millions of sperms, however, the plug is penetrated, thus opening the way for the entrance of the other sperms. Many will fail before completing the journey up the Fallopian tube to the egg. Those that reach the egg find that it is surrounded by a corona of cells which must be broken away before the egg can be reached and fertilized. Again, this requires the combined action of many sperms. The enzyme produced by the sperms dissolves away the cement that holds these cells together. A small number of sperms cannot produce sufficient enzyme to accomplish this task.

Eventually, when the cells of the corona are broken apart, a sperm will touch the covering of the egg. The head of the sperm will then fuse with the egg and the egg will immediately undergo important changes. For one thing, a **fertilization membrane** is formed around the egg that prevents the entry of more sperms. One sperm is all that is required, and more would not only be unnecessary but would completely upset the carefully balanced mechanism which provides equal inheritance from the two parents. Of course there will be times when two sperms will reach the egg at almost the same time, and studies of lower forms of life show that multiple fertilization does occur, especially in insects, but the excess spermatozoa degenerate in the egg.

After the sperm and egg have fused, the head of the sperm will absorb fluid from the cytoplasm of the egg. This causes it to increase in size until it is about the same size as the nucleus of the egg. The two nuclei move together and fuse to produce a zygote with the diploid chromosome number, forty-six. Within a short time this cell will undergo its first division and a new embryo is started on its way.

SEXUAL REPRODUCTION IN PLANTS

Alternation of Generations

Reproduction in the multicellular plants is generally somewhat more complicated than in the majority of multicellular animals because there are two distinct phases in the life cycle which differ somewhat in method of reproduction. Let us consider a fern plant as an example. Everyone is familiar with the broad spreading fronds which make up the portion of a fern visible above the ground. Yet these represent only one of the two alternating generations of the fern. The

FIG. 4–9 *A fern sporophyte growing from a gametophyte. The gametophyte is haploid and the sporophyte is diploid.*

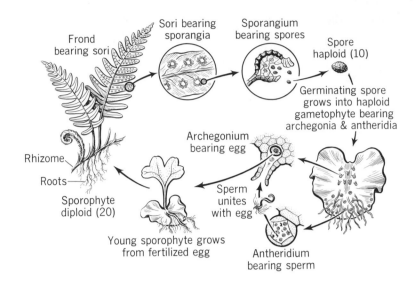

Frond
bearing sori

Sori bearing
sporangia

Sporangium
bearing spores

Spore
haploid (10)

Germinating spore
grows into haploid
gametophyte bearing
archegonia & antheridia

Rhizome

Roots

Sporophyte
diploid (20)

Archegonium
bearing egg

Sperm
unites
with egg

Young sporophyte grows
from fertilized egg

Antheridium
bearing sperm

FIG. 4–10 *The life cycle of a typical fern.*

other, smaller and less conspicuous, consists of a small, heart-shaped plant which grows flat on the ground. The fronds produce brown spore masses (sori) on the under side, and these bear hundreds of spores. When these spores are shed they do not grow into other fern plants, as one unfamiliar with botany might expect, but each spore will produce the small, heart-shaped type of plant. This plant, in turn, produces male and female gametes, two of which eventually unite to produce a zygote. The zygote grows into a typical large fern plant. This is known as an **alternation of generations** — two distinct generations alternating with one another, one producing spores that do not need to be fertilized and the other producing gametes that do have to be fertilized. The main fern plant, consisting of fronds, roots, and rhizome (underground stem), is called the **sporophyte** because it always reproduces by means of spores. The small, heart-shaped plant which grows from a spore is called the **gametophyte** because it produces gametes. The entire cycle is illustrated diagrammatically in Figure 4–10.

The primary difference between meiosis in the multicellular plants and the multicellular animals lies not in the process itself, but in the time of its occurrence. Meiosis occurs just before spore production. Hence, the spores and the gametophytes which grow from them will have the haploid number of chromosomes, while the zygote and the sporophyte which grows from it will have the diploid number of chromosomes. To illustrate with the fern, let us take a typical species which has twenty chromosomes as the diploid number. All of the cells in the sporophyte (fronds, roots, and rhizome) will have this diploid number of twenty. When the sori are formed on the under side of the frond, spore mother cells appear and undergo the two meiotic divisions and produce four **spores,** each with a **haploid** number of chromosomes. These are shed and each spore

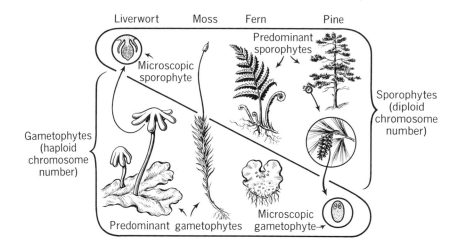

Liverwort   Moss   Fern   Pine

FIG. 4–11 *Comparison of the gametophyte and sporophyte genera-tions of liverwort, moss, fern, and pine. Note the progressive change from a predominant gametophyte to a predominant sporophyte.*

then grows into a gametophyte through repeated mitosis, so that all the cells of the entire gametophyte contain the haploid number of chromosomes. Male reproductive organs, **antheridia,** appear around the edge of the gametophyte and sperm are produced within these by ordinary mitosis. Female reproductive organs, **archegonia,** bearing eggs, form near the center of the plant. At a time when the gametes are mature and the plant is wet with dew or rain, the sperms break out of the antheridia and swim to the eggs, union occurs, and the zygote is formed. This, of course, has the **diploid** number and from it the sporophyte fern plant will grow.

The gametophyte is the predominant generation in the simpler plants, and the sporophyte is proportionately small. But as the plants become more complex, there is a progressive reduction in the size of the gametophytes and a correspond-ing increase in the size of the sporophytes, as illustrated in Figure 4-11. In the primitive liverworts, the gametophyte consists of a large thallus with its anchoring rhizoids while the sporophyte is a microscopic structure borne on the female gametophyte and nourished by it. Hence, nearly all the tissue of the plant is haploid in its chromosome number. In the mosses the gametophyte is still the predominant generation, but the sporophyte has become a sizable structure growing up from the tip of the female gametophyte though still partly dependent upon the gametophyte for nourishment. In the ferns the two genera-tions are separated from one another, but the sporophyte has become the larger generation of the two. In the **seed plants** the sporophyte comprises the primary plant and the gametophyte is a microscopic structure.

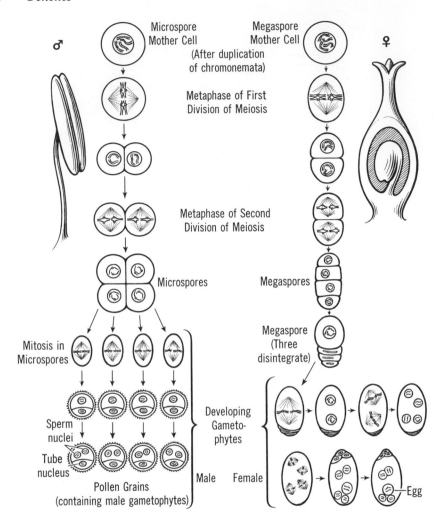

FIG. 4–12 *Microsporogenesis and megasporogenesis. Compare these diagrams with those of spermatogenesis and oogenesis in animals shown in Figure 4–6.*

Meiosis in Flowering Plants

Thus in the seed plants meiosis occurs only a short time before gamete reproduction. Since many plants which are studied genetically are flowering plants, we shall take time to study the process of meiosis and gamete formation in these plants in more detail. Two types of spores are produced, male **(microspores)** and female **(megaspores).** Their formation, known as **microsporogenesis** and **megasporogenesis,** are compared in Figure 4–12. The microspores are produced in the anthers of the flower. Cells near the inside of the anthers enlarge to become microspore **mother cells,** which we might think of as the equivalent of spermatogonial cells in animals. Each of these undergoes the two characteristic meiotic divisions to produce four microspores (pollen grains). Each of these will then undergo mitosis to produce two nuclei, which remain within the pollen grain. This is all there is to the **male gametophyte.**

The megaspores are produced within the **ovules** which, in turn, form a part of the ovary of the flower. Each ovule is a potential seed. Only one megaspore mother cell is formed in each ovule, and this cell divides by meiosis to give four cells of equal size. There is no polar body formation like that in animals. The process is very much like oogenesis, however, for three of the four cells degenerate and only the fourth enlarges to become a megaspore. In most flowering plants this cell then undergoes three mitotic divisions that produce eight nuclei. Three of these move to one end of the cell, three to the other end, and two remain in the center. Cell walls form around the cells at the ends, and this entire structure is the female gametophyte. One of the cells at one end of this structure becomes the female gamete or **egg;** the other two remain functionless as a rule. The two nuclei in the center play a part in the formation of the endosperm, which is the nutritive part of the seed. The three cells at the end of the gametophyte opposite the egg generally remain functionless.

The flowering plants are not motile organisms and must depend upon external forces to accomplish transference of the male gamete to the female. The pollen grains of most plants are adapted to be transferred by either wind or insects. The female reproductive organ of the plant, the pistil, contains a stigma at its tip to which the pollen adheres when it arrives, by whatever means of transportation. Once on the stigma a pollen grain sends out a slender pollen tube down the style of the pistil until it reaches an ovule at the base. One of the two nuclei of the pollen grain, the tube nucleus, advances with the pollen tube, but takes no part in fertilization. The other nucleus, the generative nucleus, divides by mitosis and produces two sperm nuclei; these are the male gametes. The tube penetrates the ovule and one of the sperm nuclei unites with

FIG. 4–13 *Typical fertilization in seed plants. Note how two sperm nuclei are used in fertilization; one goes to the egg, the other unites with the two nuclei of the endosperm to give a triploid condition of this portion of the seed.*

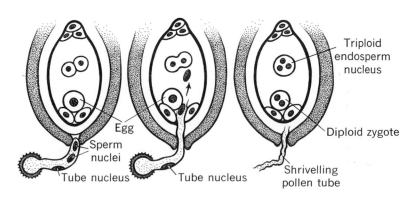

the egg to form the zygote; the latter is the beginning of the new sporophyte generation.

In most flowering plants, including corn, the one which has been investigated most extensively by geneticists, there is a very unusual second fertilization. It will be recalled that there are two nuclei in the center of the female gametophyte. There is a three-way fusion of these two nuclei and the second sperm nucleus. This produces a cell which contains three haploid sets of chromosomes; we say that it is triploid. This triple fusion is followed by successive mitotic divisions, resulting in the formation of a mass of tissue in which a large amount of food is stored. This tissue is called the **endosperm** and furnishes nourishment for the embryo until it can reach a size where it can manufacture its own food. Unusual genetic effects result from this triploid condition. As the endosperm grows the zygote also undergoes divisions to produce a young embryo. The ovule enlarges to keep pace with these growing parts within, and soon develops hardened seed coats. The entire structure is now called a **seed,** which may become enclosed in a **fruit** formed from the entire ovary of the flower.

For all practical purposes in genetic problems we may forget the complications introduced by the two alternating generations in the reproductive cycle of flowering plants and think of such problems in the same terms as we use in considering animal reproduction. In both cases we have a diploid organism which, through meiosis, produces haploid gametes; these unite and give a diploid zygote which grows into another diploid organism, and so on for generations.

PROBLEMS

1. The human egg is thousands of times larger than the human sperm, yet a child inherits equally from both parents. Explain.

2. There is evidence that there is some transmission of acquired characters from one generation to the next in certain one-celled organisms. Such transmission does not occur in the multicellular forms of life. How does the reproduction of the one-celled organisms differ from that of the multicellular organisms to make such transmission possible?

3. Could evolution take place as rapidly among animals reproducing solely by fission as among animals with sexual means of reproduction? Explain.

4. What problem would develop in future generations if meiosis did not precede the formation of gametes?

5. How many human sperms will be formed from twenty spermatids, from twenty secondary spermatocytes, from twenty primary spermatocytes?

6. How many human eggs will be formed from twenty ootids, from twenty secondary oocytes, from twenty primary oocytes?

7. In what phase of division of what type of cells do you find chromosome tetrads?

8. Basically, how does the first division of meiosis differ from mitosis?

9. How does the second division of meiosis differ from mitosis?

10. *Drosophila melanogaster* has eight chromosomes in its normal body cells. How many chromosomes would you expect to find in a sperm? primary oocyte? polar body? spermatogonium? egg? zygote?

11. Why is the egg always much larger than the sperm?

12. The common red fox has thirty-four chromosomes in its somatic cells, but another species of fox (the arctic fox) has fifty-two. These can be crossed to yield a hybrid. How many chromosomes would you expect to find in the somatic cells of a hybrid? What would happen in the prophase of the first division of meiosis when the time came for the pairing of chromosomes?

13. The hybrid between these two species of foxes is sterile. Give a reason for this sterility on the basis of the information about chromosomes.

14. What is the main difference in the relation of meiosis to gamete production in plants as compared with animals?

15. Compare the relative size of the haploid phase of the life cycle of a moss plant with the relative size of this phase of the life cycle of a flowering plant.

16. In the corn plant there are twenty chromosomes in the normal sporophyte tissue. How many chromosomes would you expect to find in the tube nucleus of the pollen grain? in the embryo within the seed? in the endosperm of the seed? in the cells of a leaf?

5

Monohybrid Inheritance

IN MOST ORGANISMS INHERITANCE is dependent upon the thousands of different kinds of genes which are transmitted through the gametes to the offspring. It would be impossible to study the influence of all of these genes at the same time and obtain any accurate picure of the way in which they work. Hence, in most experimental breeding, it is customary to use plants or animals that differ primarily in one or several clearly distinguished characteristics. For instance, if we want to know about the inheritance of kinky tail in mice we use one stock which has the kinky tail and cross it with another stock that is similar in other characteristics, but which does not have the kinky tail. This enables us to concentrate our attention on the gene which accounts for this one characteristic without worrying about other genes such as those which might make the fur wavy or straight, short or long, black or white, or present or absent. A cross of this nature is known as a monohybrid cross, and it was through such crosses that the principles of genetics were discovered. To Gregor Mendel goes the credit for the discovery of the method by which the genes are segregated and transmitted, and we shall begin our study with a survey of some of Mendel's monohybrid crosses in garden peas.

MENDEL'S FIRST STUDIES

Early Experiment

One of the characteristics which Mendel included in his first studies concerned the position of the flowers on the stem. To put it in his own words as translated from the original paper, "They are either axial, that is, distributed

68

along the main stem; or terminal, that is, bunched at the top of the stem." Here we have two definite alternative characteristics which can be clearly recognized by a glance at the mature plants. Mendel began his investigation by planting these two varieties for a number of generations, and he thereby found that the seed from flowers which were axial always grew into plants with axial flowers and that seed from flowers which were terminal always grew into plants which produced terminal flowers. After these tests Mendel felt reasonably sure that he had pure breeding varieties, for, as we pointed out in an earlier chapter, garden peas are normally self-fertilizing and there is consequently no chance for a mixing of hereditary factors between two plants. He then cross-pollinated the plants artificially in order to determine the effects of a mixture of these two pure-breeding varieties. This was done by first removing the male parts (anthers) of an immature flower of one variety before self-fertilization could take place. This flower was then carefully covered with a small paper bag to prevent possible stray pollen from reaching it. When the female portion of the flower was mature, Mendel transferred pollen from another variety to the tip of the pistil and again covered it. Seed resulting from this mixture, therefore, could be expected to bear hereditary factors from two varieties.

In this particular experiment, Mendel made thirty-four cross-pollinations, using ten different plants. For some of these crosses he took pollen from an axial flower and transferred it to the pistil of a terminal flower. For others he reversed the procedure, using pollen from terminal flowers on the pistils of axial flowers. He found that the results were the same for both types of crosses — that all of the first generation offspring bore axial flowers. He designated the first generation the F_1 (first filial) generation.

Mendel then allowed self-pollination to occur in these F_1 offspring and planted the seed from these plants to obtain a second generation (F_2). In this generation he found some terminal flowers. This indicated that the hereditary factor for terminal flowers was not lost, but merely covered up, or masked, in the first generation. He obtained 858 plants in the F_2 generation, of which 651 had axial flowers and 207 had terminal flowers. This was a ratio of approximately 3 : 1 (actually 3.14 : 1).

Dominance and Recessiveness

After a careful study of these and other results he concluded that there must be two factors for the position of flowers in each plant. Since the factor for terminal flowers reappeared in the second generation, it was obvious that it must have been present in the first generation, but there must have been another factor for axial flowers which dominated it. The factor for terminal flowers was what we call **recessive;** the factor for axial flowers was **dominant.** These two factors must have come from the original parents, one from each. Hence the gametes (pollen grains and ovules) must carry only one factor of each kind. In self-fertilization there is a random mixture of these factors and some seeds get two factors for axial flowers, some get one for axial and one for terminal, and some get two factors for terminal. This is illustrated in Figure 5–1. According to mathematical probability, the total should break down to ¼ of the first type, ½ of the second, and ¼ of the third. Since the factor for axial flowers is dominant over that for terminal flowers, however, the first two groups will produce plants with axial flowers. Hence approximately ¾ of the plants will have axial flowers and approximately ¼ will have terminal flowers — an expected ratio of 3 : 1. We see that the ratio which Mendel obtained (3.14 : 1) adheres to this very closely. Had he used smaller numbers of crosses and obtained smaller numbers in the F_2, his figures might have deviated more widely from the expected ratio. We will learn how to evaluate such deviations in Chapter 7.

To sum up, let us survey the major points of Mendel's conclusions. First, there are definite hereditary units (he called them factors, we now call them genes) which are responsible for the transmission of characteristics. Second, there are two of each type of factor in the body cells of a mature organism. Third, when these two differ, one will be expressed (will be dominant) while the other will remain latent (will be recessive). Fourth, these factors segregate unchanged into the gametes so that each gamete carries only one factor of each

FIG. 5–1 *Results of one of Mendel's crosses of garden peas bearing terminal flowers with peas bearing axial flowers, carried to the second generation.*

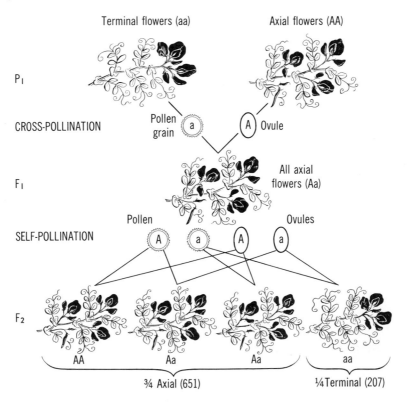

TABLE 5–1

Results of Mendel's First Experiments on Seven Pairs of Characteristics in the Garden Pea

Characteristic	F_2 Results		Ratio
Form of seed	5474 round	1850 wrinkled	2.96 : 1
Color of albumen	6022 yellow	2001 green	3.01 : 1
Color of seed-coats	705 gray-brown	224 white	3.15 : 1
Form of pods	882 inflated	299 constricted	2.95 : 1
Color of pods	428 green	152 yellow	2.82 : 1
Position of flowers	651 axial	207 terminal	3.14 : 1
Length of stem	787 long	277 short	2.84 : 1
All characteristics combined	14,889 dominant	5010 recessive	2.98 : 1

kind. Fifth, there is a random union of gametes which results in a predictable ratio of characters in the offspring.

Seven pairs of alternative characteristics were studied in Mendel's first experiments. All behaved in a manner similar to the one described. The results of all of these are shown in Table 5–1. An analysis of these results reveals a very important point relating to ratios, which we have previously mentioned. The larger numbers of offspring adhere more closely to the expected ratio than the smaller numbers. For instance, in the first cross listed there were 7,324 seeds and the ratio was 2.96 : 1. This is nearer to the expected 3 : 1 ratio than the experiment on the position of the flowers on the stem which had only 858 in the F_2. The second cross listed in this table comes even closer — 3.01 : 1 from a total of 8,023. When all of the experiments are grouped together we find a ratio of 2.98 : 1, which is almost perfect.

Mendel devised the system of using letters as symbols for the factors (the hereditary determiners). He used a small letter for a recessive factor and a capital letter of the same kind for the factor which was dominant over this recessive. Thus, he used *A* for axial and *a* for terminal flower in the monohybrid cross which we have just described. We use essentially the same system today. For instance we use *O* for the dominant gene for otosclerosis, a type of inherited deafness in man, and *o* to represent the recessive gene for normal hearing. We usually select the first letter of the word which describes the characteristic that is a deviation from normal, or which is least common.

At this point it might be well to correct an error which appears frequently in the minds of students when they study dominant and recessive genes. They sometimes get the impression that characteristics resulting from dominant genes must always be very common and those from recessive genes less common. The gene for otosclerosis is rare, yet it is dominant, while its recessive counterpart which works for normal hearing is found in nearly all people. On the other hand, the recessive gene which causes albinism (defective pigment formation in the skin, hair, and eyes) is rather rare, and its normal counterpart which functions in normal pigment production is dominant.

When we are dealing with forms of life where a comparatively large number of genes are known, it is sometimes necessary to use two or three letters for some gene symbols. For instance, in the widely studied fruit fly, *Drosophila,* we use the symbol vg for the gene which produces vestigial wings and Vg for the dominant gene which functions in the production of the more common wings of normal size.

MECHANISM OF MONOHYBRID INHERITANCE

The principle of monohybrid inheritance as worked out by Mendel with his peas applies to all forms of life, animals as well as plants. An experiment with **guinea pigs** will serve to show how it applies to animals. Let us select a black-coated male that has descended from a race which has had black coats for many generations. We can therefore assume that this is a pure line and that all the individuals in this race carry two genes for black. We use the term **homozygous** to refer to such a state and the word **heterozygous** to refer to the condition where the paired genes are different, such as genes for black and white coat in one individual. If we mate this homozygous black male with a homozygous white female, each sperm will carry one gene for black and each egg will carry one gene for white. Thus, all the offspring of this cross will be heterozygous, that is, they will carry one of each type of gene. And since all these heterozygous individuals will be black, we can conclude that black is dominant over white.

FIG. 5–2 *A monohybrid cross in guinea pigs.*

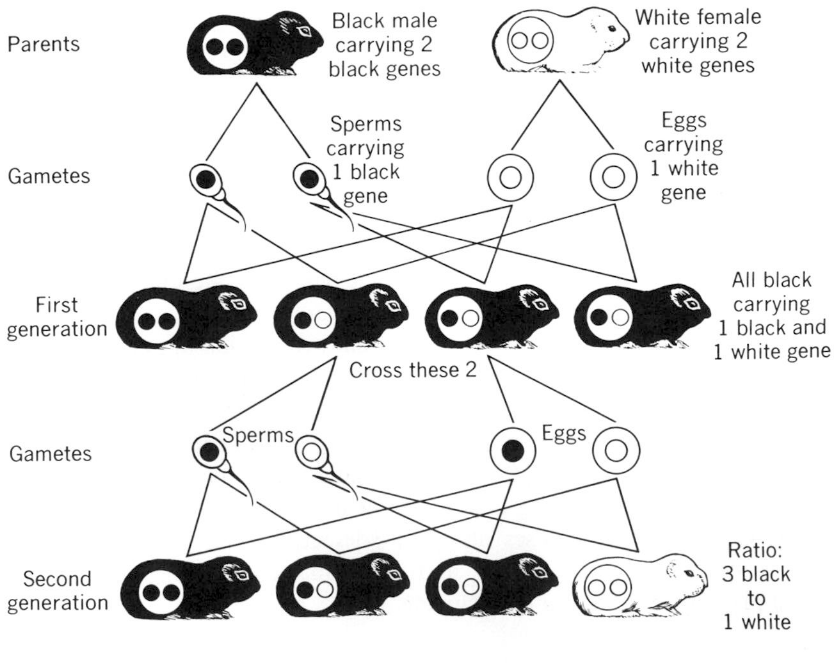

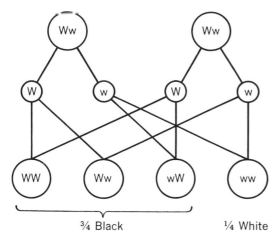

W — BLACK (dominant)

w — WHITE (recessive)

FIG. 5–3 *Method of working out a cross between two heterozygous black guinea pigs. Large circles represent the guinea pigs, small circles the gametes.*

¾ Black   ¼ White

When the cells within the testes and ovaries of these heterozygous individuals undergo meiosis, the two types of genes are separated, and one-half of the gametes will receive a gene for black coat while the other half receives a gene for white coat. There can be no such thing as self-fertilization in guinea pigs as we found in garden peas, but we can mate the first generation offspring (F₁) among themselves and achieve the same results. Again there are four possible combinations of gametes from such a cross, as illustrated in Figure 5–2. It will be noted from this diagram that each of two different kinds of sperms may unite with each of two different kinds of eggs to yield four combinations of genes in the second generation (F₂). Since white is recessive, no guinea pig can be white unless it receives two genes for white, and only one of the four types of combination contains two genes for white. This gives us the same 3 : 1 ratio as Mendel obtained with his peas. In working out this cross by diagram, we would use *w* to stand for white coat and *W* to represent the gene for black coat (Figure 5–3).

Genotypes and Phenotypes

At this point we must introduce two other very useful genetic terms which will make our future discussions much easier. The black guinea pigs are of two kinds according to the genes which they carry, but of only one kind according to the expression of the genes. When we are speaking of the different kinds of guinea pigs, it is necessary that we have special words to indicate whether we refer to the genes which are carried or the expression of the genes. When we refer to the type of genes we use the word **genotype;** when we refer to the expression of the genes we use the word **phenotype** (Gr. *phainein* = to show). The adjectives formed from these words are **genotypic** and **phenotypic.** We can say that a guinea pig has a genotype of one gene for black and one gene for white and a phenotype of black. The phenotypic expression of two genes for black is also black. The guinea pigs with a white phenotype always carry two genes for white as their genotype.

FIG. 5–4 *Vestigial wings in Drosophila. This character is due to a recessive gene.*

Alleles

Another case of monohybrid inheritance in animals will introduce still another important genetic term. In the fruit fly, *Drosophila melanogaster,* which has been used in experiments which led to many important genetic discoveries, a gene which affects wing growth is located about one-third of the way along a chromosome which is designated chromosome 2. Of course, since there are two chromosomes of this type in each cell, there will be another gene which influences wing growth at the same locus on the other chromosome 2. Some flies are homozygous for a recessive gene at this locus which produces **vestigial** wings. Such wings are greatly reduced in size, and in fact are mere stumps utterly useless for flying, so that flies with this character must hop about rather than fly. Most flies, of course, have the gene for normal-sized wings at this locus. The gene for vestigial wings and the gene for normal-sized wings are said to be **alleles.** They are both genes for the same body characteristic and they are both located at the same point on chromosome 2, but they differ somewhat in the type of wings which are produced. Thus the gene for vestigial wings is an allele of the gene for normal wings, and vice versa. Likewise the gene for axial flowers in peas is an allele of the gene for terminal flowers; and the gene for black coat in guinea pigs is an allele of the gene for white coat. The adjective of the word is **allelic.**

Determination of Genotype

The occasion often arises when one wishes to know the genotype of plants or animals with which he is working. It is easy enough to determine the genotype of the forms showing the recessive character. Such forms are always homozygous. For those which show the dominant character phenotypically, however, we cannot tell by examination whether they are homozygous or heterozygous. We can distinguish these only by crossing to the homozygous recessive.

Let us use a practical case to illustrate this point. In poultry there are two distinct types of combs which are due to two varieties of a gene. One of these types, called **rose comb,** is somewhat flattened and has multiple heads. The rose comb is produced by a dominant gene. The recessive allele produces what is called a **single comb,** which has a single head and stands up much higher. All of the fowls in the Wyandotte breed must have rose combs if they are to be considered pure bred. In some flocks, however, single-combed birds occasionally appear, even though they are descended from rose-combed parents. This indicates that there must be some genes for single comb in the flock, and, even though no single-combed birds are ever used for breeding, some single combs will still appear. Whenever heterozygous individuals are crossed, about one-fourth of the offspring will have single combs. Most Wyandotte breeders simply consider this an unavoidable occurrence and do away with the single-combed birds when they appear. It is possible, however, to eliminate this gene from a Wyandotte flock by application of the principles of monohybrid inheritance which we have learned in this chapter.

First of all, we must allow some of the single-combed birds to mature to be used as test animals. Then we will select several roosters from our flock for testing. These will be mated with the single-combed hens and a careful record kept of the results of the crosses. Should any of the roosters be heterozygous for the gene for single comb, they will produce single-combed birds in about one-half of their offspring. Such roosters should never again be used for breeding. They and their offspring can best serve mankind in the stew pot. The homozygous rose-combed roosters will produce only rose combs among their offspring.

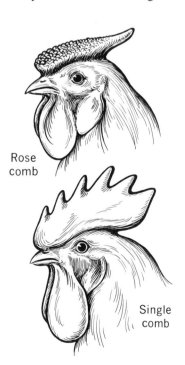

Rose
comb

Single
comb

FIG. 5–5 *Comparison of rose comb with single comb in the domestic fowl.*

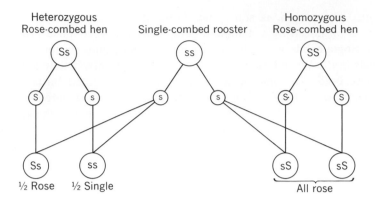

FIG. 5−6 *Test cross of two rose-combed hens (S) with a single-combed (s) rooster. The heterozygous hen yields one-half single-combed offspring, while the homozygous hen yields only rose-combed offspring. Since there is only one type of sperm, it is not necessary to show the cross of the eggs with both sperms.*

These roosters should be saved to furnish the male side of the breeding stock. The offspring of this **test cross** should be destroyed, however, for they will all be heterozygous and will perpetuate the gene for single comb if used for breeding. Hens may be tested in the same way by breeding to single-combed roosters. In this way it is possible to obtain both roosters and hens which are homozygous for rose comb, and if these are used as the foundation for a new flock of chickens, there will be no further trouble with the single comb. Figure 5−6 shows how the test crosses are made and gives the results which are obtained in the two cases.

Intermediate Genes

Whenever two athletes are matched in a test of skill and strength, one of the two is usually the stronger and wins the event. There will be some cases, however, in which both will be of about equal skill and strength and the contest will end in a draw. When we study two alleles for different expressions of a characteristic, such as rose comb and single comb, we often find that one of the two genes has greater strength than the other and dominates it. This gives us the condition of dominance and recessiveness which was first worked out by Mendel. There are some cases, however, in which the two genes seem to be of about equal strength in their determining powers and a heterozygous individual shows characteristics which are somewhat in between the two unmixed types. For instance, if one crosses a red four-o'clock with a white four-o'clock, the resulting plants produce neither red nor white flowers, but pink ones. If two of these pink-flowered plants are crossed, they produce red, pink, and white flowers in the ratio of 1 : 2 : 1, since (on the average) out of four offspring one will be homozygous for red and one for white, and two will be heterozygous. From

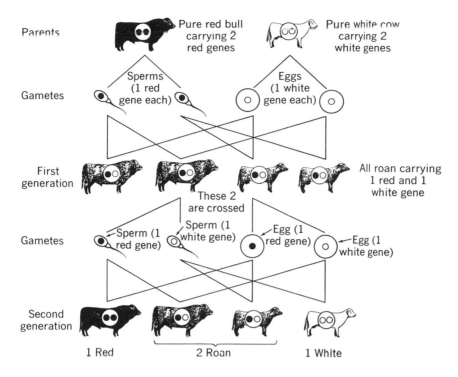

FIG. 5–7 *Cross between pure red and pure white produces roan in short horn cattle.*

these results it is evident that whenever a plant carries both the genes for red and white it will be pink. In other words, neither the gene for red nor white is dominant or recessive — this is a case where both genes express themselves partially. Such genes are called **intermediate genes.**

Another interesting example is found in short-horned cattle. If we cross a red bull with a white cow, the offspring are somewhat intermediate in their coloration — roan. A close examination of a roan animal reveals that the individual hairs are not roan — some hairs are red and some are white. The over-all effect of the combination of the two colors of hairs gives the roan appearance. A cross between two roan cattle could be expected to produce the results shown in Figure 5–8.

There are some cases where the heterozygous expression of two different genes is something new — somewhat different from either parent. For instance, if we cross a black chicken of a certain strain with a splashed white chicken (white with black splashes) the offspring are blue. When crossed together, these blue chickens yield black, blue, and white in the ratio 1 : 2 : 1. This is clearly a case of a heterozygous expression of two genes even though the offspring are not true intermediates. It is customary to speak of this condition as intermediate inheritance, however, since this term has come to refer to those cases where the heterozygote is clearly affected by both genes. The blue chickens, in

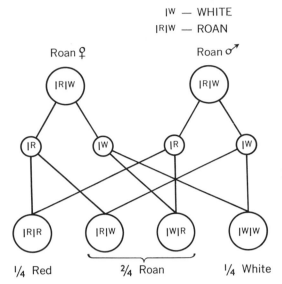

|R — RED
|W — WHITE
|R|W — ROAN

Roan ♀   Roan ♂

|R|W |R|W

|R |W |R |W

|R|R |R|W |W|R |W|W

¼ Red   2/4 Roan   ¼ White

FIG. 5–8 *How a cross between two roan cattle can result in three different colors of cattle in the offspring. I indicates intermediate inheritance.*

this case, form a special breed of domestic fowl, the **Andalusian chickens.** The rules of poultry breeders hold that no breed will be recognized unless at least one-half of the offspring breed true (are like the parents). The Andalusian breed barely gets under the wire with exactly one-half. The poultryman usually crosses Andalusians together to get more Andalusians; it seems logical that the greatest number of blue chickens would be produced by crossing two blues. Nevertheless, only one-half of the offspring breed true, and the balance must be discarded. It would be more efficient if the poultryman would get a black rooster and a group of splashed white hens and use eggs from these hens for incubation. In this way he would get 100 per cent blue Andalusians.

Incompletely Recessive Genes

In many cases it has been found that genes which have been thought to be completely recessive have some small effect when heterozygous. We cannot call it an intermediate effect because there is only a slight modification of the dominant phenotype. It is quite likely that we may find that most of the so-called recessive genes fall into this category, once the geneticists investigate all of the possible small alterations in genotype which might exist.

For human geneticists these little telltale signs of the heterozygous carriers of recessive genes are very valuable. These geneticists are often called upon to counsel prospective parents about the feasibility of having children when there is some serious hereditary abnormality in their ancestry. If the carriers of the abnormal gene can be recognized, the counselor's job is made much easier. Let us illustrate with several examples.

Friedreich's ataxia is an affliction characterized by a failure of muscular coordination and an impairment of speech. The gene causing this condition was formerly thought to be entirely recessive, but an examination of the parents of afflicted children showed that both parents had an excessive curvature of the sole of the foot, *pes cavus,* and also lacked the normal tendon reflex when the area below the knee is struck sharply. These apparently unrelated traits enable us to recognize the heterozygous carriers. If both a man and his wife show these traits, then we can tell them that each child they have will have ¼ chance of developing the ataxia.

An improper development of the mind is one of the most tragic of all human abnormalities. Nothing can bring greater heartache to parents and a greater burden to society than a child without the mental development needed to take a normal place in society. There are at least three forms of inherited idiocy which are due to recessive genes, and which, when heterozygous, have sufficient effect that we can detect them. In **juvenile amaurotic idiocy** the homozygous child is born normal and shows normal mental development until about 6 years of age. Then something goes wrong, there is a progressive decline in his mental ability, an impairment of vision leading to blindness develops, muscular weakness comes about, and death usually comes before the child is 21 years of age. In searching for some way to detect the normal carriers of this gene, geneticists found that in heterozygous persons there was an increase in the presence of vacuoles in the lymphocytes, a type of white blood cell.

Wilson's disease is due to another recessive gene with delayed action. During the second decade of life, homozygous persons show abnormal amino acid metabolism. This leads to deposits of fibrous tissue in the liver, a degeneration of brain centers, and an excretion of amino acids in the urine. Heterozygous persons can be recognized by the presence of a higher than normal concentration of amino acids in their urine.

Phenylketonuria is a condition caused by the inability of the body to utilize a single amino acid, phenylalanine. As a result, this amino acid reaches a high concentration in the blood and is excreted in the urine. It seems to act as a poison to the developing brain of a baby and leads to idiocy. Heterozygous carriers can be recognized by administering to persons to be tested an extra quantity of phenylalanine. This causes a sudden rise in the phenylalanine level of the blood. In homozygous normal persons, the blood level of phenylalanine returns to normal rapidly; however, in heterozygous persons, the level drops much more slowly. For a fuller discussion see Chapter 17.

GENE INTERRELATIONSHIPS

When we study monohybrid inheritance, it is easily possible to get the impression that there is only one gene for any one body characteristic. To refer back to the gene for **vestigial wings** in *Drosophila,* we sometimes say that the recessive gene, *vg,* produces vestigial wings and the dominant allele, *Vg,* produces the normal long wings. This does not mean that these are the only two genes which influence wing growth. A wing is much too complicated a structure to be produced by the action of one gene. We know that other genes are involved because they also show variations. One of these produces miniature wings.

These wings are perfectly formed, but are only about half the normal size. Another gene produces curly wings which curve up, and still another produces curved wings which curve down. One produces wings with notches in the end, and another produces various patterns of veins on the wings. Thus, it is evident that the wings of *Drosophila* result from the combined action of many genes, each performing a certain task. When any one of these undergoes a mutation it can produce a wing which is not of the normal type. When we wish to make a monohybrid cross, we select flies which are homozygous for all of the genes affecting wings with one exception. Thus we can investigate this single gene.

Let us illustrate further with an example in man. There is a recessive gene, *a,* which causes the condition known as **albinism.** Homozygous persons do not produce the skin pigment known as **melanin** and, as a result, they have very fair skin, their hair is almost white, and their eyes may be pink because the blood shows in the otherwise colorless iris. Some albinos have pale blue eyes because of the presence of reflective bodies in the iris but no melanin. Albinos see very poorly in bright light because they lack this pigment that normally absorbs the excess light rays. Also, they become badly sunburned if exposed to open sunlight for a short time because they lack the protection afforded by the melanin found in normal skin. They can never develop a suntan. Can we say that the dominant gene, *A,* is the gene for pigment production? Not at all. Melanin is produced through a series of enzyme-controlled reactions, and many genes are involved in its production and distribution. A common dominant gene involved causes the irregular distribution of the melanin and **freckles** result. Another dominant gene causes irregular patches of unpigmented areas, a condition known as **piebald.** Still another causes the lack of pigment only in one streak of hair at the front of the head, a trait known as a **blaze.** Still other genes regulate the amount of melanin that is produced, which accounts for the great range of color differences among the various races. Certainly, it is evident that there are a number of genes involved in skin pigmentation.

F I G . 5 – 9 *Albino brother and sister. Even though this photograph was made in subdued light, this boy and girl must squint because of the lack of pigment in the irises and the choroid coats of their eyes. The parents are first cousins with normal pigmentation and have three other children with normal pigmentation.*

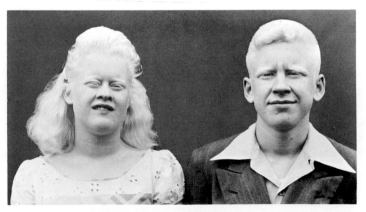

FIG. 5-10 *This young lady shows an abundance of the irregular patches of melanin known as freckles. The trait seems to be inherited as a dominant.*

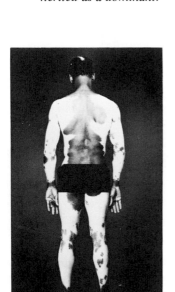

FIG. 5-11 *A piebald Negro. This trait, in which there are large unpigmented areas in the skin, is inherited as a dominant.* COURTESY CLYDE E. KEELER.

INHERITANCE IN HAPLOID ORGANISMS

Many of the haploid forms of plants are used in genetic studies since they have only one set of genes, a characteristic which has facilitated their study by geneticists. In such organisms there are no dominant, recessive, or intermediate genes, nor any need for test crosses or searches for heterozygous expressions. Since every organism expresses every gene it carries, the identification of inherited traits is greatly simplified.

Diplococcus pneumoniae is a bacterium which can cause pneumonia in man. Most of these bacteria are highly sensitive to the antibiotic known as **streptomycin** and cannot grow in its presence. A mutation of one gene, however, has produced a strain which is resistant to streptomycin. This strain produces some enzyme which breaks down streptomycin; thus it can grow on media containing this antibiotic. Every time a bacterium is divided by mitosis, the gene is passed on and all descendants of the mutant strain carry the gene for streptomycin resistance. It is very easy to detect mutations which give the pneumococcus resistance even though the mutation rate is very low. Let us say that the rate of mutation is only 1 out of each 100 million gene duplications. If we had to search for such a small number in larger plants or animals we would have little chance of finding them. For these bacteria, however, we need merely to

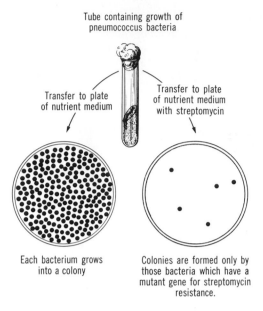

Tube containing growth of
pneumococcus bacteria

Transfer to plate
of nutrient medium

Transfer to plate
of nutrient medium
with streptomycin

FIG. 5–12 *Mutations are easy to detect in haploid organisms. When pneumonia bacteria are placed on a medium containing streptomycin, only those bacteria survive and form colonies which have a mutant gene for streptomycin resistance.*

Each bacterium grows
into a colony

Colonies are formed only by
those bacteria which have a
mutant gene for streptomycin
resistance.

place several hundred million of them on a plate of solid food media containing streptomycin. Perhaps only four of this great number would grow into colonies. These four would be the mutants, and from them we could get pure strains of streptomycin-resistant bacteria. A sheep rancher trying to establish a pure-breeding flock of a mutant type of sheep must go through many generations of breeding and selection, whereas the bacterial geneticist can obtain his pure mutant strain in about 24 hours with one transfer of a colony of bacteria.

The pink mold, *Neurospora,* exists in an albino strain which is white instead of pink. Only one gene interferes with the production of the pink pigment in the albino strain. Since this mold has sexual reproduction, we can cross a pink strain with an albino strain (see Figure 5–13). The zygote is diploid, but meiosis occurs when the zygote begins division and four haploid cells are produced. These divide again by mitosis giving eight haploid cells, each of which becomes a spore. If we carefully remove each of these spores, one by one, and transplant each into a tube of food media, we will find that four spores grow into a pink mold and four grow into an albino mold. There is no way to know whether the gene for pink would be dominant over the gene for albinism, since the pigment does not develop in the zygote, and this is the only diploid cell in the life cycle of the mold.

INHERITANCE IN MAN

It is natural for a student of genetics to have a great interest in man's inheritance, yet the genetic study of man is very difficult. If we were to choose animals on the basis of convenience and suitability for genetic analysis, man

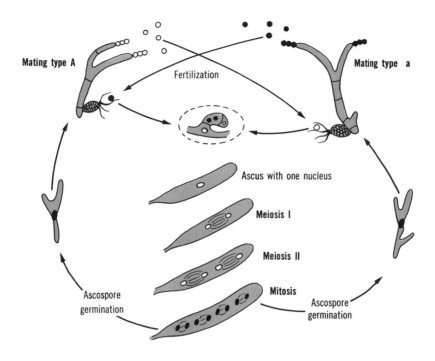

Mating type A

Fertilization

Mating type a

Ascus with one nucleus

Meiosis I

Meiosis II

Mitosis

Ascospore
germination

Ascospore
germination

FIG. 5–13 *Life cycle of the pink mold, Neurospora. There are two mating types, each of which can fertilize the other. The zygote formed by the union of gametes undergoes meiosis immediately to give four haploid nuclei. After another division by mitosis the eight nuclei form into ascospores within an ascus. These break out and each can grow into a new mold filament. Half of the spores grow into mating type A and the other half into mating type a.*

would rank near the bottom of the list. People cannot be bred like other animals; they have a very long generation cycle and a very small number of offspring. Also, they are very expensive to feed and maintain. Still, as human beings we want to know about human heredity and we must overcome all of these handicaps to learn what we can. It is surprising how much we know about human heredity in view of these handicaps. Much of this knowledge has come about by a comparison of man with some of the lower animals which we can breed easily and at little cost. Man shares the same genetic mechanism with these lower forms and much information about one form can be applied to the other. Much of our knowledge about man has come through a study of pedigrees. While we cannot tell a blue-eyed man to marry a brown-eyed woman and have many children so that we can learn about the inheritance of eye color, we can find many families where just this sort of marriage has occurred and observe their children. We can thus tabulate the parents and children in the form of a pedigree and study the results.

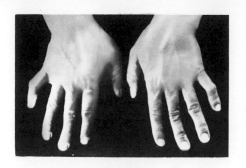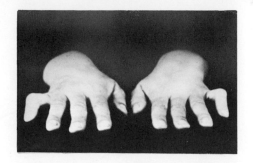

FIG. 5–14 *Crooked little fingers, streptomicrodactyly, inherited as a dominant trait.* R. A. HEFNER IN JOURNAL OF HEREDITY.

To illustrate, let us analyze an actual human pedigree and try to get some information about the method of inheritance of a particular characteristic. A woman had two sons, one of which had crooked little fingers. She had noticed the same type of fingers on her husband, but her second son had normal fingers. This led her to think that the characteristic might be inherited and she made a survey of her husband's ancestry to try to learn how the crooked fingers were inherited. She found that her husband's sister and mother both had crooked little fingers, as well as his grandfather. The characteristic also appeared in more distant relatives. The method of representing such a pedigree is shown in Figure 5–15. Through careful examination of this pedigree we can postulate that crooked little fingers are inherited through a simple dominant gene. All individuals which show this trait in this pedigree must be heterozygous, since one parent is always normal, and approximately one-half of the children from the heterozygous parent show the crooked fingers. There are no cases where a child shows the trait when neither parent shows it, as may be the case with recessive genes. Of course, there is one case on the extreme right where all (three) of the children show the character. This does not upset our conclusion by any means, however, for in the comparatively small families found in human pedigrees such deviations

FIG. 5–15 *Pedigree of a human family showing inheritance of crooked little fingers.*

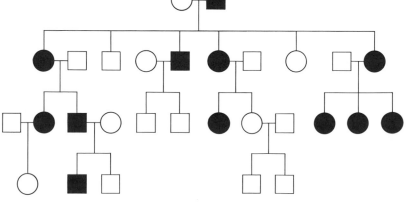

from the expected ratios arc not uncommon. When we consider all of the children in the pedigree who have one heterozygous parent, we find that the ratio is 11 : 7, which is a reasonably close approximation of the expected 1 : 1.

Of course, geneticists would not conclude from this one pedigree that this character results from a single dominant gene. But when the same characteristic is studied in many other pedigrees and the results agree, then we may feel safe in concluding that this is the correct interpretation of the method of inheritance. This is the way the principles of inheritance are applied to a study of family pedigrees. It would be well for all persons to keep a family record as complete as possible for such studies. Unfortunately, many people who have kept family records only trace the distinguished branches of the family tree and omit those gnarled and twisted offshoots which might bring disrepute to the family name. Since we inherit equally from all branches of the tree, we should include the drunkards, the horse thieves, and the morons as well as the governors, the professors, and the military geniuses. We cannot by one whit reduce the effect of the genes of the former group by ignoring their existence in our family trees. Also, many family histories are of little genetic value because they include only names, with perhaps notations of outstanding achievements for some in the group. To be of greatest value to the geneticist they should include physical, mental, and temperamental traits and aptitudes.

PROBLEMS

1. In human families it is often observed that certain characters may "skip" a generation or two and then reappear. How would you explain this in the light of the facts learned in this chapter?

2. In certain Norwegian families there is an inherited condition known as "wooly hair." Those showing this character have hair which resembles sheep's wool. A study of family pedigrees shows that a person never has wooly hair unless at least one parent also has wooly hair. How would this character most likely be inherited? Explain.

3. In some of Mendel's later crosses he found that the gene for purple flowers was dominant over the gene for white flowers. What type of offspring would you expect if you crossed a pure-line purple flower with a white flower? Show both genotype and phenotype in the first and second generations.

4. What results would you expect if you crossed one of the hybrids from the first generation, obtained in the above problem, with a plant bearing white flowers? Show one generation only, genotype and phenotype.

5. In Holstein cattle the spotting of the coat is due to a recessive gene while a solid colored coat is dominant. What types of offspring might be produced by a cross between two spotted animals? Show how you reach your conclusion.

6. A blue-eyed child has a brown-eyed mother and a brown-eyed father. From this family alone would you think that blue eyes result from a recessive gene, a dominant gene, or an intermediate gene? Give reasons for your answer.

7. In cats the gene for short hair is dominant over the gene for long hair (Angora). A short-haired tom cat is mated with an Angora female. She bears eight kittens, six short-haired and two with long hair. How do these numbers compare with the expected ratio? If you mated these same cats four more times and obtained

a total of forty offspring would you expect the results to be a closer approximation of the expected ratio? Explain.

8. Some cocker spaniel dogs have a solid body color while others are particolored (have white areas on the body). A cross between two solid colored dogs yields one particolored female in a litter of six. When this female matures she is mated with a particolored male and produces nine pups, all particolored. Diagram these two crosses and show how the results can be explained.

9. Two short-tailed (Manx) cats are bred together. They produce three kittens with long tails, five with short tails, and two without any tails. From these results, how would you think that tail length in cats is inherited? Show genotypes to support your belief.

10. The polled (hornless) condition in cattle is dominant over horned. A cattleman in Texas has a range stocked with polled cattle only, but some horned cattle occasionally appear. These are removed from the range before they can reproduce. Assuming that this man has good fences which can keep out stray bulls, how could this be explained?

11. This same cattleman wants to produce only polled cattle on his ranch. Outline a breeding program which would most efficiently eliminate the appearance of horned cattle in his herd.

12. Another cattleman has a roan bull and some white cows, but wants a pure-breeding red herd on his ranch. Outline a breeding program for him to follow which will enable him to do this without bringing in other stock.

13. In summer squash white colored fruit is dominant over yellow. If you place the pollen from a homozygous yellow-fruited plant on the pistil of a heterozygous white-fruited plant, what types of plants would you expect from the seed which come from this cross?

14. An albino man marries a normally pigmented woman who had an albino mother. Show the types of children that this couple may have and the proportions of each.

15. Refer back to the chart showing inheritance of polydactyly in Figure 1–9. According to this one pedigree, how do you think this characteristic is inherited? Give reasons for your answer.

16. Some people are afflicted with a disease known as cystic fibrosis of the pancreas, which is manifested by a faulty digestion of fats. Such persons are homozygous for a certain recessive gene which causes this condition. Today it is possible to keep them alive with careful medical treatment. A man with this affliction wishes to marry, but wonders if his children might inherit it. The object of his affections tells him that she knows there has been no such affliction in her family for at least three generations. Is there any chance they could have a child who could inherit this affliction? Explain. Give diagrams to show all of the genotypes possible for both the man and the woman, and for the children they may have.

17. Much research in human genetics has been directed toward methods of recognizing heterozygous carriers of genes which are generally recessive. What is the possible practical importance of such research?

18. In man a recessive gene is sometimes found which causes aniridia, absence of the iris of the eye. Is it correct to say that the dominant allele is the gene which produces the iris of the eye? Explain.

19. For studies of the rate of mutation resulting from treatment with radiation many genetic workers use bacteria or molds for their study. What advantage do such organisms have over larger forms, such as mice, in such investigations?

6

Dihybrid Inheritance

THE AUTHOR ONCE OVERHEARD A FOND MOTHER REMARK, "Mary has inherited her mother's blue eyes and her father's bad disposition." Without realizing it, she brought out one of the important principles of genetics — the principle of **independent segregation** of hereditary characteristics. The mother may have had blue eyes and a good disposition while the father had brown eyes and a bad disposition, but this does not mean that these characteristics will necessarily stay together in the children. It is through such independent segregation of characteristics from the two family lines that new combinations of characteristics are produced in the children.

The Principle of Independent Segregation

We can study this segregation by crossing individuals which differ with respect to two pairs of genes. This is a **dihybrid cross,** in contrast to the **monohybrid cross** in which only one genic difference is considered. Mendel worked out dihybrid crosses with his garden peas. One of these crosses involved the color and shape of the seeds after drying. Figure 6–1 illustrates this cross diagrammatically. He found that some plants produced seeds which stayed round after drying while others gave seeds which became wrinkled. A monohybrid cross had already shown him that round was the dominant character. Again, some seeds turned yellow when dried while others remained green, and in this case the yellow was due to a dominant gene. When he crossed plants producing yellow-round seed with others producing green-wrinkled seed, he obtained all yellow-round seed in the first generation, as was expected since he had already

87

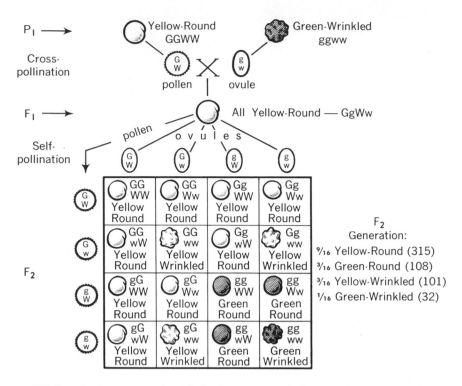

FIG. 6-1 *One of the dihybrid crosses made by Mendel, using garden peas which bore yellow-round seed crossed with peas which bore green-wrinkled seed.*

demonstrated that yellowness and roundness were the phenotypes displayed by the two dominant genes concerned. He planted these seeds and found that four distinct types of seeds were produced by the plants of the F_2 generation. From 15 second-generation plants he collected 556 seeds. Of these, 315 were yellow-round, 101 yellow-wrinkled, 108 green-round, and 32 green-wrinkled. As we examine these figures we can see that they approximate a 9 : 3 : 3 : 1 ratio — that out of the total number, 9/16 show both dominant characters, 3/16 show one dominant and one recessive, 3/16 show the other dominant and the other recessive, and 1/16 show both recessives. Thus Mendel obtained not only the two parental types but also two new types resulting from other mixtures of the characters than those found in the two parents. This showed that the two factors do not tend to stay together in the same combinations in which they are found in the original parents, P_1. This separateness of behavior among genes is called the **principle of independent segregation.** Each gene behaves exactly as it does in a monohybrid cross. Approximately ¾ of the second generation peas are yellow and ¼ are green (416 : 140); and the same holds true for round and wrinkled (423 : 133). The appearance of these 3 : 1 ratios shows us that this is not a new sort of result from genetic crosses, but that a dihybrid is only a combination of two monohybrids, each of which behaves just as it does when studied separately.

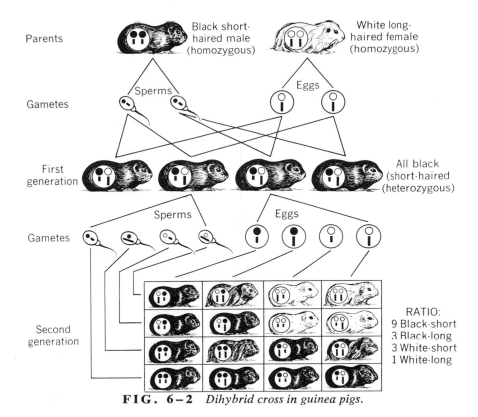

FIG. 6–2 *Dihybrid cross in guinea pigs.*

Figure 6–2 gives a graphic portrayal of dihybrid inheritance in the guinea pig. Black hair is dominant over white and short hair is dominant over long. When a homozygous, black short-haired guinea pig is mated with a homozygous, white long-haired guinea pig, the first generation offspring are all heterozygous for both pairs of alleles and show both dominants. These two pairs of alleles are on different chromosomes and there is no reason why the various paternal or the various maternal chromosomes should remain together. Hence, it is just as likely that a sperm produced by a male of the F_1 generation will receive the chromosome carrying the gene for black from the father and the chromosome carrying the gene for long from the mother as receive the genes for black and short from the father. It is equally likely that the genes for white and short or white and long will be found together. Thus, four types of sperms will be produced in approximately equal quantities, and the same will be true for eggs. Each of the four types of sperms has an equal chance of fertilizing each of the four types of eggs. This is best represented in the form of a checkerboard with four squares in either direction. If we construct such a figure, we obtain sixteen possible genotypes which may be broken down into the 9 : 3 : 3 : 1 ratio for the phenotype. Figure 6–3 is an example of this kind of diagram, showing the sixteen possible genotypes in the F_2 generation dihybrid cross of guinea pigs discussed above.

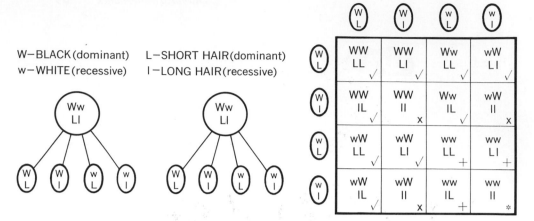

W—BLACK (dominant) L—SHORT HAIR (dominant)
w—WHITE (recessive) l—LONG HAIR (recessive)

√ 9/16 BLACK-SHORT   x 3/16 BLACK-LONG
+ 3/16 WHITE-SHORT   * 1/16 WHITE-LONG

FIG. 6–3 *The genotypic and phenotypic ratio of offspring resulting from a cross between two heterozygous black short-haired guinea pigs, worked out by the checkerboard method.*

This illustrates the principle of independent segregation. We can illustrate it further by crossing individuals differing with respect to three pairs of characteristics. In this case there will be eight types of gametes from each F_1 individual and sixty-four genotypes in the F_2 checkerboard. The phenotypic ratio will be 27 : 9 : 9 : 9 : 3 : 3 : 3 : 1. All three dominants will show in the $27/64$, but only $1/64$ of the offspring will show all three recessives. We could also make crosses considering more than three characters at a time, but the number of possible types of offspring soon becomes too great to handle conveniently in our analyses, and there are no new principles of genetics to be brought out in such crosses. Table 6–1 illustrates the infinite variety which may result in crosses between organisms which differ with respect to various numbers of genes.

TABLE 6–1

The Number of Different Kinds of Gametes and the Number of Possible Zygotic Combinations Which May Result from Hybrids of Various Numbers of Genes

Pairs of genes	Different types of gametes	Possible zygotic combinations
1 (monohybrid)	2	4
2 (dihybrid)	4	16
3 (trihybrid)	8	64
4	16	256
5	32	1,024
6	64	4,096
7	128	16,384
8	256	65,536
9	512	262,144
10	1,024	1,048,576
15	32,768	1,073,741,824
20	1,048,576	1,099,511,627,776

The Dihybrid Test Cross

We have learned that the test cross is a valuable means of determining the genotypes of the offspring of a monohybrid cross. It is equally valuable in the cases which involve more than a single pair of alleles. For instance, we can cross the heterozygous F_1 guinea pigs to the double recessives (white-long) rather than allow them to breed among themselves. This will yield a ratio of $1 : 1 : 1 : 1$ rather than $9 : 3 : 3 : 1$. This type of cross is illustrated in Figure 6–4. Since all of the gametes from the homozygous, recessive parents carry the two recessive genes, this cross brings out all the recessive genes which are present in the gametes of the opposite sex. A test cross of a trihybrid will yield a ratio of $1 : 1 : 1 : 1 : 1 : 1 : 1 : 1$.

Thus we see that the test cross is much more convenient to use where there are differences in a number of different genes because it is not necessary to use complicated checkerboards like those which must be used when individuals heterozygous for all the genes are crossed with one another. In effect, the test cross brings out the recessive genes which are present in the gametes which are produced by the heterozygous individuals.

Modified Dihybrid Ratios

Just as the monohybrid ratio is not always $3 : 1$ when the F_1 offspring are crossed among themselves, the dihybrid ratio may vary from the $9 : 3 : 3 : 1$. As an example let us use the gene for coat color in cattle in conjunction with the polled and horned condition. We have already learned that roan results as an intermediate expression of red and white. The polled (hornless) condition is due to a gene which is dominant over the gene for horned. When polled-red cattle are crossed with horned-white, the offspring are all polled-roan. Figure 6–6 shows the results of the cross of two of these F_1 individuals. Since we have

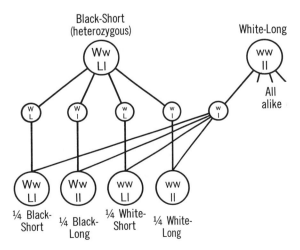

W — BLACK   L — SHORT
w — WHITE   l — LONG

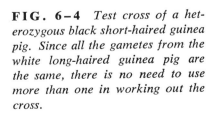

FIG. 6–4 *Test cross of a heterozygous black short-haired guinea pig. Since all the gametes from the white long-haired guinea pig are the same, there is no need to use more than one in working out the cross.*

FIG. 6–5 *Independent segregation in corn. The upper ear is from the second generation of a cross between sweet and starchy. The sweet kernels are shriveled. Note the approximate 3 : 1 ratio. The middle ear shows similar results from a cross between yellow and purple, again 3 : 1 ratio. The lower ear is from the second generation of a cross between sweet yellow and starchy purple, showing independent segregation and an approximate 9 : 3 : 3 : 1 ratio.*

three possible coat colors and two possible states of the horns, we have more than four phenotypes. The ratio comes out 6 : 3 : 3 : 2 : 1 : 1. If we consider two genes both of which exhibit intermediate inheritance, our possible number of phenotypes will become even greater. By working out such a problem one will find that the ratio will be 4 : 2 : 2 : 2 : 2 : 1 : 1 : 1 : 1. Once one understands the principle of dihybrid inheritance, however, these unusual ratios are easy to determine by the checkerboard method.

In some cases we find that certain groups are masked in their expression by the action of other genes. As an example let us consider coat color in mice. The typical wild mouse has a coat color which is called agouti (a peculiar shade of gray). Some mice, however, have black coats when they are homozygous for a recessive gene which causes the hair pigment to be black rather than agouti. Mice, like most other animals, also carry a few genes for albinism, which result in a white coat when the genes are homozygous. Albino is not an allele of the genes for agouti and black. Rather, when homozygous this gene

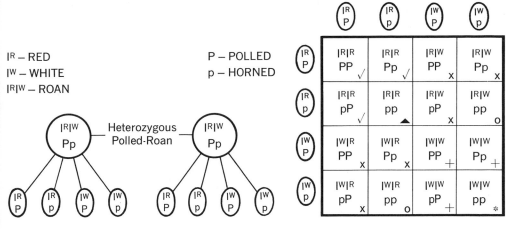

IR – RED
IW – WHITE
IRIW – ROAN

P – POLLED
p – HORNED

Heterozygous Polled-Roan

x ⁶/₁₆ POLLED-ROAN   o ²/₁₆ HORNED-ROAN
√ ³/₁₆ POLLED-RED   ▲ ¹/₁₆ HORNED-RED
+ ³/₁₆ POLLED-WHITE   * ¹/₁₆ HORNED-WHITE

FIG. 6–6 *Cross between two heterozygous polled-roan cattle to illustrate the ratio obtained when one pair of alleles is intermediate and the other pair is dominant and recessive.*

for albinism prevents the formation of pigment of any kind, regardless of the genes for agouti or black which the mouse may carry. This is due to the fact that the dominant allele of albino seems to produce an enzyme (mentioned earlier) which is necessary for the formation of pigment, and in the absence of this enzyme no pigment is produced. Now, if we cross a homozygous agouti male carrying the two dominant alleles of the gene for albinism, *BBAA,* with an albino female carrying two genes for black, *bbaa,* we will obtain all agouti mice in the offspring. When these are crossed among themselves, however, we obtain agouti, black, and albino mice in the ratio of 9 : 3 : 4. This unusual ratio results from the fact that all mice homozygous for the gene for albinism are albinos, and we cannot distinguish which ones carry the agouti genes and which ones carry the black genes. Thus, two of our classes are grouped to give the ⁴/₁₆. Figure 6–8 will help you to understand this cross.

FIG. 6–7 *How a cross between a gray mouse and a black mouse may produce a white offspring. Only one of a number of possible types of gene combinations is shown here.*

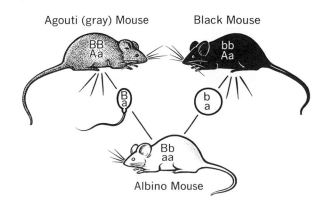

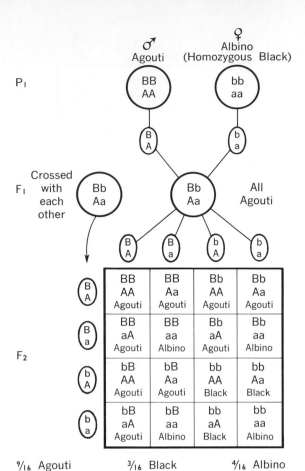

B – AGOUTI (gray)
b – BLACK
A – COLOR
a – ALBINO

P₁ Agouti ♂ ... Albino ♀ (Homozygous Black)

FIG. 6–8 *Recessive epistasis in mice. The agouti male shows the gray pigment because he carries the dominant genes which are necessary for the color of the coat. The albino female is homozygous for black, but carries two albino genes which are epistatic to the genes for color of the coat, and the mouse is white. The second generation offspring show the peculiar ratio of 9 : 3 : 4.*

$\frac{9}{16}$ Agouti $\frac{3}{16}$ Black $\frac{4}{16}$ Albino

Epistasis

The results from the experiments with the mice introduce an important principle of genetics, the principle of **epistasis.** Whenever a gene at one locus on a chromosome influences the expression of a gene at another locus, the first gene is said to be epistatic to the second. For instance, the gene for albinism is epistatic to the genes for black and agouti, because these colors cannot express themselves when the gene for albinism is homozygous. Such genes are sometimes called inhibiting genes, since they inhibit the expression of other genes, but the term epistasis (Gr., "stopping," "suppression") expresses the condition more accurately. Since the gene for albinism is recessive and two such genes must be present before any influence on the coat color is noticed, we may call this a case of **recessive epistasis.** Figure 6–9 illustrates a case of recessive epistasis in man.

Dominant epistasis results when only one gene of a pair of alleles is necessary to influence the expression of a second pair of alleles. A classical example of this type of cross may be illustrated by the White Plymouth Rock and the White Leghorn chickens. The F₁ offspring of such a cross are white, as might be expected; but the F₂ generation gives the rather startling ratio of 13 white : 3 colored chickens. On analyzing this, we might conclude that it is a case of a

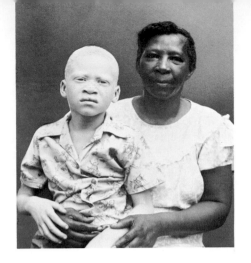

FIG. 6–9 *Recessive epistasis in man. This Negro boy, shown with his mother, has received genes for Negroid pigmentation from his Negro parents, but he is also homozygous for the gene for albinism. This gene is epistatic to genes for pigmentation of the skin, hair, and eyes.*

9 : 3 : 3 : 1 ratio with three of the classes grouped together to form the white. We can explain this result by assuming that the White Leghorns are homozygous for a dominant gene for color, *C,* but they do not show colored feathers because they are also homozygous for a dominant gene which inhibits the formation of colored pigment, *I.* The Plymouth Rocks, on the other hand, are white because they are homozygous for the recessive gene *c* which fails to produce color, but they are also homozygous for the recessive gene *i* which would allow color to form if there were any genes for color present. The F_1 of the cross are white because the dominant gene *I* inhibits the formation of color, but 3 of the F_2 receive the dominant gene for color without receiving a gene for color inhibition. A study of Figure 6–10 will clarify this picture.

As another example of a modified dihybrid ratio we might take a character which sometimes appears in human beings, the case of deaf-mutism. This unfortunate affliction produces children who are born deaf and, therefore, are

FIG. 6–10 *Colored chickens may be produced in the second generation of a cross between White Leghorns and White Plymouth Rocks.*

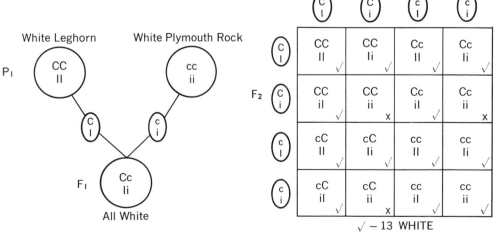

also mute unless taught to speak by special techniques. The condition may result when a person is homozygous for either of two recessive genes, let us call them *a* and *b*. Thus a person with the genotype *aaBB* would be a deaf-mute because the gene *a* is epistatic to the gene *B* for normal hearing. Similarly, a person with the genotype *AAbb* would be a deaf-mute because the gene *b* is epistatic to the gene *A* for normal hearing. This is a case of **duplicate recessive epistasis** where either of two recessive genes may be epistatic to the dominant allele of the other. When two deaf-mutes marry, one of each of the genotypes listed above, all of their children will have normal hearing and speech because both of the genes for deafness are recessive. Of course, we cannot cross the F_1 to obtain an F_2 ratio as we do in mice and guinea pigs, but, through statistical observation and comparisons with similar gene reactions in other forms of life, we can determine that the ratio of normal hearing to deaf-mutism is about 9 : 7 in the children of marriages between couples heterozygous for both genes. Figure 6–11 shows how this ratio is obtained.

It is also possible to have dominant genes which show duplicate epistasis which will give a ratio of 15 : 1. The feathers on the shanks of some chickens are inherited in this way. Most chickens have the unfeathered shanks which result when they are homozygous for two recessive genes. The presence of a single dominant gene for either of the two pairs, however, is sufficient to produce feathered shanks. In other words, if we assume that the genotype of the birds without feathers on their shanks is *aabb*, then *Aabb* or *aaBb* would result in feathered shanks. If we cross the homozygous double recessive with the homozygous double dominant, the F_1 will all have feathered shanks, and in the

FIG. 6–11 *Inheritance of deaf-mutism in man: an illustration of duplicate recessive epistasis. There are two epistatic recessive genes which can produce deaf-mutism in man. Whenever a person is homozygous for either of these two genes he will be deaf when born and, as a result, will also be mute. This diagram shows how two persons with normal hearing, but who are heterozygous for both these genes, may have both normal children and deaf mutes in the ratio of 9 : 7.*

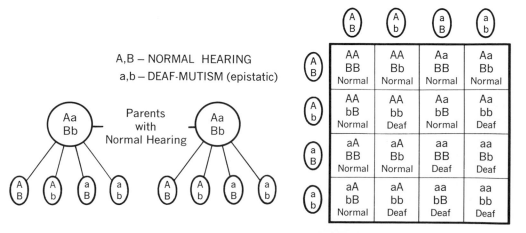

A,B – NORMAL HEARING

a,b – DEAF-MUTISM (epistatic)

	AB	Ab	aB	ab
AB	AA BB Normal	AA Bb Normal	Aa BB Normal	Aa Bb Normal
Ab	AA bB Normal	AA bb Deaf	Aa bB Normal	Aa bb Deaf
aB	aA BB Normal	aA Bb Normal	aa BB Deaf	aa Bb Deaf
ab	aA bB Normal	aA bb Deaf	aa bB Deaf	aa bb Deaf

Parents with Normal Hearing

Aa Bb — Aa Bb

$^9/_{16}$ Normal Hearing $^7/_{16}$ Deaf-mutes

FIG. 6–12 *Feathered and unfeathered shanks in chickens: an example of duplicate, dominant epistasis. The two birds on the right show the feathered condition, which may be produced by two different dominant genes.* COURTESY L. F. PAYNE, KANSAS STATE COLLEGE.

F_2 the ratio will be 15 feathered : 1 non-feathered. Only the double recessive fails to develop the feathers on the shanks.

Table 6–2 summarizes the results of the various types of dihybrid crosses which have been discussed in this chapter and shows how the typical 9 : 3 : 3 : 1 ratio is split up or combined to give the various deviations which we have studied. It should be kept in mind, however, that these deviations are the exception, and that the great majority of dihybrid crosses yield the 9 : 3 : 3 : 1 ratio. All deviations are easier to understand if we remember that a dihybrid is two monohybrids operating together.

TABLE 6–2

Modifications of the Dihybrid Ratio

Genotypes →	*AABB*	*AABb*	*AaBB*	*AaBb*	*AAbb*	*Aabb*	*aaBB*	*aaBb*	*aabb*
A & B both intermediate	1	2	2	4	1	2	1	2	1
A intermediate, B dominant	3		6		1	2	3		1
A & B both dominant (typical dihybrid)	9				3		3		1
aa epistatic to B or b (recessive epistasis)	9				3		4		
A epistatic to B or b (dominant epistasis)	13						3		1
aa epistatic to B or b bb epistatic to A or a (duplicate recessive epistasis)	9						7		
A epistatic to B or b B epistatic to A or a (duplicate dominant epistasis)	15								1

PROBLEMS

1. In *Drosophila,* vestigial wings and ebony body color are due to two separate recessive genes. The dominant alleles are normal (long) wings and normal (gray) body color. What type of offspring would you expect from a cross between a vestigial ebony female and a homozygous normal (long-winged, gray-bodied) male? If the F_1 are allowed to breed among themselves, what types of offspring would you expect in the F_2? Show complete genotype and phenotype of both generations.

2. If you made a test cross of the F_1 males of the preceding problem what results would you expect to obtain?

3. If a vestigial-winged female who is heterozygous for ebony body color is crossed with an ebony male who is heterozygous for vestigial wings, what results will be expected in the F_1?

4. In tomatoes, yellow fruit and dwarfed vine are due to recessive alleles of genes which produce the more common red fruit and tall vine. If pollen from a pure-line dwarf plant bearing red fruit is placed on the pistil of a pure-line tall plant bearing yellow fruit, what type of plant and fruit would be expected in the F_1? If these are crossed among themselves what would be expected in the F_2?

5. About 70 per cent of Americans get a bitter taste from a drug, phenyl thio-carbamide (PTC); the others do not. The ability to taste this drug results from a dominant gene while taste-blindness is recessive. A normally pigmented woman who is taste-blind has a father who is an albino-taster. She marries an albino man who is a taster, but who has a mother who is taste-blind. Show the types of children which this couple may have.

6. In the domestic swine, a white belt around the body is dominant to the uni-formly colored body. Some hogs have a fusion of the two elements of their hoof to produce the mule-footed condition (syndactyly). This also results from a dominant gene. A farmer wishes to raise only belted swine with normal feet, but there are some solid-colored and some mule-footed hogs in his pens. How would you advise him to proceed in his efforts to establish pure-breeding, belted hogs with normal feet?

7. Some dogs bark when trailing; others are silent. The barking trait is due to a dominant gene. Erect ears are dominant to drooping ears. What kind of pups would be expected from a heterozygous erect-eared barker mated to a droop-eared, silent trailer?

8. Rose comb is dominant to single comb in chickens. What type of offspring would be expected from a splashed-white rooster who is heterozygous rose-combed, crossed with a single-combed, blue Andalusian hen?

9. A young lady with normal hearing is the daughter of parents who are both deaf-mutes. She is considering marriage with a man who is a deaf-mute, but is greatly concerned about the chances of her children being deaf-mutes. If you were a marriage consultant and this young lady came to you with her problem, what would you tell her? Show by diagram just how you would explain this to her.

10. The same young lady has a second suitor who has normal hearing, but whose parents are both deaf-mutes. What could you tell her about her chances of having deaf-mutes among her children if she decided to marry this man?

11. Let F and F' stand for the two dominant genes for feathered shanks in chickens and f and f' stand for the corresponding recessives for the unfeathered condition. Show the results of a cross of a rooster with feathered shanks who is heterozygous for both genes with a hen with unfeathered shanks.

12. In *Drosophila* there is a dominant gene, *Hw,* which produces hairy wings except when a fly is homozygous for a recessive suppressor gene, *su-Hw.* Two normal-winged flies are crossed and yield all hairy-winged offspring. Show the genotype of the parents and the offspring.

13. Another dominant gene, *S,* produces a star eye in *Drosophila,* but there is a dominant suppressor gene, *Su-S,* which is epistatic to star and inhibits the action of the gene for star. A heterozygous star-eyed fly is crossed with a normal fly that is heterozygous for the *Su-S* gene and homozygous for star eye. Show the genotype and phenotype of the expected offspring.

14. In man there is a dominant gene for vitiligo, a condition where small unpigmented spots appear on the body. There is also the well-known recessive gene for albinism which causes the entire body to be unpigmented. A man with vitiligo has a mother who is an albino. He marries a woman who is an albino, but both of her parents have normal skin pigmentation. Show the genotype and phenotype of the skin of the children which this couple might have.

7

Probability and Heredity

THE STUDY OF PROBABILITY IS CLOSELY related to genetics, for a great part of the study of genetics involves a statistical analysis of the ratios of various character- istics among the offspring of parents of a known phenotype. The term "ratio," however, is often not well understood, as is clearly illustrated by the following case. A young couple with normal pigmentation produced an albino as their first child. Upon learning that such a condition was inherited, the mother visited a physician in order to learn how such a thing could have occurred and whether they should expect albinos in their future children. Since the woman seemed to possess a fair degree of intelligence, the doctor took the time to explain the simpler Mendelian principles of inheritance to her. He explained how the factor for albinism was recessive and could be carried by normal people without any expression of the character; how she and her husband apparently both carried this factor, but also carried a factor for normal pigmentation which was dominant over the albino factor; how their children would be albino only if they received both albino factors from the parents, and that this would occur in a ratio of 1 : 3 (one albino to three normal children). Suddenly the woman's face brightened as she seemed to understand the implications of this information, and she said, "I am so glad to know that, doctor; now, since we have already had the one albino, we can have three more children and they will all be normal." The great difficulty with such a line of reasoning is that the sperms and eggs which carry the genes have no way of knowing that the albino child has already been born, and gametes carrying the albino genes are just as likely to come together and produce an albino child at the second birth as at the first.

On the other hand, if we study the offspring of a large number of parents heterozygous for this trait we will find that albinos indeed do appear in only about one-fourth of the children. How, then, can we correlate the disproportionate ratios which are often found in small groups with the more nearly perfect ratios of the large samples? This is the problem which is explained by probability as it is related to heredity.

LAWS OF COINCIDENT HAPPENINGS

Two Independent Events

The laws of probability may be applied to any subject which involves chance or random happenings. For instance, if we toss a penny in the air, the chance of it falling heads is one-half. Yet we do not think it unduly noteworthy if we toss a penny four times and get heads on each toss. The expected ratio of heads to tails is 1 : 1; therefore, a perfect ratio would have been two of each out of the four tosses. If we should toss the penny four hundred times, however, we would get a ratio very close to 1 : 1; certainly, we would not get four hundred heads. This illustrates the fact that the larger the numbers, the more nearly perfect will be the ratios. We can understand why this is true if we think of each toss of the penny as an independent affair — uninfluenced by what has happened before or what is going to happen after. The chance of heads appearing on any toss is one-half, regardless of what has previously occurred. As in the case of the genes, we might say that the penny has no way of knowing whether heads appeared on the previous toss, and could not respond if it did. In any small number of tosses, it is easily possible to obtain several "heads" in succession, but as the numbers increase, the laws of probability show that the chance of continuing to obtain only heads quickly diminishes.

This is easily illustrated by a study of one of the laws of probability — that for the probability of coincident independent events. **The chance of any number of independent things happening together is equal to the product of the chances that each will happen separately.** We know that the chance of obtaining heads on any one toss is one-half; the chance of obtaining two heads on two separate tosses is: $\frac{1}{2} \times \frac{1}{2} = \frac{1}{4}$. To express it in another way — if we simultaneously

FIG. 7-1 *Probability as illustrated by tossing pennies. These pairs of photographs show the actual number of heads (left in each pair) and tails (right in each pair) which were obtained by tossing pennies. It can easily be seen that, as a larger number of pennies is tossed, the ratio more nearly approaches 1 : 1.*

FIG. 7–2 *Probability of occurrence of two events of equal chance, as illustrated by drawing poker chips. Two poker chips were drawn at once from a large mixture containing equal numbers of white and colored chips. After each drawing the chips were returned to the group and mixed thoroughly before two more were drawn. These stacks represent the various combinations which were drawn. It can be seen that the final groupings closely approximate the 1 : 2 : 1 ratio. If we think of the chips as allelic genes we can see how the ratios become more accurate as the number of offspring becomes larger.*

toss two pennies four times we can expect to get two heads in one of the four tosses. To obtain the chance of throwing four heads in a row we must multiply the chance of obtaining a head on one toss by itself four times: $\frac{1}{2} \times \frac{1}{2} \times \frac{1}{2} \times \frac{1}{2} = \frac{1}{16}$. Thus we see how the chance of obtaining only heads decreases as the number of tosses increases. If we began to calculate the chance of tossing four hundred successive heads, we would soon reach a fraction so minute that it could be considered zero.

General Applications

Now let us see how this reasoning can be applied to other things. Suppose we wanted to find out our chance of touching a blonde girl if we were blindfolded and released in the hall of a college building between classes with instructions to move about with arms extended until we touched some person. By reference to the registrar's records we could determine the ratio of girls to boys in the college; let us assume that it is 1 : 1. Then, let us assume that a survey of the general population in your section of the country has shown that the ratio of natural blondes to brunettes is about 1 : 2. We can solve our problem if we multiply the chance of a girl ($\frac{1}{2}$) by the chance for a blonde ($\frac{1}{3}$) and find a product of $\frac{1}{6}$, which is our chance of touching a blonde girl. In other words, we would have to touch six people, on the average, in order to include a blonde girl in the group.

If amateur gamblers were to apply the laws of probability to learn their chances of winning, many would, no doubt, find better ways to use their money. For instance, college campuses are sometimes invaded by the professional gambler with schemes which may seem attractive, but, when these schemes are analyzed according to the law of probability, we usually find that our chances of winning are too slight to warrant participation.

Application in Genetics: the Fraction Method

Now that we understand how this law of probability works, let us see how it may be applied to genetic problems. To return to the case of the young couple who had produced an albino child, we can now see that the chance of their producing an albino child at any birth is one-fourth. The fact that the first child is albino, however, will in no way influence the heredity of future children in this respect. If they have a second child, its chance of being albino is also one-fourth. Let us suppose that they have decided to have three more children and want to know the chance that all three will be albino. The answer would be: $\frac{1}{4} \times \frac{1}{4} \times \frac{1}{4} = \frac{1}{64}$; not a very great chance. If they should want to know the chance of all three being normal the answer would be: $\frac{3}{4} \times \frac{3}{4} \times \frac{3}{4} = \frac{27}{64}$. Such an application of the laws of probability can be very valuable to a couple who know that they carry some serious hereditary defect (such as deaf-mutism) when they try to decide on the advisability of having children.

One may often note the working of the laws of probability in the proportion of the sexes in human families. Statistics show that the chance that a child will be a boy is slightly greater than one-half, but for practical purposes we can assume that it is just one-half. In many families, however, there is a preponderance of one sex or the other. Does this mean that there is something wrong with the law of coincident probabilities? Not at all. In fact there would be something wrong if we did not get a certain proportion of families with all boys or all girls. Let us consider a family with four children — all boys. Using the same method we used for the pennies, we find that such families would be expected in one out of every sixteen families with four children. In other words, if we surveyed a large number of families in which there were four children, we would find about one-sixteenth of them with four boys. There would, in fact, be something wrong with our assumptions if we did not find such a proportion in a large survey. In such a survey we would find families with all possible combinations of the sexes. The product method of determining the probability of a coincidence can easily be applied to determine the chance of finding all girls or all boys in such families, but it is more difficult to use to determine various mixtures of the sexes, such as the chances of finding three boys and one girl.

The Binomial Method

It will be easier to obtain the answer to such a problem by using a bit of simple algebra, so simple, in fact, that it requires no previous knowledge of this subject. The method involves the expansion of the binomial, $(a + b)^n$, in which we allow a to equal the chance of a certain thing happening, b the chance of an alternative happening, and n the total number of events to be considered. For instance, let a equal the chance that a child will be a girl, which is one-half; b the chance that it will be a boy, which is also one-half; and n the total number of children in the family, which is four.

Before proceeding further we shall digress long enough to gain an understanding of the method of expanding the binomial $(a + b)^4$ for the benefit of those who may not remember the method from their high school algebra. Of course, we can multiply it out algebraically, but there is a quicker way. The first term in the expression is always the first letter with an exponent the same as n. This would be a^4. The exponents of the second term will be a^3b, the third a^2b^2, the fourth ab^3, and the fifth b^4. The exponents of a drop one with each progression and the exponents of b increase one, but the two combined always total 4. In addition to the exponents, there will be numbers in front of the terms which we call coefficients. The coefficient of the first term (a^4) is always 1 and is usually not written out; we merely write a^4. To determine the coefficient of the second term we can multiply the coefficient of the first by the exponent of the first and divide by 1. Thus $1 \times 4 \div 1 = 4$. The second term then is $4a^3b$. To get the coefficient of the third term we multiply the coefficient of the second by the exponent of a and divide by 2. $4 \times 3 \div 2 = 6$. The fourth coefficient may be obtained from the third term in the same way, but this time we divide by 3. $6 \times 2 \div 3 = 4$. Note that the fifth coefficient turns out to be 1 by the same method of calculation. The complete expansion of the binomial now appears as follows:

$$a^4 + 4a^3b + 6a^2b^2 + 4ab^3 + b^4$$

From this expansion we must choose the particular term which will give us the probability of three boys and one girl. We have already indicated that the exponent indicates the number of children or other individuals or events being considered. As we look over the terms we find one which fulfills our require-

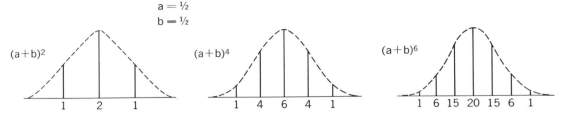

$a = \frac{1}{2}$
$b = \frac{1}{2}$

$(a+b)^2$ 1   2   1

$(a+b)^4$ 1   4   6   4   1

$(a+b)^6$ 1   6   15   20   15   6   1

FIG. 7–3 *Probability curves formed by expansion of the binomial a + b to the second, fourth, and sixth powers, where the chances of a and b are both equal to one-half. Note the bell-shaped nature of the curve in each case.*

ments, $4ab^3$, which indicated one girl (a^1) and three boys (b^3). To get our answer we substitute the known probabilities for a and b and multiply:

$$4 \times \frac{1}{2} \times (\frac{1}{2})^3 = \frac{4}{16} = \frac{1}{4}$$

This means that there is one chance in four of having one girl and three boys.

To find the chance of two children of each sex we take the term $6a^2b^2$, which gives $\frac{6}{16}$ or $\frac{3}{8}$. If we figure the chances for each of the possible combinations of the sexes, beginning with girls only and ending with boys only, we get the following results:

$$\frac{1}{16} + \frac{4}{16} + \frac{6}{16} + \frac{4}{16} + \frac{1}{16} = \frac{16}{16} \text{ or } 1$$

It will be noted that these chances form a bell-shaped curve when plotted on a graph as shown in Figure 7–3. This will be true for any group of probabilities where the chances for each of two alternatives is one-half. In those cases where the chances are not one-half a skewed curve will be produced.

Figure 7–4 shows the expansion of the binomial up to $(a + b)^5$. There is another way of obtaining the coefficients through the use of Pascal's triangle, which is shown in Figure 7–5. This works nicely for the expansions where the number of events is small, but becomes cumbersome when they become larger.

FIG. 7–4 *An expansion of the binomial a + b through the fifth power. These simple algebraic terms are of great value in studies of probability.*

$(a + b)^1$	$a + b$
$(a + b)^2$	$a^2 + 2ab + b^2$
$(a + b)^3$	$a^3 + 3a^2b + 3ab^2 + b^3$
$(a + b)^4$	$a^4 + 4a^3b + 6a^2b^2 + 4ab^3 + b^4$
$(a + b)^5$	$a^5 + 5a^4b + 10a^3b^2 + 10a^2b^3 + 5ab^4 + b^5$

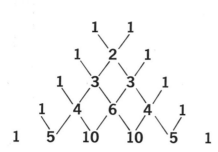

FIG. 7–5 *The Pascal triangle. This triangle can be used to obtain the coefficients of binomial equations. Note that each number after the first is obtained by adding the numbers above it. The second number in each line indicates the power of the equation. Thus, the last line has 5 as its second number and is used in expanding* $(a + b)^5$.

Let us now go back to our original problem of the parents heterozygous for albinism. What are the chances that they will have two normal children and one albino in their future family, assuming that they plan to have three additional children? To answer this question we would use the expansion of the binomial $(a + b)^3$.

Let us allow a to represent albinism, which has a chance of ¼, and b to represent normal, which has a chance of ¾. We select the term $3ab^2$ for one albino and two normal children. If we substitute the probabilities and multiply this out, we get ²⁷⁄₆₄. Figure 7–6 shows the type of curve which results when all the probabilities are worked out for a family of four children from such heterozygous parents.

Combination of Two Methods

In some types of problems we can combine the two methods which we have studied. For instance, if we wish to determine the chances that the heterozygous couple will have two albino boys and one normal girl in their future family, before we use the binomial $a + b$, we must first obtain the probabilities of the two

FIG. 7–6 *A probability curve formed by an expansion of the binomial $a + b$ to the fourth power, where a has a chance of one-fourth, and b has a chance of three-fourths. Note the skewed nature of the curve as compared with the curves in Figure 7–3.*

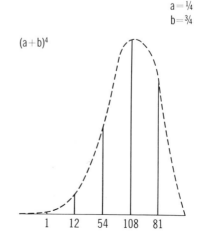

$a = ¼$
$b = ¾$

$(a+b)^4$

1   12   54   108   81

events. We know that the chance for an albino is ¼ and the chance for a boy is ½, so we multiply these two fractions to obtain ⅛ as the chance for an albino boy. To obtain the chance for a normal girl we multiply ¾ by ½, which gives ⅜. We can now let a represent the chance for an albino boy (⅛) and b represent the chance for a normal girl (⅜). We expand the binomial $(a + b)^3$, since there are three children involved, and choose the expression $3a^2b$ to solve our problem. When we substitute the fractions and multiply, we obtain %₁₂ as the answer to our problem.

Reverse Application of Laws of Probability

It is possible to apply the principles of probability in reverse to determine the chances of independent events when one knows the chance of two such events happening together. For instance, we learned that, in tossing pennies, the chance of tossing two heads in succession is ¼. Now, suppose that from this figure we want to determine the chance of tossing one head in one toss. We take the square root of ¼ and obtain ½. We can thus say, **the chance that either of two independent events of equal frequency will occur separately is equal to the square root of the chance that the two will occur together.** Please note that the events must be of equal frequency. Thus while the chance that heterozygous parents will have an albino boy is ⅛, the chance that they will have an albino child of either sex is not the square root of ⅛. This is true because the chance of having a boy and the chance of having an albino are different.

Now let us see how this principle could be applied in a genetics problem. Suppose that one baby out of every 100,000 which is born in the United States dies at birth because of an inherited lung deformity which results from a recessive autosomal lethal gene. How many such lethal genes are there in the living population? Each baby which dies because it is homozygous represents the coincident occurrence of independent events of equal frequency: the union of germ cells both of which carry the gene. Therefore the chance that any one germ cell in the population will carry the gene is equal to the square root of 1/100,000 or about 1/316. A more extensive application of this principle is discussed in Chapter 23 under the Hardy-Weinberg law.

PROBABILITY IN SOLVING DIHYBRID RATIOS

The principles of probability can be used to obtain the phenotypic ratios to be expected from dihybrid, trihybrid, and even higher hybrid crosses, thus obviating the necessity of using extensive checkerboards when only phenotypic ratios are desired. It is possible to obtain these results in this manner because each of these multiple hybrids is actually only two or more monohybrids segregating at the same time and we can use the multiplication method to obtain the chance of both happening together.

As an example, in the guinea pig when homozygous black is crossed with homozygous white, we obtain a ratio of three black to one white in the second generation. Likewise, a cross between homozygous short-haired guinea pigs and homozygous long-haired guinea pigs yields a second generation of three short to one long. Now suppose both crosses are conducted simultaneously. We can obtain the number of black-short to be expected by multiplying the expectation of each trait independently:

$$\text{¾ black} \times \text{¾ short} = \text{⁹⁄₁₆ black-short}$$

To find the number of black-long to be expected, we again multiply the chance of each trait appearing independently:

$$\text{¾ black} \times \text{¼ long} = \text{³⁄₁₆ black-long}$$

To get the entire dihybrid results we can use algebraic multiplication:

$$\frac{\begin{array}{r}\text{¾ black} + \text{¼ white} \\ \times\ \text{¾ short}\ + \text{¼ long}\end{array}}{\text{⁹⁄₁₆ black-short} + \text{³⁄₁₆ white-short} + \text{³⁄₁₆ black-long} + \text{¹⁄₁₆ white-long}}$$

If we wish to add a third trait to get a trihybrid ratio, we can multiply the phenotypic ratio of a third cross with the above results.

We can even use this method in cases of epistasis if we keep in mind which gene is epistatic to the other. For instance, if we use the duplicate recessive epistasis involving two separate genes which can cause deaf-mutism in man, we obtain the ratio as follows:

$$\frac{\begin{array}{r}\text{¾ normal } (A) + \text{¼ deaf } (aa) \\ \times\ \text{¾ normal } (B) + \text{¼ deaf } (bb)\end{array}}{\text{⁹⁄₁₆ normal } (AB) + \text{³⁄₁₆ deaf } (aaB) + \text{³⁄₁₆ deaf } (Abb) + \text{¹⁄₁₆ deaf } (aabb)}$$

Ratio: 9 normal : 7 deaf-mutes

This method can be used for test crosses or any combination of genes in the parents. Just remember to obtain the monohybrid ratio for each trait first and then multiply the two ratios together. As a final example, let us assume that a man heterozygous for the dominant gene for piebaldness (unpigmented spots on the body) is also heterozygous for the recessive gene for albinism. He marries an albino woman with normal parents. From such a marriage we would expect a ratio of one piebald to one normal. We would also expect a ratio of one albino to one normal. We must keep in mind, however, that an albino person cannot show the dominant gene for piebaldness because albinism causes a complete absence of pigment all over the skin. Thus, the results would be:

$$\frac{\text{¼ albino (piebald-masked)} + \text{¼ albino} + \text{¼ piebald} + \text{¼ normal}}{\begin{array}{c}\text{½ piebald} + \text{½ non-piebald} \\ \times\ \text{½ albino}\ + \text{½ non-albino}\end{array}}$$

Ratio: 2 albino : 1 piebald : 1 normal

ANALYSIS OF RESULTS

When we compare the actual results of a genetic cross with the expected results which have been calculated by the laws of probability, it is obvious that they will seldom come out in exact agreement. In fact, the laws of probability themselves tell us that we will expect deviations from the ratio because of chance. If we flip a coin one hundred times and obtain 46 heads and 54 tails, we do not think such an occurrence to be unusual. We understand that chance variation could easily account for this deviation from the mathematical ratio of 1:1. But suppose we get 40 heads and 60 tails — is this deviation from the ratio great enough to indicate that perhaps something other than chance is operating to cause tails to come up more often? Perhaps the metal is heavier on the head side of the coin. What if we should get 25 heads and 75 tails? This would surely seem to indicate that something other than chance is operating. But where do we draw the line? Theoretically, we could actually get 100 tails and no heads from pure chance. The chance of obtaining this unusual distribution, however, would only be one in billions, so for practical purposes we can consider such a distribution as impossible unless something other than chance is intervening. As an arbitrary cut-off point, we generally say that something other than chance seems to be involved when the distribution would occur less than one in twenty times, or 5 per cent of the total cases. There are several ways in which we can determine when deviations from the mathematical ratio are so great as to exceed this cut-off point. One of these methods, which is very widely used, is known as **chi-square.**

Determining Goodness of Fit by Chi-Square

When we use a mathematical test to determine how close an obtained ratio fits the "expected" ratio, we say we are determining the "goodness of fit," that is, how close the obtained ratio fits the "expected" ratio. The equation for determining chi-square (χ^2) is as follows:

$$\chi^2 = \frac{(x_1 - m_1)^2}{m_1} + \frac{(x_2 - m_2)^2}{m_2}$$

In this equation we allow x_1 to represent the observed number of one group; x_2 to represent the observed number of the other group; and m_1 or m_2 to represent the "expected" number of each group.

To illustrate how this equation is used, let us take the results obtained from a cross between a black male and a gray female that is heterozygous for black body color in *Drosophila*. The gene for black body color is recessive. Among the offspring there are twenty-six black flies and thirty-six gray flies. In a cross of this nature the expected ratio is 1 : 1, ½ gray and ½ black. A perfect 1 : 1 ratio in a total of sixty-two flies would be thirty-one of each. Each group obtained deviates from the "expected" value by 5. Is this a significant deviation? It might be that the gene for black reduces the viability of the flies; or there may be epistatic genes which mask this color in some cases, or environmental factors may play a part in the expression of the gene for black. Chi-square gives us an objective means of evaluating the degree of deviation.

Total number of offspring	62
Expected numbers of black	31 (m_1)
" " " gray	31 (m_2)
Obtained numbers of black	26 (x_1)
" " " gray	36 (x_2)
Deviation of both groups from expected	5

$$\chi^2 = \frac{(26-31)^2}{31} + \frac{(36-31)^2}{31} = \frac{25}{31} + \frac{25}{31} = 1.6$$

Although we now know the chi-square value for this set of numbers, what does this tell us about the goodness of fit of our results? To determine this we must consult a chart of chi-square values (Table 7–1). In the left-hand column you will note numbers which indicate the degree of freedom. In this problem where we have two classes of offspring, there is one degree of freedom. The degrees of freedom of choice are one less than the total number of classes involved. If there are three classes of offspring, there are two degrees of freedom; four classes would mean three degrees of freedom, etc. This reasoning is predicated on the fact that there is no choice for the final class. If you are judging a beauty contest with only four contestants and are to give four prizes, you have a freedom of choice for first prize, second prize, and third prize, but no freedom of choice for the fourth prize. The girl remaining must get the fourth prize, since she is the only one left. Likewise if you have sixty-two offspring in the above problem and twenty-six have black body color, the remaining thirty-six must be gray, if only these two alternatives exist. Hence, there is only one degree of freedom.

TABLE 7–1

Chi-Square Numbers Arranged According to Chance Occurrence in Percentage

Degrees of freedom	Possibility of chance occurrence in percentage (5% or less considered significant)								
	90%	80%	70%	50%	30%	20%	10%	5% (sig.)	1%
1	0.016	0.064	0.148	0.455	1.074	1.642	2.706	3.841	6.635
2	0.211	0.446	0.713	1.386	2.408	3.219	4.605	5.991	9.210
3	0.584	1.005	1.424	2.366	3.665	4.642	6.251	7.815	11.341
4	1.064	1.649	2.195	3.357	4.878	5.989	7.779	9.488	13.277
5	1.610	2.343	3.000	4.351	6.064	7.289	9.236	11.070	15.086
6	2.204	3.070	3.828	5.348	7.231	8.558	10.645	12.592	16.812
7	2.833	3.822	4.671	6.346	8.383	9.803	12.017	14.067	18.475
8	3.490	4.594	5.527	7.344	9.524	11.030	13.362	15.507	20.090
9	4.168	5.380	6.393	8.343	10.656	12.242	14.684	16.919	21.666

Having settled this point, we select the top horizontal row of figures corresponding to one degree of freedom. For a different problem with three degrees of freedom, we would choose the third horizontal row. We find the values of chi-square on this table which most closely match the value which we obtained. Our obtained number of 1.6 falls between 1.074 and 1.642 in the table, which indicates probabilities of 30 per cent and 20 per cent, respectively. The obtained chi-square is actually very close to the latter. This means that the deviation which we obtained in our *Drosophila* cross would be expected to be this great or greater in about 20 per cent of the cases when such a cross was made, provided that only chance is operating. Had the chi-square which we obtained been greater than 3.841, we would have considered that something other than chance was affecting our results. This is the number on the chi-square scale which represents only a 5 per cent chance of occurrence, which we have chosen as the level of significance. In this particular cross, therefore, we can conclude that the deviation we obtained was not at all unusual and that it can be ascribed to pure chance.

Since the chi-square we obtained does not exactly agree with the closest number in the table, we can be more precise by indicating that the chi-square lies between the 20 and the 30 per cent columns. We do this as follows:

$$30\% > P > 20\%$$

This expression means that 30 per cent is greater than the probability of chance occurrence and 20 per cent is less than such a probability. We can be still more specific by interpolation, that is by calculating the exact point between the two percentages in the table where the observed chi-square falls. When we do this we can say that $P = 20.08\%$.

When we first began studying ratios in connection with Mendel's crosses in garden peas, we emphasized the fact that ratios are more likely to adhere to the "expected" numbers when the total numbers were large. We can now show that this is true by considering an example from *Drosophila* in which we have a larger number of flies than in our previous problem.

Assume that in another cross of black males by gray (heterozygous black) females we obtain five hundred and sixty flies instead of only sixty-two. These are distributed as follows:

Total number of offspring	560
Expected numbers of black	280 (m_1)
” ” ” gray	280 (m_2)
Obtained numbers of black	270 (x_1)
” ” ” gray	290 (x_2)
Deviation of both groups from expected	10

$$\chi^2 = \frac{(270 - 280)^2}{280} + \frac{(290 - 280)^2}{280} = 0.72; P = 43\%$$

When we compare these results with the first set of results, it will be noted that this deviation is greater in absolute magnitude (10 as compared to 5), but the percentage deviation is smaller —10 is only 1.8 per cent of the total number of flies (560), while in the earlier cross 5 was 7.9 per cent of the total (62). This relation is reflected in the smaller chi-square for the second set of results and in the fact that a deviation of the magnitude seen in the second set is expected to occur in 43 per cent of the cases. Neither of the results in these two crosses shows a significant deviation from what might be expected on the basis of the laws of chance, but the odds against the occurrence of so large a deviation are greater in the first experiment than in the second experiment. These results show why experimental biologists like to obtain as large numbers as feasible whenever they are dealing with studies involving statistical analysis.

As an example of how chi-square may be applied to human populations, consider a family pedigree which shows a rather heavy incidence of cancer. Does heredity influence a person's chance of dying of cancer? Perhaps we can gain a clue by a study of a particular family pedigree. Suppose this pedigree shows that thirteen out of sixty-four persons have cancer listed as the cause of death. Now, assume that statistical studies of deaths in the United States show that, during the past 20 years, one death in each ten has been the result of cancer. Is the number of deaths in this one pedigree greater than would be expected by the laws of chance? Chi-square can help us to find out.

Total number of deaths in pedigree	64
Expected deaths from cancer (10%)	6.4 (m_1)
Expected deaths from other causes (90%)	57.6 (m_2)
Observed deaths from cancer	13 (x_1)
Observed deaths from other causes	51 (x_2)
Deviation of observed from expected	6.6

$$\chi^2 = \frac{(13 - 6.4)^2}{6.4} + \frac{(51 - 57.6)^2}{57.6} = 7.5; P < 1\%$$

This chi-square value with a probability of chance occurrence below 1 per cent shows rather definitely that the number of cases of cancer in this pedigree must be due to something other than chance. We might presume that heredity is involved, but this certainly does not exclude other factors which might be more commonly found in the environment of this particular family than in that of the people of the United States as a whole.

It is possible to work out chi-square problems by a tabular method and some prefer this procedure. Table 7–2 shows how this method can be used with some of Mendel's dihybrid results as well as how chi-square can be used when there are more than two classes in the results. In the latter case there are four classes and thus three degrees of freedom. We must keep this in mind and use the third horizontal column of figures in Table 7–1 when determining the value of *P* in this problem. We obtain a chi-square of 0.5102, which indicates that the

TABLE 7–2

**Tabular Method of Working Out the Chi-Square Value
for One of Mendel's Dihybrid Crosses With Peas**

Phenotype of peas	Observed frequency x	Expected frequency m	Deviation squared $(x - m)^2$	$\dfrac{(x - m)^2}{m}$
Yellow-round	315	313	4	0.0128
Yellow-wrinkled	101	104	9	0.0865
Green-round	108	104	16	0.1538
Green-wrinkled	32	35	9	0.2571
Σ	556	556		0.5102

$\chi^2 = 0.5102$ Upper limits of chi-square at the
5 per cent level of significance 7.815
$P > 90\%$

deviations obtained by Mendel would occur by chance in more than 90 per cent of the time. We should not conclude this discussion of the use of chi-square without a note of caution. The minimal number in any group observed or expected should be at least 5 before chi-square is a valid measure. It is better if the smallest number is larger, but 5 should be considered the absolute minimum if the use of this method is to be reliable.

In this chapter we have learned how certain mathematical procedures can be used to obtain an objective evaluation of the results of genetic crosses. Many other procedures are also applicable to genetics, and we shall take up some of these as we continue our study.

PROBLEMS

1. In tossing pennies, what are the chances of tossing eight successive tails? Would this be the same as the chances of obtaining eight tails if eight pennies were tossed simultaneously? If not, explain.

2. Suppose you toss a penny three times and it has come up heads each time. What are the chances that it will come up heads on the fourth toss?

3. A certain couple have three boys and are expecting their fourth child. What are the chances that it will be a boy?

4. Another couple has just married. They plan to have four children and would like to have three girls and a boy. What are the chances that their desires will be fulfilled?

5. Some dogs bark while they are trailing an animal; others are silent. The barking trait is inherited as a dominant. A barker, who had a silent mother, is mated to a silent trailer. The female bears four puppies. What are the chances that there will be two of each kind in the group?

6. In man, brachyphalangy (shortened fingers) is a heterozygous expression of a gene which causes death when homozygous. Two persons with brachyphalangy

married during the last century and have three grown children. What are the chances that two of these will be normal and one display brachyphalangy?

7. Two heterozygous black, short-haired guinea pigs are crossed. Both carry the genes for white and long hair in the recessive state. They bear only two offspring. What are the chances that one of these offspring will be white-long and the other will be black-short?

8. A poultryman has all blue Andalusian hens and roosters in his flock. He incubates three eggs and hopes to obtain two white-splashed hens and one black rooster from the chicks which hatch from the eggs. What are the chances that his wishes will be fulfilled?

9. In summer squash, disc fruit is dominant over spherical fruit. A farmer plants seeds which are from heterozygous parents and plants three seeds to a hill. Assuming that they all grow, what proportion of the hills will bear two plants with spherical fruit and one plant with disc fruit? What proportion of the hills will bear all spherical fruit?

10. The chance of throwing two sixes with a pair of dice is $\frac{1}{36}$. What is the chance of throwing one six with one die?

11. Use the probability method of multiplying to obtain the phenotypic ratio to be expected from a cross of heterozygous black short-haired guinea pigs with guinea pigs which are white (homozygous) and heterozygous short.

12. In domestic chickens a dominant gene produces rose comb and the recessive allele produces single comb. The blue Andalusian condition is an intermediate expression of the genes for black and white-splashed. Suppose you cross two blue Andalusians, both of which have rose combs, but which carry the gene for single comb (are heterozygous). Use the probability method of determining the phenotypic ratio to be expected among the offspring of these two chickens.

13. About 1 person in each 20,000 in Europe is an albino. What proportion of the gene pool of Europeans includes the gene for albinism? In other words, what fraction of the germ cells of Europeans contains the gene for albinism?

14. Dr. H. Bentley Glass of The State University of New York at Stony Brook studied an isolated group of people in Pennsylvania known as the Dunkers. He found that about 60 per cent of them have type A blood whereas only 40 per cent of the people in the United States have type A blood. There were 228 people in this sample of the Dunkers. Is this difference greater than what would be expected on the basis of chance alone?

15. In a certain family pedigree it was found that among the living children produced there were 36 girls and 20 boys. Some old members of the family say that a witch put a curse on the family many years ago which causes fewer boys to be born. Use chi-square to obtain an objective evaluation of the ratio of the sexes in this family and to determine whether factors other than chance appear to affect the ratio. Use a 1 : 1 ratio for the expected distribution of the two sexes.

16. On one pea plant which grew from a seed which was heterozygous round, Mendel obtained 26 round seeds and 6 wrinkled seeds. Do these results fall within the normal probabilities of a chance distribution of the expected 3 : 1 ratio?

17. Two blue Andalusian chickens are crossed. Eggs taken from the hen are incubated and chicks with the following traits are hatched out: 10 white-splashed, 18 Andalusian blue, and 6 black. Do these numbers agree with the expected genetic ratio of 1 : 2 : 1?

8

The Determination of Sex

MANY PERSONS ASSUME THAT THE FUNDAMENTAL biological function performed by sex is reproduction. A brief study of biology, however, soon reveals the fact that reproduction is possible without sex — most plants and many of the simpler animals have asexual means of reproduction which are fully as efficient in accomplishing the propagation of the species; and some few species reproduce only by asexual means. Sex, however, has a second biological function in addition to reproduction, and one which asexual reproduction does not perform. This function consists of providing within families, populations, and species that genetic variety without which long evolutionary success is improbable. The contrast between the evolutionary value of sexual and asexual reproduction is well illustrated by Figure 4–2, which shows that the chance of securing favorable recombinations of genes is much greater under sexual reproduction. Hence, we find that sexual reproduction has become established in the great majority of the forms of life which exist on the earth, including bacteria, as noted in Chapter 4.

Sexual reproduction, as the term is used by biologists, does not necessarily imply a clear-cut distinction into male and female sexes. Any situation in which there is a genetic mingling of nuclear material from two different cells will come under the heading of sexual reproduction. In the one-celled protozoan, *Paramecium,* there is a form of sexual union called **conjugation,** in which two organisms unite and exchange nuclear material. Morphologically, we cannot distinguish between the two conjugants, but experimental evidence shows that they are different physiologically. In other words, conjugation cannot take place between just any two *Paramecia;* the two individuals must have some sexual distinction of a physiological nature before the process can occur. These animals also differ from higher forms of animal life in that there are not necessarily just

115

two sexes (or mating types) as we find in higher animal life. In *P. bursaria,* Jennings found eight different mating types. An individual of any one of these types can mate with one of any of the other mating types, but not with other animals of the same mating type as itself. In plants, the bread-mold, *Mucor,* is another good example of an organism in which conjugation occurs between physiologically different individuals and which has a number of different mating types. Such mating types are possibly due to variations in the alleles of genes related to the physiological distinctions of sex, and would be determined on the same genetic basis as applies to other characteristics. Among the more advanced forms of life, the mating types are reduced to two, called sexes, possibly through an evolutionary adaptation because of the greater efficiency of having only two types rather than as many as could be produced by all the combinations of all the mutant alleles involved.

In higher animals these two types — male and female — exhibit a great variety of differences from one another, differences far too extensive to be accounted for merely by the principles of inheritance as we have presented them up to this point. Let us consider two children of the same parents — one a boy and the other a girl. These two young persons will show a great variety of distinctions which extend far beyond the obvious characteristics associated with reproduction and lactation. Every part of their bodies is affected — the skin, the muscles, the blood, the hair, the bones; in fact there is hardly any part of the human body which does not show some distinction according to the sex of the individual. How can two persons who are so different be the offspring of the same parents? We certainly cannot explain these distinctions on the basis of a difference in a single gene, such as would result in variation in the color of the eyes among children of the same parents. The differences are far too extensive for that. Neither could we find a satisfactory possible explanation in the differences of a number of individual genes, such as those involved in differences of body height. Such multiple gene variations result in quantitative rather than qualitative effects. People are not either tall or short; there are countless gradations between the extremes. With regard to sex, however, such gradations do not normally exist. People are either male or female, with no gradations from one to the other such as would be expected if we should seek to explain sex determination on the basis of multiple gene inheritance. (We will postpone a consideration of those very rare cases of mixed sex until we have gained a better understanding of our subject.) Sexual characteristics therefore must depend upon some genic reaction which is different from the method of inheritance of characteristics unrelated to sex. This distinctive mechanism of inheritance will be the subject for consideration in this chapter. The mechanism is not the same for all living organisms, so we must learn several different methods.

SEXUAL DIFFERENTIATION IN MONOECIOUS ORGANISMS

The word **monoecious** is a term used to describe a plant or animal which produces both types of gametes within itself. In other words, in any monoecious

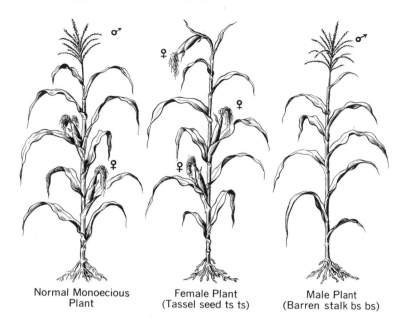

Normal Monoecious
Plant

Female Plant
(Tassel seed ts ts)

Male Plant
(Barren stalk bs bs)

FIG. 8–1 *Genetically, the line of distinction between monoecious and dioecious organisms is thin. The corn plant normally bears both male sex organs (tassels) and female sex organs (ears) on one stalk. One gene causes an ear to be formed in place of tassels, thus producing a plant bearing only female organs. Another gene causes a barren stalk, producing a plant with only male organs.*

species there is only one kind of organism with reference to reproduction; no distinction between males and females exists. Sex determination in this type of organism requires no unusual genic mechanism. In most higher plants, for example, the sexual differentiation is limited to male and female reproductive organs within the flowers, which appear only after the plant is mature and ready for seed production. Genes for stamens and genes for pistils express themselves at the proper time and place just as the genes for leaves, roots, bark, and other parts of the plant express themselves at the proper time and place. The genes for these sex organs are present in every cell of the plant body — just as the genes to produce leaves are in every cell — yet these genes express themselves only at a certain place during flower formation, just as the genes for leaves express themselves only at the proper place on the stem.

This inherent ability of all cells to produce both types of sex organs is well illustrated by maize (corn). In this plant the ears (female flowers) normally develop along the sides of the stalk, and the tassel (male flowers) at the top of the plant. There is a mutant gene, however, which may convert the male flowers into female flowers, so that grains of corn may actually be produced at the tip of the plant where the tassels are normally found. This shows that the cells in this region have the power to produce both sex organs and that under the stimulation of the mutant gene they will produce female rather than male flowers. Another mutant gene suppresses the development of the ears, and when affected by this gene a plant bears male flowers only. When affected by both these mutant genes, a plant bears female flowers only, but these are at the upper tip rather than in the normal place for female flowers along the side of the stalk. This shows us not only that cells in different parts of the plant possess genes for the production of both sex organs, but also indicates the rather minor genetic difference between the monoecious and dioecious organisms. Variations in two genes, in this case, convert a normally monoecious organism into a dioecious one.

In some of the lower animals there is also a monoecious type of sexual reproduction. *Hydra* produces male and female gonads only after it has reached maturity, and their appearance at a definite place requires no more complicated genic basis than the appearance of tentacles at the proper place during the embryonic development of the animal. Sponges have a similar method of producing gonads and gametes. In some monoecious animals the differentiation into male and female tissue occurs much earlier in the life of the individual, but the same principle of formation holds true. An earthworm, for instance, begins to produce the tissue to form both testes and ovaries early in its embryonic life. Their origin, however, is no different from the origin of the crop, the gizzard, the esophagus, or the heart, and we need no special mechanism of sex determination to account for sexual differentiation early in the development of monoecious individuals.

SEXUAL DIFFERENTIATION IN DIOECIOUS ORGANISMS

When we begin to study **dioecious** organisms (those with distinct male and female sexes) we face the problems of sex determination discussed at the beginning of this chapter. Even though the appearance of sex organs comes late in the life of the organism, there must be some mechanism for determining which of the two types of sex organs will make their appearance. For instance, cedar trees may all look alike as they grow and mature, yet when they produce cones, some trees will produce only female cones and others only male cones. There must be some distinction in the tissue which extends to all parts of the plant, for new plants which may arise from a portion of the roots will continue to produce the same type of cones as were produced by the original plant.

Sometimes a hopeful gardener will set out a young holly tree, brought in from the forest, with visions of bright green foliage and beautiful red berries for Christmas ornamentation. Unfortunately, however, the holly tree is dioecious and the gardener may get foliage without berries should he happen to set out a tree of the male sex. A retired businessman in Florida may wish to grow papayas, a delicious tropical fruit. He may plant seed or set out young plants; yet he will find that only about one-half of the plants bear fruit. The others are males and bear no ovules from which the fruit may arise.

In most dioecious animals, however, the sexual distinctions appear much earlier in the life of the individual. In some simpler forms of animal life these distinctions are restricted to the gonads. Starfish are dioecious, yet one cannot detect any external indication of sex at any stage of their life cycle. Internally, only the gonads show any differences according to the sex of the animal. In the great majority of animals, however, there is a distinct **sexual dimorphism** (sexual differences which are clearly evident from external observation). Some male butterflies are so different from the females that they would be classified as a different species had they not been observed mating with the females. The plumage and head furnishings of birds provide another well-known case of extreme sexual dimorphism.

In spite of these variations in the time when sexual differences appear and in the extent of these differences in dioecious forms of life, the problem of sex determination in such organisms is similar to that in the preceding examples cited. As in monoecious organisms, we have ample evidence to indicate that in dioecious ones all cells, and therefore all organisms, with a diploid set of chromosomes are **bipotential** as to sex. That is, all have the necessary genes to produce the primary and secondary characteristics associated with both sexes as well as the less tangible chemical differences between gametes. The cells in a woman's body contain all of the genes for the production of a fully formed beard, yet a beard normally does not develop. A man, on the other hand, has all the genes necessary for fully formed feminine breasts, yet his breasts normally remain rudimentary and do not realize the potentialities of these genes.

With these facts in mind, the problem of understanding sex determination narrows itself down to discovering a mechanism for the stimulation of one set of genes to full expression while, at the same time, inhibiting the set of genes for the opposite sex so that they will not be expressed. The problem thus becomes much simpler than it would appear to be without the concept of sexual bipotentiality of all diploid cells. The next step of our study logically concerns the means by which this simultaneous stimulation and inhibition of genes are accomplished. Since the mechanism is not the same in all dioecious forms of life, we shall consider various mechanisms separately.

Sex Determination by Environment

In a few forms of life the male or the female genes are brought into expression by environmental influences. One of the most interesting cases of this occurs in the marine worm, *Bonellia*. This is a creature which lives a rather sedentary existence in the shallow water near the shore of the ocean in many parts of the world. *Bonellia* exhibits a most extreme sexual dimorphism, for the female is several inches long and has a long proboscis extending out from the anterior end of the body, while the male is a minute, degenerate creature which lives within the reproductive tract of the female. The male is thus conveniently located for fertilization of the eggs produced by the female, but this seems to be about the only activity in life for which he is fitted.

Because of such extreme differences between the sexes, *Bonellia* furnishes excellent material for a study of sex determination. After considerable experimentation, Baltzer found that if a single worm is reared from the egg in isolation

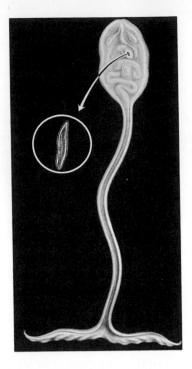

FIG. 8–2 *The marine worm, Bonellia. A mature female is shown at the left, bearing the tiny parasitic male in her uterus. An enlarged view of the male is shown in the circle at right.*

from all others of the same species, it invariably develops into a female. On the other hand, if newly hatched worms are released in water containing mature females, some of the young worms attach themselves to the proboscis of the female and exist parasitically by sucking nourishment from the female host. Such young worms mature rather quickly into the tiny males, which then migrate down to the reproductive tract of the female, enter it, and take up their abode there. From these results, it would seem as if there is something in the secretions absorbed by each male from the proboscis of the female which causes the genes for maleness to be expressed. Such a substance would have an effect like the hormones of vertebrates, which have the power to stimulate the expression of genes related to one sex and inhibit genes related to the opposite sex.

Another well-known case of sex determination by environment is found in the marine mollusk, *Crepidula*. If the young animals are reared alone they always become females, but if reared in close proximity to an adult they become males. Again, this would seem to be a case of hormone-like effect from some secretion of the adult animals.

These cases of sex determination by environment achieve a result which is genetically beneficial. The very nature of the process insures that there will be some females and some males. This method is far from ideal, however, because of the likelihood of disproportion in the production of the two sexes. It is not surprising to find that this somewhat inefficient mechanism for sex determination should be supplanted by more stable ones in most dioecious organisms.

There are only a very few organisms which are thus influenced by their environment. Since chromosomes follow a regular pattern of distribution in meiosis, they offer a much more stable mechanism upon which sex determination can be based.

Sex Determination by Chromosomes: the XY Method

In the great majority of dioecious organisms, the chromosomes play a major role in the determination of sex. We can use our old standby, *Drosophila melanogaster,* to illustrate the most common method of sex determination by chromosomes which has come to be known as the *XY* method. We have previously brought out the fact that this insect has eight chromosomes in its diploid cells. In female cells these are represented by four like pairs of chromosomes — "like" in the sense that for each chromosome in the cell there is an identical mate. In a metaphase plate these may be seen as one pair of short, dot-like chromosomes, one pair of long rod-like chromosomes, and two pairs of V-shaped chromosomes. Examination of male diploid cells, however, reveals a difference. For these contain three like pairs and one unlike pair. The male cells have the two pairs of V-shaped chromosomes and the one pair of dot-like chromosomes, but there is only one long rod-like chromosome and, to mate with it, a straight chromosome with a hook on the end of it (Figure 8–3). Here, then, we find a chromosomal difference between the sexes which extends to every cell of the fly's body. It is obvious that this chromosomal difference must be related to sex determination.

For convenience, the long, rod-like chromosomes are called **X-chromosomes,** while the unlike mate with the hook on it is called the **Y-chromosome.** Both types are called **sex chromosomes.** For distinction in discussing them, the rest of the chromosomes are called **autosomes.** Thus, *Drosophila melanogaster* has three pairs of autosomes and one pair of sex chromosomes in each diploid cell. The gametes, being haploid, each carry only three autosomes and one sex chromosome.

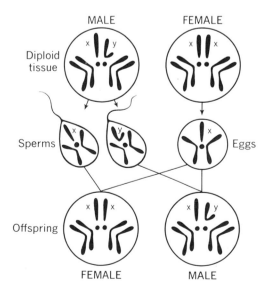

FIG. 8–3 *The sex chromosome differences in Drosophila. Male diploid tissue contains both an X- and a Y-chromosome. Female diploid tissue contains two X-chromosomes. Sperms are of two types, one carrying the X-chromosome and one the Y-chromosome. The sex of the offspring depends upon which of the two types of sperm fertilizes the egg.*

In oogenesis the four pairs of chromosomes line up in the metaphase and separate and go to the two poles. The egg and the polar bodies always receive the same kind of chromosomes — one *X*-chromosome and three autosomes. Thus, all eggs carry this **genome,** that is, haploid set of chromosomes. In spermatogenesis the chromosomes also pair in meiosis, but here the two members of one pair are unlike. As a result, the secondary spermatocytes (and later the spermatids and sperms) will be of two distinct types — one carrying three autosomes and an *X*-chromosome, and one carrying three autosomes and a *Y*-chromosome. These two types will be produced in equal quantities, for, every time a secondary spermatocyte is formed with an *X*-chromosome, another secondary spermatocyte will be formed with a *Y*-chromosome. Sex determination, therefore, depends upon which of the two types of male reproductive cells fertilizes the egg. Eggs fertilized by the *X*-bearing sperms will produce females, while those fertilized by *Y*-bearing sperms will result in males. Since these two types of sperms are produced in equal quantities, the proportion of the sexes would normally be expected to show a ratio of 1 : 1. This is a much more reliable mechanism for sex determination than the rather haphazard environmental one.

From these observed facts alone, one might think that the *Y*-chromosome, appearing only in males, was the important factor in sex determination. It would seem quite simple to assume that the *Y*-chromosome contains some factors which stimulate genes for maleness, present in all the body cells, to develop. In the absence of the *Y*-chromosome, on the other hand, it would seem that the genes for femaleness were free to develop, with the genes for maleness held in abeyance. Further investigation, however, reveals that the *Y*-chromosome plays no part in sex determination; in fact, it plays very little part in heredity at all. It is possible, in certain rare cases of abnormal meiosis to be discussed later, to obtain flies which contain the normal three pairs of autosomes, but only one *X*-chromosome and no *Y*-chromosome. Such flies will mature into perfectly normal males so far as appearance is concerned. They will be sterile, however, a fact which indicates that the *Y*-chromosome has something to do with fertility in the male. It is also possible to obtain flies with three pairs of autosomes, two *X*-chromosomes, and one *Y*-chromosome. Such a fly will be a perfectly normal and fully fertile female — so it is clear that the *Y*-chromosome in no way interferes with the expression of female characteristics.

This neutral state of the *Y*-chromosome in sex determination may be better understood if we learn more about the nature of this odd chromosome. Studies show that it is almost entirely devoid of genes. In an anthropomorphic flight of fancy we might surmise that the *Y*-chromosome is present in the group just to keep the *X*-chromosome from getting lonesome; we may imagine that, with each of the other chromosomes neatly paired with a mate at meiosis, the *X*-chromosome would feel neglected without a partner. Certain it is that the *Y*-chromosome is present as a mate even though it seems to play little part in the activities of the cell. If we compare chromosomes to sausages and the genes to the meat within the outer skin, we could think of the *Y*-chromosome as an empty skin with only a few bits of meat at one end, but puffed up to look like the other sausages. We shall see in Chapter 10 that this lack of genes has an important bearing on the method of inheritance for characteristics resulting from genes on the X-chromosome.

With the Y-chromosome eliminated as a factor in sex determination, we find that the only distinction remaining between male and female cells lies in the number of X-chromosomes they contain: females have two X-chromosomes, whereas males have only one. The mechanism of sex determination, therefore, must be related to this distinction. Yet how can the fact that a cell contains one or two of a given chromosome affect the sex of the individual? In an attempt to answer this question, Dr. C. B. Bridges worked out a very interesting and most convincing **ratio theory** of sex determination in *Drosophila*.

To understand this theory, let us keep in mind that there are genes for both sexes in all the body cells and that we search only for a trigger to trip the balance in favor of one or the other when we study any means of sex determination. It is as if we had a double-barreled shotgun, fully loaded and ready to shoot. One barrel will deliver a female charge, the other a male charge. Which charge is delivered depends upon which of the two triggers is pulled. Bridges found that the ratio of the X-chromosomes to the autosomes appears to be the trigger-pulling mechanism. It seems that the X-chromosome contains genes which tend to pull the female trigger while the autosomes contain opposing genes which tend to pull the male trigger. A single X-chromosome seems to contain female-stimulating genes with a relative strength of 1.5, while a complete haploid set of autosomes (three in *Drosophila*) contains male-stimulating genes with a relative strength of 1.0. Since we have introduced the word genome for a haploid set of chromosomes, we can use the term **autosomal genome** to refer to such a haploid set of autosomes. The Y-chromosome contains no sex-influencing genes and would be rated at 0 in such a scheme.

Considered in this light, an egg bearing three autosomes and an X-chromosome, fertilized by a sperm carrying three autosomes and an X-chromosome, would have two autosomal genomes with a strength of 2 in male determination and two X-chromosomes with a strength of 3 in female determination. Since 3 is greater than 2, the fly would be a female. On the other hand, if the egg were fertilized by a Y sperm, the fly would have two autosomal genomes with a male-determining value of 2, a single X-chromosome with a female-determining value of 1.5, and a Y-chromosome with a value of 0. Such a fly would be a male, since 2 is greater than 1.5.

Considerable support to this theory is furnished by studies of flies arising after an abnormal distribution of chromosomes. In rare cases the X-chromosomes of *Drosophila* fail to separate during the first meiosis of oogenesis, and both X-chromosomes go to one pole of the meiotic spindle. This gives some eggs with the normal single autosomal genome, but with two X-chromosomes; and some eggs with the normal single autosomal genome, but with no sex chromosomes at all. Such an unusual occurrence is called **non-disjunction** and is illus-

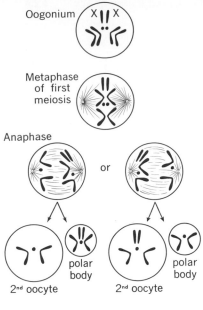

FIG. 8–4 *Non-disjunction in Drosophila melanogaster. In rare cases the two X-chromosomes adhere during the first meiosis and both go to the secondary oocyte or to the polar body. This produces some eggs with two X-chromosomes and others without any X-chromosomes.*

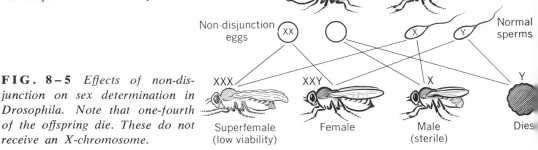

FIG. 8–5 *Effects of non-disjunction on sex determination in Drosophila. Note that one-fourth of the offspring die. These do not receive an X-chromosome.*

trated in Figure 8–4. The results of fertilization with the two types of sperm are shown in Figure 8–5. It can be seen that normal females and normal males may be produced, but the females will carry an extra Y-chromosome, although it does not influence their sex in any way. Also, the males will be sterile, since they lack the Y-chromosome, but will be perfectly normal in appearance and reactions toward the females. The fly receiving the Y-chromosome and no X-chromosome naturally dies, because it fails to receive the genes which lie on the X-chromosome and a complete body cannot be produced without them. The fly receiving the three X-chromosomes is a so-called "superfemale," for the ratio here is 4.5 for femaleness and 2 for maleness, which is greater than the 3 : 2 ratio characteristic of normal females. Interestingly enough, the superfemale does not, as the term might imply, show any exaggeration of the female characteristics. Rather, she is sterile and somewhat underdeveloped in her sexual characteristics.

Further evidence in support of the ratio theory may be found in other instances of abnormal meiosis where not only the X-chromosomes but also the autosomes are diploid in the gametes. Those flies receiving three autosomal genomes and three X-chromosomes have the same ratio of autosomes to X-chromosomes as is found in the normal females and are females with a triploid rather than a diploid chromosome number. The most interesting fly of all, however, is the one receiving three autosomal genomes and two X-chromosomes. The sex balance ratio in this case is 3 : 3 (equality) and the fly is said to be an intersex — its sexual characteristics are approximately intermediate between the two sexes. Table 8–1 summarizes the various effects which can be produced through different ratios of autosomes and sex chromosomes such as occur naturally in very rare instances and may be produced more abundantly by special genetic techniques. It will be noted from this table that it is possible to produce a supermale as well as a superfemale.

TABLE 8–1

Ratio Theory of Sex Determination

Chromosome combination	Ratio of male : female determiners	Sex
AAXX	2 : 3	female
AAXY	2 : 1.5	male
AAAXX	3 : 3	intersex
AAAXY	3 : 1.5	supermale
AAXXX	2 : 4.5	superfemale

A — one haploid set of autosomes with a male-determining value of 1.

X — one *X*-chromosome with a female-determining value of 1.5.

Y — one *Y*-chromosome with a value of 0.

The *XY* method of sex determination has been described primarily as it has been worked out in *Drosophila melanogaster,* but the majority of dioecious animals, including that very important animal, *Homo sapiens,* share this method of sex determination. It is also found in most dioecious plants, but there seem to be certain important differences in the ratios of the chromosomes related to sex determination. This is discussed later in this chapter.

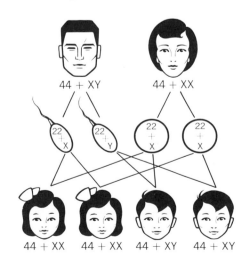

FIG. 8–6 *Sex determination in man. All eggs carry an X-chromosome, but about one-half of the sperms carry an X-chromosome and one-half carry a Y-chromosome. The sex of the children depends upon which type of sperms fertilize the eggs.*

The XO Method

Although the *XY* method of sex determination is by far the most common among animals, it was not the first to be discovered. In 1902, C. E. McClung, while studying meiosis in the testes of the grasshopper, noted that there were eleven pairs of chromosomes and an odd chromosome with no mate. He reasoned that this odd chromosome was an accessory one which was associated with sex determination. The true relationship was later worked out by E. B. Wilson on the bugs, *Protenor* and *Anasa,* and Nettie Stevens on the plant louse, *Aphis*. In the squash bug, *Anasa tristis,* Wilson found that the females have twenty-two chromosomes in somatic cells which form eleven pairs at meiosis. All eggs, therefore, bear the haploid set of eleven. Males, on the other hand, have only twenty-one chromosomes in their somatic cells and at meiosis there are only ten pairs and the odd accessory chromosome. He suggested that this be called the *X*-chromosome. About one-half of the sperms carry this *X*-chromosome and one-half do not. Fertilization of an egg by a sperm carrying the full eleven chromosomes will result in a female, while fertilization by a sperm with ten chromosomes will result in a male. This is known as the *XO* method of sex determination. It was not long after this, in 1905, that both Wilson and Stevens found the *Y*-chromosome as they extended their studies to other forms of insects.

The *XO* method is found in many *Orthoptera* and *Heteroptera* among the insects. It differs from the *XY* method only in the absence of a *Y*-chromosome. Since we have learned that the *Y*-chromosome plays no part in sex determination in *Drosophila,* no new principles are introduced by the *XO* method. We can easily see how it could have arisen from the *XY* method through a loss of the *Y*-chromosome in evolutionary development.

An unusual variation of the *XO* method of sex determination is found in the aphids (plant lice). These insects have parthenogenic reproduction through the spring and summer, when the females lay unfertilized eggs and these hatch into females only. In the fall, however, both sexes come from the eggs; these mate, and eggs are laid which remain dormant throughout the winter and hatch into females the following spring.

It might appear that such a peculiar set of circumstances would require some very distinctive method of sex determination, but, from the intensive study of the chromosomes by Boveri and Morgan, we can see that the pattern is not greatly different from that found in the grasshopper. These studies show that the winter eggs all have six chromosomes. The females which hatch from these mature and produce eggs which also have six chromosomes each. There is no reduction of chromosomes in their oogenesis, only a mitotic type of division. In the fall, however, there is a variation of this — one of the chromosomes (the *X*-chromosome) lags behind on the metaphase plate. Thus one of the daughter nuclei would receive six chromosomes, including two *X*-chromosomes, while the other would receive only five, with only one *X*-chromosome. Females hatch from those with six chromosomes and males from those with five chromosomes. Oogenesis in the females is normal and the eggs have three chromosomes each.

Spermatogenesis in the male is also normal, but spermatids that do not receive an *X* degenerate. Thus, all eggs are fertilized with *X* sperm and all zygotes hatch into females.

The ZW Method

A number of animals display a method of sex determination by which — as if to confuse beginning students in genetics — the relation of the sex chromosomes is exactly the reverse of the method which we have just studied. It is found in a comparatively large group of animals — butterflies, moths, caddis flies, birds, and some fishes. In these animals the male-determining genes are on the chromosome which corresponds to the *X*-chromosome of animals with the *XY* method. In order to avoid the confusion which would surely result if we used the *X* and *Y* designation for the sex chromosomes, it is common practice to use *Z* and *W* instead. Thus a rooster would have two like sex chromosomes which we would designate *ZZ*, while a hen would have two unlike sex chromosomes, *ZW*. Sperms would all carry a *Z*-chromosome, but the eggs would carry either *Z*- or *W*-chromosomes in approximately equal numbers. Sex determination, therefore, depends upon which of the two types of eggs is fertilized.

The Honeybee Method

The honeybee and certain other hymenopterans possess a peculiar type of sex determination which from some points of view might make them a subject of envy. These industrious insects, which are of such great commercial value, can produce either sex at will (if we assume for the moment that they have a will). Examination of the brood cells in a hive reveals two distinct sizes of cells. The smaller of the two are reserved for the development of the workers, which are females, while the larger cells are for drones, which are males. The cells are made by the workers, however, before the queen deposits the eggs in them, so it is imperative that she lay an egg for the proper sex in a cell designed for the development of that sex. To understand how she accomplishes this feat we must go back to her nuptial flight. Most of you will recall how the queen leaves the hive on her nuptial flight and flies high in the air, pursued by drones from her own and surrounding hives. She finally allows herself to be caught by a drone and is inseminated as the pair drops to the earth. Sperms from the male are stored in a seminal receptacle within her body and remain available for fertilization when eggs are laid throughout the remainder of her life. When she lays an egg in a worker cell, sperms are emitted from the seminal receptacle to fertilize the egg and it will develop into a female; all fertilized eggs form females. (Incidentally, the workers can make this egg into a queen by enlarging the cell and feeding the developing larva on a richer diet, but both workers and queens are females.) When the queen comes to a drone cell she exerts some sort of pressure on the ducts leading from the seminal receptacles so that sperm cannot pass out and fertilize the egg as it passes down the oviduct. Thus an unfertilized egg will be laid. Normally an infertile egg of most species of animals will not hatch. We all know that we have to keep a rooster around the barnyard if we

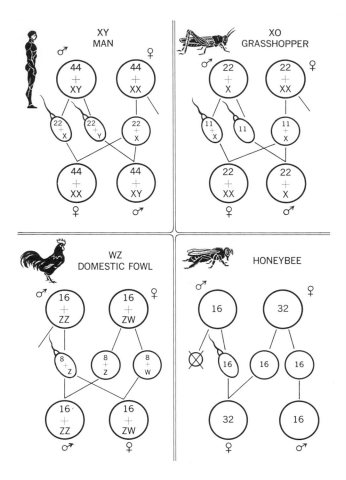

FIG. 8-7 *Comparison of the four types of sex determination by chromosomes in animals: man, the grasshopper, the domestic fowl, and the honeybee.*

expect the hens to obtain any results from their efforts to hatch their eggs. It should be mentioned, however, that infertile eggs of sea urchins, rabbits, and other animals have been induced to begin development without fertilization by experimental treatment with such things as ice water, salt water, and the point of a fine needle. Biologists at Beltsville, Maryland, even found that some eggs from a strain of white turkeys would hatch without fertilization. In the honeybee, however, it is normal for infertile eggs to hatch, and they always produce males. This gives us a method of sex determination which is quite distinct from the other chromosomal methods because the ratio of the chromosomes to one another is the same in both cases. We cannot say that one of these is a sex chromosome and that the others are autosomes.

The explanation for such an unusual method of sex determination has been brought out by the brilliant work of Whiting at the University of Pennsylvania. He has worked very extensively with the small parasitic wasp *Habrobracon juglandis* which, like the honeybee, is a hymenopteran and has the same method of sex determination. In this form there is a gene for sex which bears a large number of multiple alleles. We might call them x^a, x^b, x^c, x^d, and so on. Femaleness is caused when any two of these alleles are heterozygous. The haploid males, of course, since they develop from eggs which have not been fertilized, can carry only one of the alleles, which is equivalent to homozygosity of that allele. Whiting found that through close inbreeding of this wasp it was possible to produce diploid males, a result which would be expected whenever one of the alleles became homozygous in the diploid zygote. Such males are highly sterile, however, and would not function in reproduction in those rare cases where they might be found in natural populations. To illustrate, suppose we had a female heterozygous for the sex genes, x^a and x^z. When crossed with a male carrying x^m, the females would be x^a/x^m and x^z/x^m. Now, if the same male was crossed to his daughters, one-half of the diploid offspring would be homozygous x^m/x^m in both cases. These would be diploid males. Of course, at the same time the females would lay some infertile eggs which would hatch into the normal, fertile, haploid males.

SEX CHROMOSOMES IN DIOECIOUS PLANTS

We have already pointed out that the majority of the plants are monoecious and therefore have no problem of sex determination. There are a number, however, which are dioecious and require some mechanism to determine which type of reproductive organ is to be produced. Most of the dioecious plants which have been studied have the *XY* type of sex determination, but plants have also been discovered with the *XO* and the *ZW* methods. A few cases have been studied in an effort to determine how the ratio theory may be applied to plants, and the results, admittedly based on a limited number of studies, so far indicate that the male determiners are located in the *Y*-chromosome.

This was beautifully demonstrated by Westergaard and Warmke in a study of *Melandrium,* a seed plant in the pink family. In *Melandrium album,* the diploid female plants have twenty-four chromosomes — twenty-two autosomes and two *X*-chromosomes. Diploid males have the same chromosome number, but show a cytological distinction between one of the pairs into an *X*- and a *Y*-chromosome. In this species there are some plants which are tetraploid — have four haploid sets of chromosomes. These consist of three types — females which have forty-four autosomes and four *X*-chromosomes; males with forty-four autosomes, two *X*-chromosomes and two *Y*-chromosomes; and males with forty-four autosomes, three *X*-chromosomes and one *Y*-chromosome. This indicates that the *Y*-chromosome is the sex-determining element and carries the genes to produce maleness, that one *Y*-chromosome, even in the presence of three *X*-chromosomes, is sufficient to produce a male. Triploid plants may be obtained by crossing the diploid with the tetraploid. This results in thirty-three plus three *X* females, thirty-three plus two *X* plus *Y* males, and thirty-three plus *X* plus two *Y* males, a distribution in the males which further bears out the theory of the sex-determining power of the *Y*-chromosome. The final proof is furnished in studies involving fragmentation of the *Y*-chromosome, which are similar to the work of Dobzhansky and Schultz on *Drosophila*. Plants bearing both types of sex organs (hermaphrodites) have been obtained when about one-half of the *Y*-chromosome is missing from a plant without any other *Y*-chromosome. This implies at least two sex genes on the *Y*-chromosome.

Sex chromosomes in plants were first studied by Allen in the liverworts. It will be recalled from our study in Chapter 4 that in the liverworts the sex organs are borne on a haploid gametophyte. Allen found that male gametophytes had seven autosomes and a *Y*-chromosome while the female gametophyte had seven autosomes and an *X*-chromosome. After union of the sperm and egg a diploid zygote is formed with fourteen autosomes and one *X*- and one *Y*-chromosome. This forms the diploid sporophyte which, through reduction division, produces the haploid spores carrying only one of the sex chromosomes. These mature into male and female gametophytes, according to which of the two sex chromosomes they receive. This method of sex determination is illustrated in Figure 8–9.

FIG. 8–8 *The chromosomes of the male and female plants of the liverwort, Sphaerocarpos. These are haploid cells. Plant with the Y-chromosome is male, and plant with the X-chromosome is female.*

FIG. 8–9 *Sex determination in the liverwort, Sphaerocarpos. Note that meiosis comes before the production of male and female organisms rather than after, as in the higher plants and animals.*

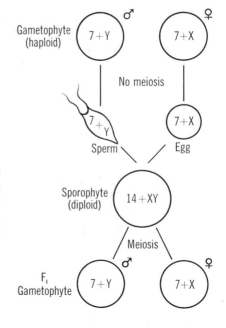

SEX DETERMINATION IN BACTERIA

After the discovery of sexual reproduction in bacteria, much research was directed toward the exact method by which genes pass from one cell to another. Electron photomicrographs revealed conjugation tubes connecting different bacterial cells, and it was first thought that there might be an exchange of genes somewhat on the order of conjugation of protozoans, such as occurs in *Paramecium*. Investigation of sexual reproduction in the bacterium, *Escherichia coli,* however, showed that this was not the case. Instead, it was found that in conjugation one of the two cells acted as a male and passed genes to a recipient or female cell. In Chapter 14 we shall learn more about this process, but at this point we shall confine our interest to the way in which one cell acquires the properties of maleness while another cell acquires femaleness.

Careful study has shown that the male or donor cells always possess an **episome,** which is distinct from the chromatin of the cell. This episome must be the male-determiner. In sexual reproduction one of these male cells joins onto one of the female cells and transfers a portion of its single chromosome into the female cell. The male cell now lacks some of its genes and, since the bacteria are haploid, it dies shortly after contributing genes to the female cell. Since the episome itself is not transferred in conjugation, as a general rule, it might appear as if there would be a gradual reduction in the male population to the point of extinction. Such an impression, however, fails to take into account the fact that asexual fission is much more common than sexual reproduction in these bacteria.

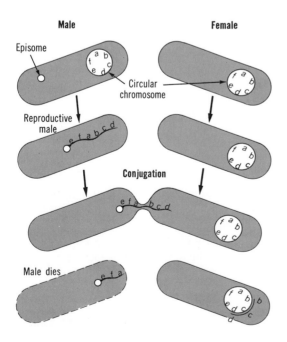

FIG. 8–10 *Sexual reproduction in the bacterium, Escherichia coli. The male cell has an extrachromosomal body, the episome which becomes attached to the circular chromosome and the chromosome opens out with the episome at one end. During conjugation a portion of the chromosome is injected into the female cell, generally leaving the episome and part of the chromosome behind in the male cell. This makes the female diploid temporarily for some of the genes, but the duplicate genes, either from the male or the female, are cast off in cell divisions which follow.*

The episome duplicates and is passed on to the daughter cells formed by fission of male cells. Hence, fission of male cells always produces more male cells, and this insures a plentiful supply of male cells when the time for sexual reproduction arrives. Also, it is known that there are occasions when the entire chromosome and the episome are transferred from the male to the female and thus convert the recipient female into a male. In addition, there are occasions when conjugation occurs between male and female and only the episome is transferred, and thus the female is converted into a male.

There is one other problem which remains. If the female cell receives a portion of a chromosome from a male, there would be a diploid condition for the genes of this injected portion of a chromosome. The female would have alleles of these genes on its own chromosome. The haploid condition is restored, however, as this female cell undergoes divisions. One gene of each of those which are diploid is thrown off from the cells and there is only one gene of each kind remaining. The gene that is retained may be either the original allele or the introduced allele.

CHROMOSOMES AND SEX IN MAN

Until comparatively recent years chromosome studies in man have been difficult. The chromosome number was large and the cells were small. Also, it was difficult to get fresh human tissue which is needed for chromosome studies. In spite of these difficulties, some chromosome studies were made and it was found that man has the *XY* method of sex determination. It was presumed that the ratio of *X*-chromosomes to autosomes was the determining mechanism, as had been worked out so well for *Drosophila* and some other insects. As the techniques of tissue culture developed, however, human cells could be grown outside of the body and a plentiful supply of fresh cells became available. Early tissue cultures were prepared from bone marrow taken from the breast bone, but today such cultures can be made from cells obtained more easily. A tiny bit of skin snipped painlessly from the body can serve; or, a few ml. of blood can be withdrawn and treated with an extract of various plants, such as the kidney bean, *Phaseolus vulgaris*. This extract agglutinates the red blood cells so that they can be easily removed and also stimulates mitosis of the white blood cells which remain. When placed in a suitable nutrient medium, these white blood cells soon grow into a mass of tissue which can be smeared and stained on a slide for microscopic examination. This technique, which is much superior to that of cutting thin slices of cells for study, permits a study of the entire chromosome complex in the cells. Soon it became apparent that there were differences between the human sex-determining mechanism and that found in *Drosophila,* even though both are *XY*. Figure 8–11 shows the sex chromosomes of a man obtained from a squash preparation of a cell.

Since 1942 geneticists and physicians had been familiar with a human abnormality of sex known as **Klinefelter's syndrome** (see Figure 8–12). About one male child out of every 5000 who are born expresses the symptoms char-

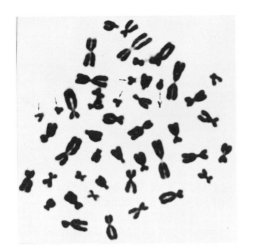 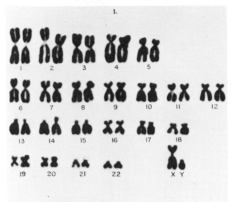

FIG. 8–11 *Human chromosomes showing the sex chromosome dif-*
ference in a man. The photograph at left shows the appearance of a
squash preparation of a cell, on the right is a karyotype; each chromo-
some has been cut out of a photograph and the pairs matched together.
There are 22 pairs of equal size and one pair which includes the long
X-chromosome and the short Y-chromosome. COURTESY MURRAY BARR.

acterizing this syndrome. Such children have typical male sex organs, but, as
they grow, the testes do not grow proportionately to the rest of the body and in
the adult are only about one-half normal size. Also, the fat deposits and growth
of face and hair are somewhat feminine in nature and some degree of breast
enlargement may occur. Men who have Klinefelter's syndrome are always sterile
and about 25 per cent have some degree of mental retardation. About 1 per cent

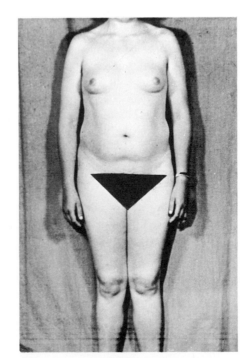

FIG. 8–12 *Klinefelter's syn-*
drome. The person pictured here
is a male, but the reproductive
organs are underdeveloped and
there is some development of fe-
male characteristics. Chromosome
studies show this person to have the
XXY sex chromosomes. COURTESY
POVL RIIS, COPENHAGEN.

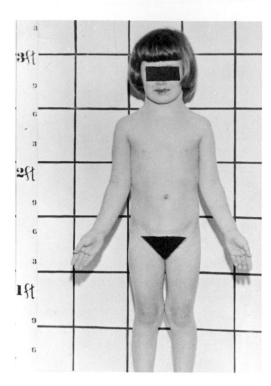

FIG. 8–13 *Turner's syndrome. This child is a female, but her reproductive organs will show no maturation and she will remain an undeveloped female. Note the extreme angle of the arm formed at the elbow. Injection of female hormones will help, but cannot completely correct the condition. She has only a single X-chromosome.* COURTESY HENRY K. SILVER.

of all men in mental institutions have this syndrome. Their blood is low in androgens (male hormones) and sometimes they can benefit from injections of androgenic hormones which give them the appearance and psychological attitude of more normal males.

When cells from a person with Klinefelter's syndrome were examined, it was found that each cell had forty-seven instead of the normal forty-six chromosomes. An analysis of the individual chromosomes showed that there were two X-chromosomes and one Y-chromosome in each somatic cell in addition to the forty-four autosomes. In *Drosophila* such a combination would produce a female, normal in appearance and reproduction, but in man it produces a male, even though an abnormal one. In man it appears as if the Y-chromosome possesses some genes which promote the emergence of male characteristics. Still, since these male characteristics are not expressed as fully as in normal males, we know that the two-X combination is related to the expression of female characteristics or the inhibition of male characteristics.

Further clues as to the nature of the sex-determining mechanism in human beings has come from studies of another sex abnormality, **Turner's syndrome.** About one out of every 3,000 female births results in a child with this abnormality. These are phenotypic females, but at adolescence the adult characteristics do not develop normally and they never reach functional maturity. Persons afflicted with Turner's syndrome are also dwarfed physically, averaging only about 4 feet, 10 inches in stature when adult, and often show mental retardation. Many of them also have a characteristic "webbing" of the skin on the side of the neck and wide-spaced nipples of the mammary glands, although these

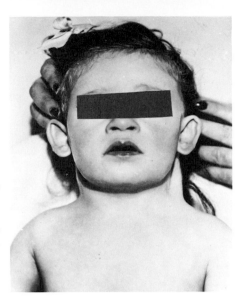

FIG. 8–14 *Webbing of the neck in Turner's syndrome. Many persons with this syndrome have the folds of skin on the side of the neck as shown here.* COURTESY HENRY K. SILVER.

glands do not enlarge as in normal women. Autopsies performed on persons with this syndrome show that their ovaries are composed mainly of masses of connective tissue, so that there is little or no secretion of estrogenic hormones. Injection with estrogens brings about marked changes toward a more normal female phenotype.

Studies of the cells show only forty-five chromosomes in persons with Turner's syndrome; there is only one X-chromosome and no Y-chromosome. This sex chromosome combination would produce a perfect phenotypic male in *Drosophila,* although it would be sterile. In human beings it appears as if a single X-chromosome can bring about the development of a female body, but that two X-chromosomes are required for the development of normal ovarian tissue and the secretion of the hormones required for maturity of the female organs and reproduction. A small percentage of the persons with Turner's syndrome have been found with forty-six chromosomes, but on close examination of the chromosomes only a part of one of the X-chromosomes is found present. The portion which was lost evidently contained the genes needed for complete femaleness.

Non-disjunction during meiosis of either the male or the female can produce gametes with odd sex chromosome complements and can result in these sexual abnormalities. We should also expect cases of XXX individuals. Such cases have been found to occur about once in each 830 female births and thus are somewhat more common than the XO females. *Drosophila XXX* females are weak and sterile, but human XXX females are normal in appearance and may be fully fertile although most are mentally retarded.

RECOGNITION OF SEX FROM ISOLATED INTERPHASE CELLS

The preparation of a tissue culture is an exacting and time-consuming procedure. It would be a great aid to the diagnosis of sexual abnormalities if there were some way to remove cells from a person's body and determine the sex

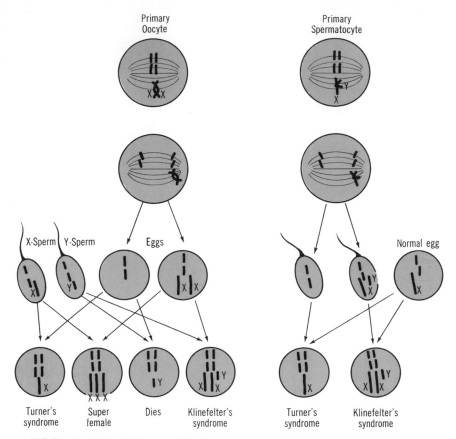

FIG. 8–15 *Different kinds of abnormalities which can arise as a result of non-disjunction of the sex chromosomes in human oogenesis and spermatogenesis.*

chromosome constitution of those cells without the difficulties of tissue culture. Such a method has been found somewhat by accident. Murray Barr, geneticist at the University of Western Ontario, was conducting an experiment on the effect of fatigue on the cell structure in cats. He found that, although the nerve cells looked the same when at rest as they did after repeated stimulations, some of them had a small body in the nucleus which stained heavily with stains having an affinity for DNA. Other cells did not have these bodies. Upon checking his records, Barr noted that the cells with the stained bodies were always from female cats. Further investigation showed that not only nerve cells but many types of cells from female cats had these bodies, now known as **sex chromatin** or **Barr bodies.** Figure 8–16 shows Barr bodies in cells obtained from various parts of a human female's body. It was soon learned that these bodies can be found in females of many classes of mammals, including the primates (the order which includes man).

Epithelial cells can easily be obtained from the lining of the mouth, vagina, or urethra. In women the Barr body lies against the nuclear membrane and appears as a rounded disc. It appears that there is regularly one less Barr body than the number of *X*-chromosomes. A normal woman has two *X*-chromosomes and one Barr body; a woman with Turner's syndrome, although a female, has one *X*-chromosome and no Barr body; and a female with three *X*-chromosomes has two Barr bodies. Normal males have no Barr body, but males with Klinefelter's syndrome have one Barr body. Thus, many abnormalities of sex resulting from chromosome unbalance can be diagnosed from a few easily obtained cells.

It has also been discovered that it is possible to distinguish sex by studies of simple blood smears. The neutrophils, which are the most common of the white blood cells, have a nucleus that is commonly divided into two or three lobes. Female neutrophils may also have a small "drumstick" extending out from one of the lobes. In only a few of these cells, as seen on a microscope slide, will the drumstick be turned in such a way as to be visible; however, its presence is a definite indication of the female chromosome complement in the cells.

FIG. 8–16 *Barr bodies in cells from the human female. These are: A. cells scraped from inside the mouth; only the nuclei are visible. B. outer epithelial cells from the skin. C. cell from a vaginal smear. D. white blood cell showing drumstick.* COURTESY MURRAY BARR.

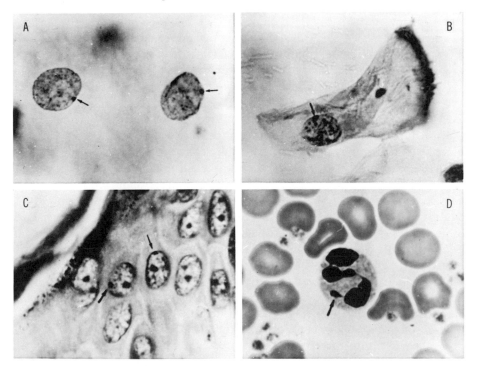

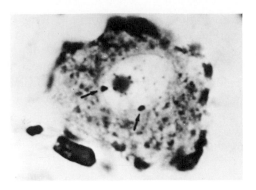

FIG. 8–17   *Cell from human superfemale (XXX) showing two Barr bodies against the nuclear membrane.*

With the discovery of the sex chromatin, some workers turned to an application of this technique to ways of determining the sex of human embryos before birth. The amniotic fluid, in which the embryo floats, contains many epithelial cells which have sloughed off from the skin of the embryo. A sample of this fluid can be removed, either by abdominal puncture or through the cervix of the uterus, and the sex determined by the presence or absence of Barr bodies in the cells. Such knowledge of sex before birth could have a value much greater than mere satisfaction of curiosity. Since some inherited traits are related to sex, it might be possible to deduce whether a serious abnormality was present in the embryo, if its sex could be determined. Unfortunately, however, experiments on the withdrawal of the amniotic fluid from the amnionic sac around embryos of experimental animals such as rats show an increased incidence of harelip and other abnormalities in the young at birth. The embryo is so sensitive during its early development that very small changes in its environment can cause serious results.

Studies of the cells from aborted embryos show that the Barr bodies can be distinguished at about 15 to 16 days after conception, which is several weeks before there will be any differentiation of the gonads or appearance of any other distinguishing features of sex. Hence, it is evident that the Barr body is not produced as a result of the action of female hormones upon the cells. No one can say definitely just what material forms the Barr body — we know that it contains nucleic acid because it stains with the nucleic acid stains like the chromosomes of a dividing cell. It has been suggested that it may be one of the X-chromosomes which fails to uncoil completely as the cell goes into the interphase. Yet, some investigators have reported seeing bodies in the cells from female birds which appear to be of the same nature as the Barr bodies of mammals. Since birds have the WZ method of sex determination, the female is the heterogametic sex and has only one Z-chromosome, which is the homologue of the X-chromosome. Male birds do not show these bodies even though they have two Z-chromosomes. More research is needed to unravel this mystery.

PROBLEMS

1. Why is it not possible to designate the mating types of *Paramecium* as male and female sexes?

2. Would you expect to find sex chromosomes in monoecious animals such as earthworms? Explain your answer.

3. Use corn as an example to show that the distinction between monoecious and dioecious plants is not great.

4. If you collected a large number of *Crepidula* from the seashore and a large number of *Drosophila* from a banana stalk, in which group would you expect the greatest variation in the sex ratio? Explain.

5. It is sometimes possible to produce tetraploid *Drosophila*. What would be the sex of the following: $AAAA$ plus $XXXY$, $AAAA$ plus $XXYY$, $AAAA$ plus YYY? Show how you derive your answer.

6. It is sometimes said that a drone honeybee has no father but has a grandfather. Explain how this can be.

7. Would you expect to find a greater variation in the sex ratio of the honeybees in a hive or in a group of monarch butterflies? Explain.

8. In one breed of white turkeys infertile eggs sometimes hatch. What sex would you expect these to be? Do you think they would be fertile? Explain.

9. A *Habrobracon* female with sex genes x^d/x^p is mated to a male with the sex gene x^s. She bears both male and female offspring. Show the possible sex genes which may be carried by the offspring of both sexes.

10. Choose one of the males and one of the females from problem 9 and show the sex genes and the sex of their offspring.

11. In both *Drosophila* and *Melandrium*, XX produces a female and XY a male in normal diploid organisms. In what way do their methods of sex determination differ?

12. The haploid male and female liverwort plants produce gametes which unite on the female plant to produce a small diploid sporophyte. Each spore mother cell (it is diploid) produces four spores through the two divisions of meiosis. These adhere together and can be removed in a group. What would be the sex of the four plants which grow from these spores? Would this be definite or only an expected ratio?

13. In the bacterium, *Escherichia coli*, the male cells die after transferring genes to female cells. Since the offspring of the female are female cells, how is the supply of male cells maintained?

14. Bacteria are commonly said to be haploid in their gene number, yet a female cell becomes diploid for certain genes after receiving a portion of a chromosome from a male cell. How is the haploid gene number restored?

15. Compare *Drosophila* with man with respect to the part played by the Y-chromosome in sex determination.

16. Suppose that non-disjunction of the sex chromosomes takes place during meiosis of one primary spermatocyte of a man. Show by diagram the kinds of sperms which will be produced and the sexual abnormalities which will result if these sperms fertilize normal eggs.

17. Outline a plan of research which might reveal the nature of the Barr body.

18. Shortly after the Barr bodies were discovered, some physicians would determine the sex of unborn embryos by removing a small amount of the amniotic fluid. This is seldom, if ever, done today. Explain.

Other Factors Related to Sex Determination

IN THE LAST CHAPTER WE FOUND that chromosomes and chromosome ratios furnish the trigger which sets in motion the mechanism of sex determination in the great majority of organisms with separate sexes. But these are not the only factors related to the development of sexual characteristics. The expression of these characteristics may be modified by various other agents, some of which will be studied in this chapter.

HORMONES AND SEX DETERMINATION

Functions of the Gonads

Hormones are relatively recent scientific discoveries, yet it has long been known that there is a rather close relationship between the sex glands of the male and the sexual characteristics which appear in all parts of his body. A boy castrated before adolescence fails to develop the beard, the voice, the musculature, and other characteristics of a man. In past centuries, choir boys were castrated to prevent the loss of the fine quality of their soprano voices as they reached the age of adolescence. Some religious sects required castration of men who joined their orders so that they might not be subject to the temptations that would accompany normal masculinity. Some oriental rulers had many eunuchs as slaves so they could feel assured that none would usurp their conjugal rights.

Castration of the female is a much more serious and difficult operation, and until recently little was known about the relationship of the ovaries to the sexual characteristics of women. Today, however, largely as a result of modern surgery, we know that the relationship is similar to that in men. Removal of the ovaries from an immature girl results in a person devoid of the many feminine characteristics which normally come with womanhood. The gonads, in short, are dual in function. First, they produce the reproductive cells, the sperms or eggs; and second, they produce the male or female hormones which are distributed throughout the body by the blood stream and which influence the various sexual characteristics in all regions of the body. Such a dual function seems to be characteristic of the gonads of all vertebrates.

Partial Sex Reversal

Striking evidence that the gonads perform this second function comes from an apparent change in the sex of some chickens. Almost all poultrymen have noticed that occasionally old hens which have been good egg-layers for years will begin to show characteristics suggestive of the male. They may develop the spurs and head furnishings of the rooster, begin to crow, and attempt to mate with other hens. One investigator (Crew) even reported one that produced viable sperms and became the father of two chicks after several years of egg-laying as a hen. Something happens within such hens to release the inhibition which has so long held the genes for maleness in check and they produce the characteristics normally found only in the males.

Upon autopsy we find the answer. In birds there is only one functional ovary, the left, and in the center of this ovary there is a small mass of tissue similar to the male testis. The ovary normally produces the female hormones which allow the full expression of the female characters and inhibit the analogous characteristics of the male. In some hens, however, there may be a tubercular infection of the ovary which largely destroys it. In such cases, the testis-like tissue, released from the inhibition of the female hormones, may enlarge and produce the male hormone in sufficient quantity to cause the apparent reversal of sex. Similar effects may be produced by surgical removal of the entire ovary except the testis-like inner portion or, more rapidly, by complete removal of the ovary and the implantation of mature testes from a rooster. Within a few months after such an operation, a female bird will be superficially indistinguishable from a normal male. Removal of the testes and transplantation of an ovary into an immature male will likewise produce a bird with all the characteristics of a normal hen.

Steinach successfully transplanted the gonads of rats from one sex to the other and discovered similar effects, so we know that the same relationship exists in mammals. Such experiments cannot of course be conducted on human beings, but there is every evidence for assuming that the effects would be the same. Although such a thing would never be done, it is entirely possible that the ovaries could be removed from a baby girl and that, through treatment with male hormone, she would develop into a person who would be completely masculine in appearance and who could compete physically with any man. Of course, there

FIG. 9–1 *Precocious sexual development and reversal of sexual characteristics in chicks. The chick at the left is a 6-week-old female which received an injection of male hormone, testosterone, at 1 week of age. It has not only developed many male characteristics, but has done so much earlier than would be the case in a male not receiving any injection. The center photograph shows an uninjected female of the same age as the one at the left and in the right photograph is shown a male of the same age.*

could not be a complete change-over of sexual organs because of the advanced state of development of these organs at birth, but there would be rather extensive changes in them even so. Transplantation of human testes might also be possible in the light of recent success in transplanting kidneys from one person to another.

The Freemartin in Cattle

Further light on the relation of hormones to sexual development is brought out by the freemartin condition in cattle. For many years cattlemen have known that, when twin calves of opposite sexes are born, the female is usually somewhat abnormal and is sterile. Such a calf is called a **freemartin.** An extensive study of embryonic twin calves was made by Frank Lillie at the University of Chicago, using the abundant material from the Chicago stockyards. He found that, in most cases, there was a junction between the blood vessels of the two calves within the uterus. When calves of opposite sexes have this junction, the hormones can pass freely from one to the other. Since the male hormone appears earlier than the female hormone, the effect exerted is primarily on the female. A partial reversal of sex takes place and the calf is born with incompletely developed female organs and partially developed male organs. Although released from the influence of the male hormone at birth, these changes have already been so extensive that the female calf cannot grow into a normal fertile cow. It remains a freemartin.

Early Sex Differentiation in Man

All these facts seem to point to hormones as the sex-determining factor in the vertebrates. How can we correlate such findings with the indisputable evidence of chromosome determination studied in Chapter 8? A study of gonad development in embryology will help. During the early stages of human embryonic development there is no indication of the eventual sex of the individual being formed. The various layers of body tissues are formed and differentiate into various body organs. Sexual organs make their appearance, though in their early stages they are the same regardless of the future sex of the embryo. A pair of gonads is formed, sometimes called the ovotestes, which have an inner portion of testicular material and an outer shell of ovarian tissue. Then one or the other gains the advantage in growth; let us assume that it is the testicular tissue. It enlarges in size and secretes the male hormone which stimulates the further growth of the testicular tissue and inhibits the growth of the ovarian portions of the ovotestes. This process begins to have its effects on the indeterminate sexual organs at about the end of the second month of human embryogenesis. The genital tubercle enlarges, surrounds the urethral opening, and becomes the male penis. The vaginal opening closes, and the tiny uterus remains small and undeveloped. The folds of skin on either side of the vagina enlarge and form the scrotum, which is to receive the testes when they descend from the body cavity shortly before birth. It is interesting to note at this point that the testes may not descend if there is a deficiency of male hormone, and a child may then be born

FIG. 9–2 *The human male and female sex organs develop from a common embryonic beginning. The undifferentiated sex organs, shown at left, consist of a genital tubercle, a urogenital opening, and a pair of genital folds on either side. If the embryo becomes a male (upper sequence), the genital tubercle enlarges to produce the penis and encloses the urogenital opening, while the genital folds form the scrotum which contains the testes after birth. If the embryo becomes a female (lower sequence), the genital tubercle remains small and forms the clitoris, the urogenital opening divides to form the openings of the urethra and the vagina, while the genital folds form the outer lips, labia majora.*

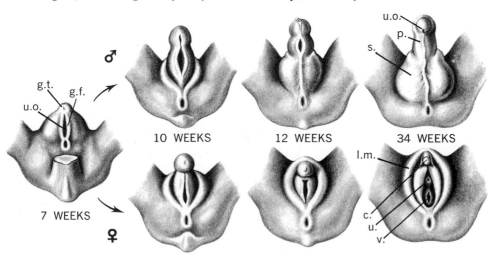

with the testes inside the body cavity — a condition known as *cryptorchism.* Treatment with male hormone may induce the testes to descend in such cases.

Should the ovarian tissue gain the upper hand in the embryological battle, events would be quite different. It will enlarge and inhibit the testicular portions, but these remain as rudimentary bodies in the center of the ovaries. (We have seen how they may become active when released from the female domination in the case of the hens discussed earlier in this chapter.) Under the influence of female hormone the genital tubercle remains rudimentary and forms the clitoris, which does not surround the urethral opening as in the male. The vagina and uterus continue development and reach functional maturity at puberty. The folds of skin on either side of the vagina remain to form the lips (labia) of the female genitalia.

Early Sex Differentiation in the Opossum

Carl R. Moore of The University of Chicago has done some work on the opossum which indicates that the original differentiation and early development of sex organs in mammals come from a hormone-like secretion from the cells of the body as a whole. The opossum is an ideal mammal upon which to make such studies. In most other mammals the embryo is so enclosed that it is not possible to reach it for experimental purposes until it is born. In the opossum, however, which is a marsupial, the embryos are born only 13 days after fertilization, when they are in a very early stage of embryonic development. They will be no more than one-half inch in length, but are born with sharp claws by means of which they climb up the mother's body and enter the pouch on the ventral surface. There they find nipples to which they become attached through a fusion of the skin of their mouths with the skin of the nipples. It is possible to open this pouch and experiment on the tiny animals inside as they complete their embryonic development.

Moore worked out a technique of removing the gonads from these embryos and applying the hormone of the opposite sex, and he found that there was no trend toward a reversal of sex after such treatment when the embryos were still

FIG. 9–3 *Young opossums in the pouch of their mother. These are in a very early stage of embryogenesis, yet they have already been born and have climbed up into the pouch where each has attached itself to a nipple to complete its embryonic development. They furnish excellent material for studies on the effect of hormones on sex in mammals.* COURTESY CARL R. MOORE, UNIVERSITY OF CHICAGO.

quite young. Female organs would continue to develop even though the ovaries had been removed, and male hormone was administered daily, and vice versa. It was only when the embryos were 63 days old, or older, that the hormones caused any tendency toward sex reversal. From these results, Moore concluded that it must be a hormone-like substance secreted by the body cells rather than by the gonads which influences the early sexual differentiation of the embryo, but that at a certain stage in embryonic development this is supplanted by the hormones from the gonads, which take over from that point on. According to this view we could assume that the sex chromosome balance in the body cells determines whether these cells shall secrete male- or female-stimulating substances. These substances, in turn, stimulate the early development of the sex organs, and when the gonads reach sufficient maturity they begin secreting the hormones to continue the job. If such a condition were found in other mammals, it would supplement rather than contradict the other hormone studies which we have described.

Relation of Chromosomes and Hormones

The apparent overlapping of chromosomes and hormone mechanisms in sex determination in vertebrates is greatly simplified when we understand that the chromosomes play their most important role during the early embryonic development of the individual. The sex chromosomes influence the early development of sex organs and determine which of the sex hormones will be produced. Then, as the hormones begin to function, the sex differentiation becomes influenced by the hormones. It appears as if the female sex organs are the basic type because they will develop if no sex hormones are present. These organs will not mature and function, however, without the presence of the female hormones. The male hormones bring about an alteration of the basic female organs and the male organs are formed. Thus, we can see how a hen, with cells having the female chromosome complex, can show a reversal of characteristics and become a phenotypic male.

SEX INTERGRADES

The combination of chromosomal and hormonal factors in dioecious animals results in normal males and females in most instances. There are times, however, when either the chromosomes or the hormones become unbalanced and various degrees of sex intergrades result. We have already learned how an abnormal balance of the sex chromosomes can result in intersexes in *Drosophila* and sex abnormalities, such as Klinefelter's and Turner's syndromes in man. Also, it is possible to produce intersexes in *Bonellia* by removing larvae from the proboscis of the adult females before these larvae have become fully developed males. The partially developed males, suddenly deprived of the hormone from the females, cease developing male characteristics and form female ones. The result is a mixture of both sexes — an intersex.

Human Sex Mosaics

Since the discovery of a way to identify the sex chromosomes of individual cells in different parts of the body, many cases have been found which show a variation in the sex chromosomes in these different areas. **Mosaicism** is the term used to describe the condition where a person's cells consist of two or more populations, each with a different chromosome complement. As an example, Murray Barr found a girl in which both buccal (mouth) and vaginal smears showed two Barr bodies, thus indicating the XXX-chromosome complement. A study of the blood, however, was negative with respect to sex chromatin, thus indicating an XO complement. Phenotypically, the girl showed the symptoms of Turner's syndrome, thus indicating gonadal dysgenesis and presumably an XO condition of the gonads.

These mosaics arise as a result of errors in mitosis in the early stages of embryonic development. As an example, let us start with a normal XX zygote. The X-chromosomes sometimes fail to disjoin in mitosis, as we have already learned in *Drosophila* studies. If this occurs in the early embryo, one cell is formed with XXX and another with XO. This will result in an individual with the XXX/XO type of sex chromosome mosaicism. The effect on the individual depends upon which of the two types of tissue forms the gonads. Non-disjunction in XY cells can also lead to various types of mosaics. If the Y-chromosomes fail to disjoin and both pass to one pole of the spindle, the cells produced will be XXY and XO. This could result in a person showing symptoms of Klinefelter's or Turner's syndrome, depending upon which of the two types of cells formed the gonads.

Likewise, a sex chromosome mosaic can be produced if one of the sex chromosomes lags behind at anaphase and fails to become included in the nucleus as the daughter cells are formed. If an X-chromosome lags behind in one of the nuclei being formed from an XX cell, a mosaic will result which will be XO/XX. An XY cell can result in an XO/XY mosaic if there is a lagging of the Y-chromosome in one of the nuclei being formed in the telophase of mitosis. Such sex mosaics are more common than has previously been realized, because many people with such mosaicism are perfectly normal, as would be expected when the gonads receive a normal sex chromosome complement.

The Gynander in Insects

In those forms of life which do not have sex hormones which affect all of the cells of the body equally, a mosaic condition of the sex chromosomes can lead to a phenotypic sex mosaic. Such sex mosaics, known as **gynanders,** will have male tissue in one part of the body and female tissue in another. These have been observed in many different species of insects, but have been studied in detail especially in *Drosophila*. The most commonly seen is the bilateral type of gynander, in which one side of the fly is male and the other side is female. There is no dilution of the characteristics — the male side is fully male and the female side fully female. Dr. J. T. Patterson of The University of Texas reported an amusing case involving the sexual reaction of a gynander. During courtship,

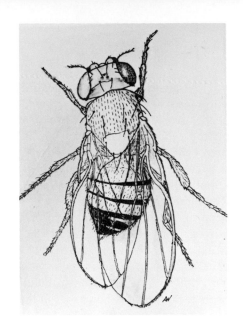

FIG. 9-4 *A Drosophila gynander. Male tissue is on the left and female tissue is on the right.*

male flies hold their wings over their backs and vibrate them rapidly while the females spread their wings to the side. This particular gynander, when placed in a vial with both sexes of flies, vibrated the wing on the male side and spread the wing on the female side. A typical bilateral gynander is shown in Figure 9–4.

Gynanders start their life cycle as females with a normal pair of *X*-chromosomes. In the first mitotic division, however, it is possible for one of the *X*-chromosomes to be lost in the cytoplasm and so fail to be included in the nucleus of one of the two cells formed. Thus, in the two-cell stage, one normal female cell is matched with a cell having only one *X*-chromosome, a fact which makes this cell male tissue. As these two cells continue to divide, the bilateral type of gynander is produced. Should the mishap occur at the second mitosis of embryogenesis, only one cell out of four would produce male tissue. This would yield a gynander with one-fourth of the body male. Even smaller portions of male tissue may be produced by the loss of an *X*-chromosome at yet later divisions, but such small regions are difficult to detect.

FIG. 9-5 *Possible method of formation of a bilateral gynander in Drosophila. Starting with an XX female zygote, one X-chromosome lags behind in the first mitosis. This yields one nucleus with the male-determining chromosome ratio and the other nucleus with the female-determining ratio. Through successive divisions this may produce a fly with male tissue on one side and female tissue on the other side.*

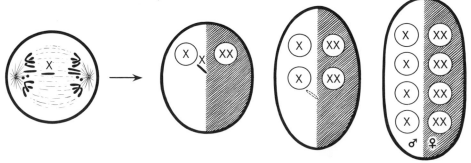

At circus side-shows and carnivals one may sometimes see a person who is purported to be a human *gynander* — "half man and half woman," as the barker says. With a hairy, muscular leg on one side and a delicate silk-clad leg on the other; with a masculine chest on one side and the apparent feminine breast development on the other; with a stubby beard on one side and a smooth, delicately rouged cheek on the other; with a crew haircut on one side and feminine curls on the other — such an individual may appear quite convincing. The author has even seen one who would talk out of one side of the mouth in a deep masculine voice and then switch to a higher-pitched feminine voice when talking from the other side of the mouth. Such cases are amusing sources of wonder to the gullible and naïve, but they are not true gynanders. They are usually men who have a somewhat underdeveloped musculature on one side, and, with that as a basis, go on to build up the illusion of a bilateral sexual division of their bodies. Sex hormones in the vertebrates tend to give even distribution of sex characteristics all over the body.

The Pseudo-Hermaphrodite

It is possible, however, for human beings to show the intersexual condition with characteristics of both sexes in one body. Such an individual is often called a "morfadite," which is a common corruption of the word *hermaphrodite* (which, in turn, is a combination of the names of two classical deities, Hermes and Aphrodite). The latter word is used in biology to refer to animals which normally have both sexes in one body — those which produce both sperms and eggs. This is a perfectly normal state in many of the lower animals. It is therefore probably best to use the word **pseudo-hermaphrodite** to refer to the cases of mixed sex in human beings and other higher animals which normally have separate sexes, since such individuals do not produce viable gametes of both sexes. Such cases are very rare in human beings, although there are many under-sexed individuals who show some development of sex characteristics resembling those typical of the opposite sex. These individuals are either male or female, but, owing to some deficiency of the sex hormones, they fail to show the characteristics of their sex as fully as more normal individuals do. The pseudo-hermaphrodite, on the other hand, has the complete sex organs of both sexes and shows secondary sexual characteristics about half way between the two sexes.

A typical example was a child born to a couple living in a rural community in Oklahoma near the turn of the century. The attending physician told the couple that they were the parents of a baby girl; they took his word for it and dressed and treated the child accordingly. At adolescence, however, this young person decided that she — or he — was not a girl and began dressing and acting like a boy. Masculine characters developed, although not to the extent which would be expected in a normal man. At the age of 24 this person married a normal woman and lived a normal married life as a man. At the age of 42, he was examined by the author and was found to have fully developed sex organs of both sexes. The breast development was that of a girl of about 15, and the musculature was about intermediate between the norm for the two sexes.

The voice was also intermediate in its tonal quality. The person shaved regularly, but the beard growth was not heavy. Such individuals probably result from hormonal rather than chromosomal unbalance.

Some light has been shed on this type of abnormality by some interesting work which has been done by Crew with pigs. Pseudo-hermaphrodites occur with comparative frequency among pigs, and these animals therefore offer an excellent avenue of investigation. It was found that the differentiation of the ovotestes of the early embryo into testes or ovaries proceeds from the anterior to the posterior end. In pseudo-hermaphrodites it was discovered that the ovarian tissue begins its differentiation at the anterior end in the usual fashion, but, before the differentiation has been completed, there is a change which causes the testicular portion to attain dominance at the posterior end.

These two portions of the gonads sometimes separate to produce a pig with both testes and ovaries. Under the influence of both hormones it is understandable that the animal shows characteristics of both sexes. But how can this unusual condition result on the basis of our previous study of genes for sex determination? Crew postulated that the genes for sexual differentiation of the ovotestes vary somewhat in the time at which they begin their work. If a genetic male (XY) inherits quickly functioning female genes and slowly elaborating male genes, the female genes will get in their work for a time before the male genes become active and inhibit them. This will result in the condition described above.

A few years ago the medical journals reported a case that substantiates the application of this principle to human beings. A human hermaphrodite was discovered who, quite naturally, found it very difficult to fit into a society which is divided into males and females. Such minor matters, for instance, as a trip to a rest room became a major problem. In an effort to improve this person's psychological adjustment to life, a group of doctors asked — for reasons of grammar, let us say "him" — to decide whether he would prefer to be male or female. He decided on the former, and the doctors then performed an operation to remove the ovarian tissue which was producing the female hormone. Within 2 years the person was a well-adjusted male member of human society, although the nonfunctional parts of the female organs remained.

SEX REVERSAL

We have just reported cases of partial reversal of sex owing to the changes in the hormonal balance after sex determination has already begun to produce the sex characteristics. On some occasions the reversal has gone so far as to produce individuals indistinguishable from those who have the opposite sex chromosome balance and phenotype. When Murray Barr developed the sex chromatin method which facilitated recognition of the X-chromosome complement of human cells, many persons with various sexual abnormalities came to him for examination. In the many thousands of these he found a few who were sex chromatin-negative, that is, had no Barr bodies in their cells, and yet were perfectly normal phenotypic females with the exception that they were sterile. An examination of their chromosomes showed that the cells had the XY-chromosome balance. These persons should have been males, yet somehow a reversal of sex had taken place and they became females. How can these cases be explained? We do not know for sure, but one theory holds that there may be gene mutations which bring about the development of the ovarian portion of the embryonic ovotestes, in spite of the presence of the Y-chromosome which usually inhibits the development of this tissue. Other theories involve the metabolic rate in the early embryo and other factors.

Regardless of the factor involved, we know that a reversal of the hormone output in the early embryo can result in a complete phenotypic reversal of sex. Breneman of Indiana University has demonstrated this rather conclusively by experiments with domestic chickens. He injected a female hormone into the air chambers of incubating eggs. When these eggs hatched, there were over twice as many females as males, while the control eggs, which were injected with salt water, gave the expected 1 : 1 ratio of the sexes. Since there were over two hundred eggs in the experimental group, it appears that the female hormone causes a reversal of sex in some of the eggs with the ZZ sex chromosome complement, eggs which normally would have produced males.

Sex reversal has also been reported among plants. Hemp is a plant which is dioecious, probably with an XY mechanism of sex determination. It has been found, however, that plants which normally would function as males may be made to function as females if the environmental conditions are varied. For instance, hemp planted from May 1 to July 15 gave the expected 1 : 1 ratio of the sexes, whereas somewhat later plantings resulted in a greater percentage of females, and November plantings produced 100 per cent females. The variation in the length of the day is thought to be the factor responsible for the reversal of sex among those late-started plants which received the chromosome complex that normally produces males.

All these cases of sex reversal, as well as the partial reversals, serve to re-emphasize our basic contention that all individuals are bipotential as to sex.

THE PROPORTION OF THE SEXES IN MAN

Ratio of Sexes at Birth

It is probable that many a person approaching marriageable age has wondered if there are enough members of the opposite sex to go around. In our own country the proportion varies somewhat according to location — there is some excess of males on the west coast and some deficiency of males in the New England States. These variations, however, are due to population migration and not to any differences in the over-all proportion of the sexes in the country as a whole. (Apparently too many young men have taken Horace Greeley's advice to go west.) Statistics indicate that the sexes are almost exactly equal among persons 20 years of age in the United States. So, you may be thinking, is this not exactly what would be expected in view of the XY method of sex determination found in man? Indeed it is, but when we make a statistical

study of the proportion of the two sexes at birth, we find an astonishing thing —
the sexes are not born in an exact 1 : 1 ratio as our chromosome studies would
lead us to expect. About one hundred and six boys are born for every one hundred girls. How can this be, when we have a mechanism that produces male-
and female-determining sperms in equal quantities? At first thought, one might
assume that the female embryos are weaker than the male embryos and, therefore, that a smaller proportion of them survives embryonic development. We
know that from 16 to 20 per cent of all conceptions terminate in death to the
fetus before or at the time of birth. Perhaps, one might assume, there is a greater
proportion of girl than boy babies among these. A study of abortions, mis-
carriages, and stillbirths does not bear out this assumption, however. A survey
of the great collection of human embryos at the Carnegie Institution of Washington has shown that there is a greater number of males among the embryos
of all ages, from the earliest stages in which sex can be distinguished on to the
time of normal birth.

Reasons for Variation in Ratio

To reconcile these statistical results with the theoretical ratio of 1 : 1, we
must assume that the male-determining sperms possess some superiority over
the female-determining sperms in the race to reach the egg. To such a small
body as a sperm, with its limited food supply, the journey up the uterus and the
Fallopian tube must be arduous indeed. To reduce it to human terms we might
compare it to a race of about 64 miles. We obtain this approximation by considering the length of a sperm in relation to the distance which it must travel
from the cervix to the Fallopian tube where fertilization takes place, as compared to the height of a man and a proportional distance to travel. If a group
of men started on such a grueling race, even a slight advantage would make
a tremendous difference in the chances of winning. But what advantage could
the Y-carrying sperms possess over the others? For one thing, they might be
lighter; true, the difference could only be very minute, but they possess the
smaller Y-chromosome in place of the larger X-chromosome which must weigh
somewhat more. A few pounds of weight makes a great difference in a horse
race; perhaps this small variation makes a difference in the sperm race. Also, it
is conceivable that the Y sperms may have a slight advantage in penetrating the
egg after they have completed the journey. Perhaps they are very slightly smaller
in diameter. Recently, some very accurate measurements have been made on
sperm size. When these were tabulated and plotted on a graph, it was found
that they formed a bimodal curve, which seems to indicate that there are two
sizes of sperms. It is possible that the two sizes represent the X and Y sperms,
respectively. At any rate, we know that a few more eggs are fertilized by Y
sperms than X sperms.

Ratio of Sexes by Age Groups

This excess of boy babies would seem to give girls an advantage in the
marriage mart. They should have a greater number of possible mates to choose
from. We have already learned, however, that at about the time for marriage the

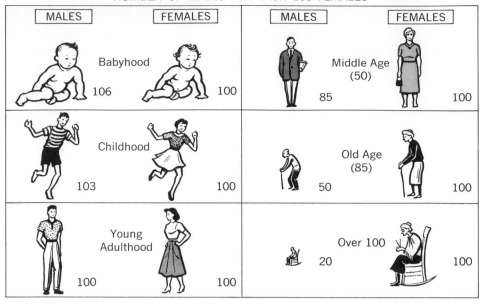

FIG. 9–6 *The weaker sex? A graphic comparison of the proportion of the sexes at different ages. Although the males start with an advantage, they are in the minority in later life.*

proportion of the two sexes is approximately equal. This is because of a differential viability which begins operating early in life and which results in numerical equality between the sexes at a time when such equality is most important. Boys just do not survive the diseases and dangers of childhood as well as girls. This predestined weakness in the male continues to manifest itself throughout life. The insurance companies recognize it; a woman receiving an annuity upon retirement will receive smaller monthly payments than a man of the same age because of her greater chances for a longer life. At 50 years of age we find that the male population has shrunk from the equality at age 20 to the point where there will be only about eighty-five men to every hundred women. At age 85, the women outnumber the men almost 2 : 1 and at age 100, there are about five women for each man. In terms of muscular power certainly men are superior, but in terms of physical qualifications for survival we must concede the advantage to the women.

ATTEMPTS TO CONTROL SEX DETERMINATION

This question of sex determination has intrigued man since the dawn of history. The human mind is such that it seeks an explanation for observed facts — man wants to know why things happen as they do and, if possible, to control the forces of nature and to shape them to his own ends. It is quite natural, therefore, that he should speculate on such an important phenomenon as sex determination and attempt to direct the course of events which produces one sex or the other. In some parts of the earth, where women occupy inferior positions in society, the birth of a boy baby is an occasion for great rejoicing, while that of a girl baby brings sadness. Such people greatly desire a method of insuring the former. How the population could be maintained should they discover such a method would be another problem. Even in our own society where we are generally satisfied with whichever sex our babies happen to be, many persons would like to have a method of determining sex and have a family according to a definite plan or sequence of sexes. The animal breeder would welcome any method that would also apply to domestic animals. How convenient it would be to produce a predominance of females when the breeder wanted egg-laying hens or milk-producing cows! The breeder of race horses, on the other hand, might prefer an excess of stallions.

It is therefore not at all surprising to find that the experimental modification of the sex ratio in man and domestic animals should have been the subject of conjecture and investigation since the time of our earliest historical records. Over 200 years ago Drelincourt collected and tabulated the various methods which had been proposed to predetermine the sex of unconceived as well as conceived but still unborn children. He found 262 theories, all of which have failed to stand the test of time. It is rather obvious that any method proposed for producing one sex rather than the other will succeed about 50 per cent of the time so that advocates of any technique can produce results to substantiate their viewpoint. In ancient times ridiculous concoctions such as the boiled comb and wattles of a rooster, blood from the heart of a male lion, and tissue fluid squeezed from the testes of a bull have been given to prospective mothers in attempts to produce male offspring.

In more recent times such things as diet, acid or alkali pre-coital douches, and the timing of conception within the monthly cycle have been advocated as methods to control the sex of the offspring. All have failed to stand the test of time.

Theoretically, predetermination of sex is possible, but it must be based upon some means of separation of the two kinds of sperms and this appears to offer an almost insurmountable obstacle. Two methods of approach might be used. If the Y sperms are a little lighter than the X sperms, it is conceivable that a delicate centrifuge might separate the two. The difference in weight would be so slight, however, that there seems little likelihood that there can be any successful separation. Also, it is not impossible that there could be some antigenic difference between the two kinds of sperms and some serum might possibly be able to immobilize one without affecting the other. Again, however, the prospects for success are extremely minute.

PROBLEMS

1. Crew discovered a hen which became transformed into a functional rooster and actually produced chicks after mating with a normal hen. Show the ratio of the sexes which would be expected among the chicks produced as the result of such a mating.

2. Steinach transplanted testes of a male rat into a castrated female rat and achieved a reversal of sexual characteristics. If such a transformed female could transfer sperm to a normal female, what sex ratio would be expected among the offspring? Explain.

3. Whenever human twins of opposite sexes are born, the female is usually fertile, but the freemartin of cattle is sterile. Twin calves usually have a junction of the blood vessels during their embryonic existence, but human twins do not. Why is the freemartin usually sterile while the girl twin is not?

4. What evidence indicates that the earliest differentiation of the sex organs in the mammals is not influenced by the sex hormones?

5. A certain species of frog has the *XY* method of sex determination. When raised at about 70°F. they produce approximately half males and half females as would be expected. However, when the tadpoles are kept in water which is about 90°F., they produce all males. From the discussions given in this chapter what theory would you propose to account for this?

6. One research worker proposed the theory that homosexual men and women are cases of sex reversal. He used the statistics of sex ratios among the brothers and sisters of these people to substantiate the theory. According to the facts brought out in this chapter how could this theory be definitely proved or disproved?

7. A study of sex ratios according to order of birth in the United States shows 106.32 males per 100 females for first born children, but only 105.10 to 100 for the fifth born. What could be a possible explanation for this?

8. Among the United States Whites the sex ratio of live births is almost 106 to 100, among the Negroes of the United States it is 102.6 to 100, and among the Negroes of Cuba it is 101.1 to 100. What explanation can you offer for this difference?

9. In a certain area in India the population is very dense and the people are nearly all very destitute and live on the verge of starvation. It has been found that there are about 120 girls of marriageable age for each 100 males of such an age group. There is little migration either into or out of this region. Give a possible explanation for this sex ratio divergence.

10. Individuals who have mosaicism involving the sex chromosomes are to be found in insects and in man. The phenotypic expressions of the mosaicism are quite different in the two. Explain.

10

Inheritance Related to Sex

SOME YEARS AGO THE CROWN PRINCE OF SPAIN, Prince Alfonso, exiled from his **native** land because of the turn of political fortunes, was living in Miami, Florida. One night he was driving down one of the main boulevards of the city in a high-powered automobile. A sudden emergency arose as another car pulled in from a side road. There was a screeching of brakes, a sudden crash, and the shattering of glass as the cars collided. It was not a serious collision, as automobile collisions go, but Prince Alfonso was cut by the flying glass. The cuts were not of a serious nature, but they were bleeding profusely and continued bleeding long after the normal time for blood to coagulate. The prince was rushed to a hospital, but too late. The loss of blood had been too great and he died before any treatment could be given.

This event recalled to mind the unusual disease of **hemophilia,** "bleeder's disease," that has plagued the royal families of Europe for many years. Persons with this disease have blood that does not clot properly; even small injuries assume major importance as blood continues to pour from wounds for hours after an injury. There is something lacking in the blood of such persons which is necessary for normal clotting. Whereas normal blood clots within 2 to 8 minutes after flowing from a blood vessel, in persons with hemophilia the time may be prolonged to 30 minutes up to 24 hours.

This condition is inherited and was perpetuated in the royal families of Europe because of the rather close intermarriage necessitated by the custom that royalty could marry only royalty. Poor little Alexis of the ill-fated Romanoff dynasty of Russia had many a close brush with death during his brief life because of this disorder — although it was the revolution rather than the disorder which finally ended his life. Rasputin, the mad monk of Russia, obtained his strong

155

hold on the government of Russia partly because of his apparently successful treatment of the little Tsarevitch during one of these bleeding attacks. Rasputin's unscrupulous use of his power was a notable factor in the precipitation of the Russian revolution. When we consider how the recent history of the world has been influenced by this revolution, we may ponder how far-reaching can be the effects of a single gene.

Hemophilia is an ancient human affliction. Among the Hebrews of antiquity there were cases reported of male infants who bled to death after circumcision. An Arabian surgeon of the 11th century, Albucasis, described boys in a village who would bleed to death if their gums were rubbed harshly and men who bled to death from slight wounds. Strangely, however, in all of the accounts of the disease it seems to have appeared only in the male members of a family, and pedigrees show that it follows an unusual "skip generation" pattern of inheritance. What is there about the gene which results in hemophilia that is different from other genes inherited in the usual manner and affecting both sexes with equal frequency?

SEX-LINKED GENES

Hemophilia

Hemophilia in man results from what is known as a sex-linked gene. This simply means that the gene is located on the *X*-chromosome. In our discussions of the relation of chromosomes to sex determination, we have emphasized the fact that the *X*-chromosome contains genes which tend to influence the expression of the genes for femaleness. Outside of this distinction, however, the *X*-chromosome is no different from other chromosomes in its nature — it contains genes for characters in all parts of the body which may have nothing to do with sex. It happens that one of these is associated with the production of some substance in the blood which is necessary for the normal clotting process. A mutant form of this gene causes this substance (substance VIII) not to be produced in proper quantity and hemophilia results. Since the *Y*-chromosome contains few genes which are homologous with those of the *X*-chromosome, it is evident that such

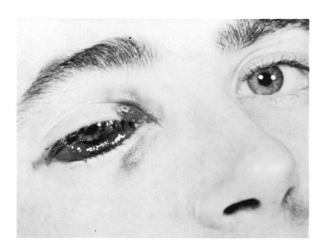

FIG. 10–1 *Hemophilia, a sex-linked trait. The rupture of a small blood vessel under the skin of the eyelid of this boy has resulted in the extensive bleeding shown by the accumulation of blood under the skin in this region.* COURTESY J. V. NEEL.

FIG. 10–2 *A white-eyed Drosophila male. This was the first character which was discovered by T. H. Morgan to show sex-linked inheritance. This fly also shows miniature wings, another character due to a sex-linked gene.*

sex-linked genes must be inherited in a manner somewhat different from the genes on the autosomes.

Discovery of Sex Linkage

Sex linkage was first discovered by Thomas H. Morgan in his work with *Drosophila melanogaster* at Columbia University. In the course of examining thousands of flies in connection with other experiments, Morgan found one with **white** eyes instead of the normal **red** color. This arose as a mutation. When this fly was crossed with others, strange to say, it was found that the method by which the eye variation was inherited followed an unusual pattern. A white-eyed female, when crossed with a red-eyed male, produced only white-eyed males and red-eyed females in the first generation. This was different from any kind of ratio that had been studied up to that time, and it seemed that this character was transmitted from female to male and male to female exclusively. When the first generation flies were crossed among themselves, however, the offspring included both white- and red-eyed flies among both sexes and in approximately equal proportions. Morgan reasoned correctly that such a means of inheritance would indicate a gene located on the *X*-chromosome. Figure 10–3 illustrates how these results are obtained.

One important fact should be emphasized in connection with the study of sex linkage. Since the male has only one *X*-chromosome, he can carry only one haploid set of sex-linked genes. Some of these will be dominant and some recessive, but all will express themselves since there will be no dominant alleles to overcome the recessives. We have already learned that dominance is relative; it depends upon the relative strengths of two genes for somewhat different expressions of a character. In most cases, when two different alleles are present, the more potent one dominates the weaker and expresses itself. Without a dominant allele to prevent its expression, however, a recessive gene will express itself. A person can always win a race if he is the only one entered in the contest. Thus all sex-linked genes in individuals of that sex which carry only a single set of such genes will be expressed, dominant or recessive.

In the cross of the flies just described, white is recessive to the dominant red, but the white shows in the males, even though there is only one white gene present,

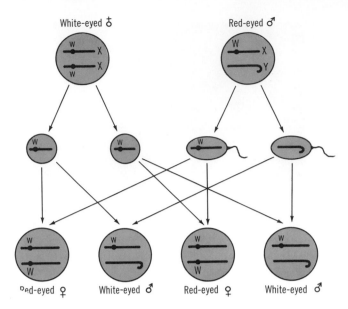

FIG. 10-3 *Sex-linked inheritance in Drosophila. A cross between a white-eyed (w) female and a red-eyed (W) male yields red-eyed females and white-eyed males in the first generation.*

since it has no normal allele to dominate over it. The word **hemizygous** is used in reference to such genes in the sex where they are haploid. Hemophilia in man is also due to a recessive gene. The hemophilia gene which became so widespread in the royal families of Europe seems to have originated with Queen Victoria of England, probably as a mutation, since there is no record of it in her ancestry. Figure 10–4 shows how this gene was spread throughout Europe by the descendants of this queen.

FIG. 10-4 *The inheritance of hemophilia in the royal families of Europe. The gene probably arose as a mutation in Queen Victoria or her immediate ancestors, since she bore one afflicted son and two carrier daughters and there was no record of hemophilia in her ancestry. The gene had a great influence in shaping the history of the world.*

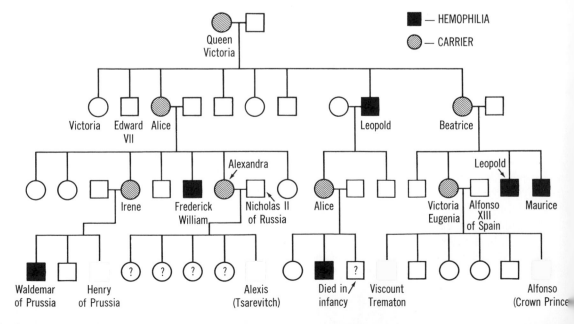

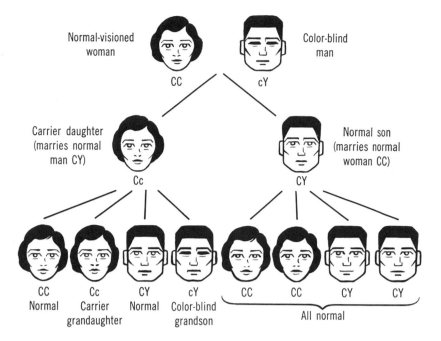

FIG. 10–5 *"Skip generation" transmission of a sex-linked character. The children of a color-blind man and a normal woman will all be normal, but the daughters will be carriers of the gene for color blindness. The character will then be expressed in one-half the grandsons through these daughters. All grandchildren through the sons will be free of the gene provided it is not brought back into the family through marriage.*

Color Blindness

Another sex-linked gene in man, one which is much more widespread, is the gene for red-green color blindness. There is a gene on the *X*-chromosome which plays a part in the formation of the color-sensitive cells in the retina which are necessary for the distinction of red and green. One recessive form of this gene fails to do its job properly and results in red-green color blindness. It is difficult for a person with normal vision to realize what the world looks like to a color-blind person; since colors cannot be described, we cannot get a word picture of this condition. It has been suggested, however, that normal vision at twilight is somewhat like the vision of a color-blind person. As darkness settles on the earth one may notice that reds and greens gradually become less distinct, while blues and yellows retain their characteristic hues for a while longer. Brilliant reds and greens can be distinguished longer than lower intensities of the same colors. In a similar manner, color-blind persons can often distinguish these two colors when they are intense and in good light, but the paler shades and poor light cause confusion. See Chapter 25 for a more detailed discussion of color blindness and its other forms brought about by other genes.

There are a great many more color-blind men than women. Because of this predominance of color blindness among men it is sometimes thought that sex-linked genes show only in males. It is possible for a woman to be color-blind, however, if she receives the gene from both parents. It is estimated that 8 per cent of the men in the United States are color-blind to some degree, but only 0.5 per cent of the women. It is unfortunate that red and green have been chosen as stop and go signals for traffic lights in view of the possible confusion of these colors by so large a proportion of our population.

False Concept of Sex Linkage

Another popular misconception which should be corrected at this point is that boys tend to inherit most of their characteristics from the mother while girls inherit most of theirs from the father. This is an excellent illustration of how a scientific fact may be misinterpreted and expanded into a belief which is quite false. It is true that a man receives all of his sex-linked genes from his mother and none from his father, but this represents genes on only one of twenty-three pairs of chromosomes — certainly not a significant proportion. Since girls receive exactly the same number of chromosomes and genes from each of their parents, no distinction of this kind can exist in their inheritance.

Hemophilia in Women

Hemophilia in females is very rare and it is obvious that it could not occur with the same degree of severity as found in some males. A girl with severe bleeding would be doomed to death by the time she reached adolescence if she survived that long. The chance of a homozygous girl being born in random matings is very slight, for the frequency of hemophilic male births is only about 1 in 10,000. There should be an equal probability of female carriers and, if the male survival is not affected, the mathematical frequency of homozygous females would be equal to the square of one-ten-thousandth which would give a chance of only one in 100,000,000. Even so, this probability is too high, for it is reduced by the fact that about three-fourths of the male bleeders die before reaching adulthood and some of the survivors may never marry because of their condition. But there are many women on the earth, and especially in isolated regions, where the gene may be of higher than average frequency, there could be cases where it could become homozygous in a female child. In England three such cases were reported in a single pedigree where there were frequent cousin marriages. Not only did all of these survive beyond adolescence, but they actually bore children.

A more thorough study of hemophilia indicates how such an event could be possible. There is variation in the time of blood clotting in different lines of hemophilia inheritance. Normal blood, when removed into a container, will clot in about 5 minutes. In one pedigree there were nine bleeders, but the blood clotting time did not exceed 16 minutes for any of these. In another pedigree the time ranged between 20 and 50 minutes, while in a third it ran up to 2 hours. This suggests that there may be several varieties of this gene (multiple alleles) which vary in the degree of hemophilia which they produce. Multiple alleles will be discussed in the next chapter. If this is the true explanation of the variation then it would be plausible that a woman could survive when homozygous for one of the genes which produces only a slight increase in the blood-clotting time. There is also a possibility that this pedigree from England is a case of a different kind of blood clotting disorder resulting from genes at other loci and is not true hemophilia. Recent discoveries have uncovered a number of these, such as Christmas disease and parahemophilia.

It would be well to mention that there is now good evidence to show that the blood of women heterozygous for the gene for hemphilia has a somewhat longer blood-clotting time than the normal. This slowness is not great enough to cause any serious danger in time of injury, but it represents a measurable increase above the normal nevertheless. Further study may make it possible to identify carriers by this characteristic.

Dominant Sex-Linked Genes

The two sex-linked genes which we have so far discussed, those for hemophilia and for color blindness, are expressed much more commonly in men, but there are some which are actually more prevalent in women. Those of the latter type are the dominant genes which lie on the X-chromosomes. Most genes that deviate from the normal condition are recessive, but a few are dominant and will express themselves even though a normal allele may also be present. For instance, one dominant sex-linked gene produces defective enamel of the teeth, and persons having this condition wear their teeth down so that they are usually just stubs protruding from the gums. A woman, having two X-chromosomes, has twice as great a chance of receiving this gene as a man with his single X-chromosome. Thus women show this condition about twice as frequently as men.

SEX LINKAGE IN XO AND ZW ANIMALS

The method of inheritance for sex-linked genes among those comparatively few animals with XO sex determination is identical with the method we have studied for the XY method. With a very few exceptions to be mentioned later in this chapter, the Y-chromosome does not bear genes homologous with those on the X-chromosome; hence, it plays no part in the inheritance of sex-linked genes. It is to be expected, therefore, that the complete absence of a Y-chromosome will have no bearing on the way in which sex-linked genes are inherited.

In those animals with the ZW method of sex determination, however, we find an exact reversal of the method of inheritance for sex-linked genes. Males carry the diploid and females the haploid number of such genes. An example from the domestic fowl will serve to illustrate. The **barred** condition of the feathers, found in Plymouth Rock chickens, is due to a dominant gene located on the Z-chromosome. The **non-barred** allele is recessive. When a non-barred

bb   BW

Bb   bW

FIG. 10–6 *The inheritance of barred (B) and non-barred (b) feather pattern.*

rooster is mated with a barred hen, the male chicks will all be barred and the female chicks will all be non-barred. This is exactly like the white-eyed by red-eyed cross in *Drosophila*, except that in the expression of the sex-linked character the sexes are reversed.

Y-CHROMOSOME INHERITANCE

Incompletely Sex-Linked Genes

As we have studied the methods of sex-determination we have emphasized the fact that the *Y*-chromosome does not carry a large number of genes. We did note, however, that male fruit flies which did not receive the *Y*-chromosome were sterile, thus indicating that there must be a gene or genes for male fertility on the *Y*-chromosome. Any such gene would seem to have no normal allele on the *X*-chromosome, since the sterile males have an *X*-chromosome. To understand how this may be, let us consider the relationship of the *X*- and *Y*-chromosomes more fully. Figure 10–7 shows better than words can explain just how these two

FIG. 10–7 *The classification of genes on the sex chromosomes of animals with the XY method of sex determination. Only the incompletely sex-linked genes have alleles on both X- and Y-chromosomes.*

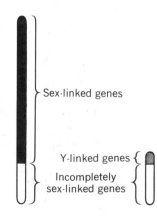

Sex-linked genes

Y-linked genes

Incompletely sex-linked genes

chromosomes are related. Most of the genes are located in that portion of the *X*-chromosome for which the *Y*-chromosome has no homologous portion. These are the sex-linked genes we have been considering. The homologous portions of the *X*- and *Y*-chromosomes are largely devoid of genes, although one gene for bobbed bristles has been discovered in this region. Such a mutant gene may be present on either the *X*- or the *Y*-chromosome and will have an allele in both males and females. The method of inheritance of such genes is somewhat similar to that for autosomal genes, and in the case of recessive genes it is only through special linkage tests that their location on sex chromosomes can be determined. Such genes are said to be **incompletely sex-linked.**

Y-Linked Genes

There are also a few genes in the non-homologous portion of the *Y*-chromosome of the male (known as **Y-linked genes**) which present a still different pattern of inheritance. There seem to be two genes for male fertility in *Drosophila* which are located on this portion of the *Y*-chromosome.

Y-linked inheritance of seventeen human genes has been suggested by various workers, but further investigation has shown that the majority of these must be ruled out. A dominant autosomal gene, sex-limited to men, could give pedigrees in some cases which would appear to show *Y*-linked inheritance. The characteristic which appears most likely to be *Y*-linked is *hypertrichosis* (long hair growth) of the ears.

SEX-LIMITED AND SEX-INFLUENCED GENES

Sex-limited genes are those which produce characteristics that are expressed in only one of the sexes. They are often confused with sex-linked genes, but are entirely different in mode of inheritance. Whereas sex-linked genes are those which are located on the *X*- or *Z*-chromosomes, sex-limited genes may be located on any of the chromosomes. Sex-linked genes may be expressed in both sexes, although they usually show more abundantly in one sex, while sex-limited genes show in only one sex. Sex-limited genes are responsible for what we commonly call secondary sexual characteristics as well as primary sexual characters. The antlers in deer are good examples.

In Man

Let us consider a well-known example in man. The beard is definitely produced by such sex-limited genes. A woman, normally, does not have a beard, yet she surely carries all the genes necessary to produce a beard, and a son can inherit his type of beard through dominant genes which come to him from his mother. In rare cases abnormalities in hormone secretion may occur in a woman which allow these genes to express themselves, and the result is a bearded lady. Similarly, breast development is normally limited to women, but hormone unbalance may cause feminine breast development in a man. All the evidence indicates that the expression of sex-limited characteristics in the vertebrates depends upon

FIG. 10–8 *A bearded lady. This person shows all the feminine characteristics of a normal woman except for the growth of masculine hair on the face and chest. This condition serves very well to illustrate the fact that all people carry the genes necessary for the characteristics of both sexes and in rare cases (usually due to hormone disturbance) some of the characteristics of the opposite sex may be expressed.* UNITED PRESS PHOTO.

the presence or absence of one of the sex hormones. For instance, genes for the deep masculine voice and masculine musculature in man will express themselves only when the male hormone is present. Genes for the feminine voice and feminine muscular development, on the other hand, express themselves in the absence of the male hormone — they do not require the presence of the female hormones. A castrated male will show these characteristics even though no female hormones are present. Genes for feminine breast development, however, require the presence of the female hormones to express themselves rather than mere absence of the male hormone. It can easily be seen that certain sex-limited characteristics are expressed in the absence of certain hormones while others appear only in the presence of sex hormones.

Milk Production in Cattle

An excellent illustration of sex-limited inheritance is provided by the transmission of high milk-producing qualities in cattle. The genes of the bull have just as much to do with the milk production of the offspring as do the genes of the cow which is the female parent. Of course, there is no way to determine by observation of the bull whether he carries genes for high milk production, but it is possible to tell by progeny tests. For instance, a bull may be of the finest quality by all the show-ring standards, and yet may cause a lowering of the milk production of all his female offspring. Another bull may lack the fine show points of our first specimen, and yet his offspring may consistently show a higher milk yield than the cows which bore them. If this should happen we could conclude that the latter bull would carry genes for high milk production and that the former bull did not.

Examples in Birds

The number and variety of sex-limited characteristics in birds produce a sexual dimorphism in many species which causes remarkable differences between the sexes. The brilliant plumage of the male pheasant or peacock stands out in bold contrast to the drab pattern characteristic of the female. There is an interesting case in the domestic fowl, however, which shows that the male may show the female feather pattern if he fails to receive the gene combination necessary to produce the male plumage. In some breeds, males as well as females may show the hen-feathered condition. It has been found that this condition in the male is due to a dominant gene, *H*. Any male that receives at least one *H* will be hen-feathered, but those males which are homozygous for the recessive, *hh,* will show the cock-feathered condition. Females are all hen-feathered regardless of genotype. Figure 10–9 illustrates the inheritance of this sex-limited character.

Sex-Limited Genes in Invertebrates

In invertebrate animals, the sex-limited factors seem to respond to the chromosomal make-up of the individual cells. Gynanders of butterflies with distinct sexual dimorphism clearly show the color pattern of the male on the male side and that of the female on the female side. An interesting example of sex-limited genes in invertebrates is found in the clover butterfly. In this species the males are always yellow, but the females may be either yellow or white. There

FIG. 10–9 *Sex-limited inheritance in the domestic fowl. The dominant gene H causes the males to have hen-feathering, but all females are hen-feathered regardless of the presence or absence of this gene. In the cross diagrammed here, one (one-half) of the male offspring has hen-feathering although he has the spurs and head furnishings of the normal cock.* h = *cock-feathering in males.*

H — HEN-FEATHERING IN MALES
h — COCK-FEATHERING IN MALES

hh   Hh

Hh   hh   Hh   hh

Hen-feathered   Cock-feathered   Hen-feathered   Hen-feathered
cock   cock   hen   hen

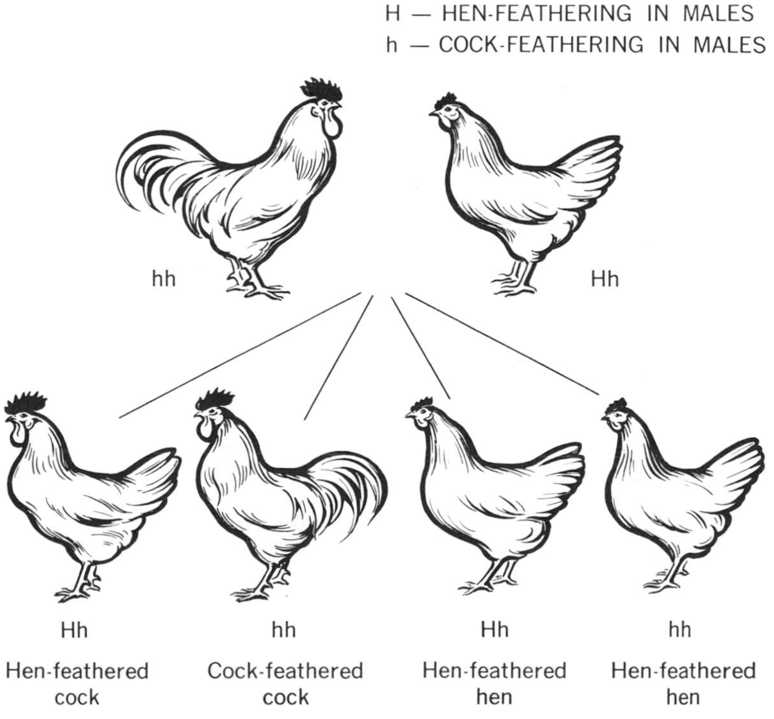

is a dominant gene for white, but it can show only in the females. If we use the letter Y for white then the females with genotype YY or Yy will be white; those with yy will be yellow. All three genotypes will be yellow in the male. Gynanders carrying the gene for white will be white on the female side and yellow on the male side.

Baldness in Man

As one sits in the balcony of a theater and looks out over the audience below one may be impressed by the rather large number of bald heads which are so prominent in a gathering of this nature. Various environmental factors are often mentioned as possible causes of baldness, but as a matter of fact the great majority of bald people are that way because they inherit a genic combination which produces baldness. The character is certainly more prevalent in men, although a greater number of women are bald than is commonly believed — for they can hide it more easily. At first sight, such a condition would suggest recessive sex-linked inheritance, such as we find for color blindness. Investigation of the inheritance of baldness, however, definitely shows that it does not follow the pattern of sex-linked genes, for sons often inherit the condition from their fathers. Further study shows that it is due to one of those peculiar genes which are called

FIG. 10–10 *Pattern baldness, a sex-influenced trait in man. This diagram shows how a man will develop the baldness if he receives at least one gene for it, but a woman must receive two before she will show the baldness.*

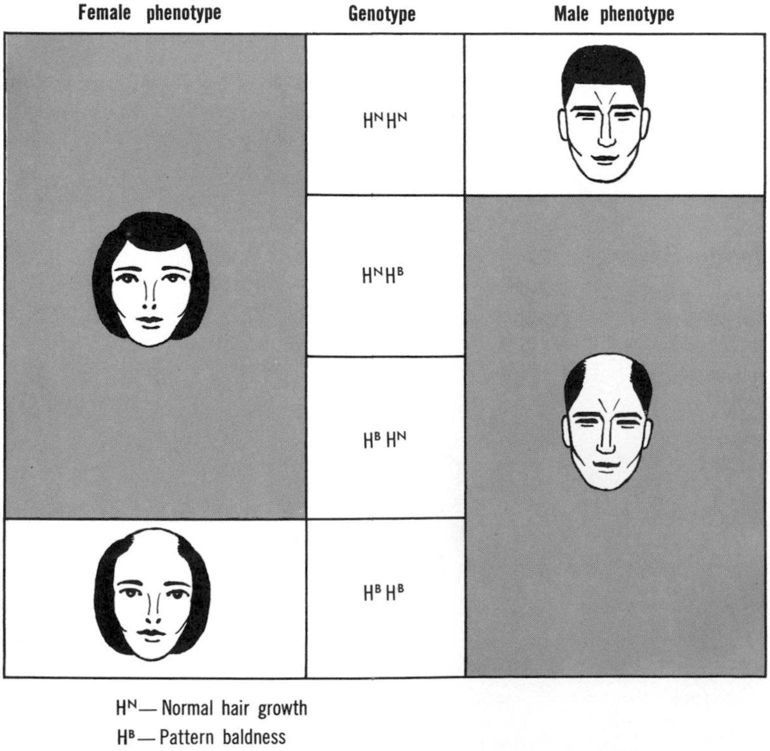

Female phenotype	Genotype	Male phenotype
	$H^N H^N$	
	$H^N H^B$	
	$H^B H^N$	
	$H^B H^B$	

H^N — Normal hair growth
H^B — Pattern baldness

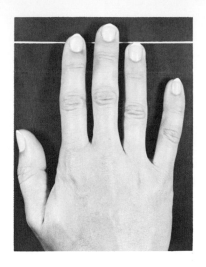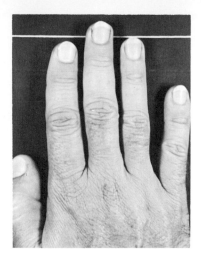

FIG. 10-11 *Sex-influenced inheritance of length of index finger. When the fourth, or ring, finger is placed on a line it will be found that in some persons the second, or index, finger will reach or extend beyond this line as shown in the photo at the left. In others it will be shorter, as shown at right. The short index finger seems to be inherited as a dominant in men and as a recessive in women.*

sex-influenced genes. This particular trait is dominant in men and recessive in women. This accounts for the preponderance of the expression of this character in men as compared to women, for a man is bald if he has only one gene for baldness, whereas a woman must receive two such genes to be bald. It seems that a single gene can operate only in the presence of the male hormone. One interesting case which indicated that this was true was that of an undersexed man who began taking treatment with male hormone. He obtained the desired masculinity, but he lost his hair. This man was evidently heterozygous for baldness, but previously had not had enough male hormone so that the single gene would express itself. Not only is the baldness itself inherited, but the particular pattern of exposed epidermis on the scalp is determined by the genes which influence this condition. In any studies of the inheritance of baldness, however, we must keep in mind the fact that the condition may be induced by purely environmental agents. Excessive exposure to high energy radiation, such as is emitted when the energy of the atom is released, will cause a loss of the hair in both sexes. Also, such diseases as syphilis, seborrhea, and thyroid disease may cause baldness. Hair growth may be restored when it is due to these environmental agents, if the cause is removed, but there seems to be little hope for the man with the gene.

Length of Index Finger

There is another interesting sex-influenced gene in man which affects the length of the index finger. When the hand is placed so that the tip of the fourth finger just touches a horizontal line, it will be noted that the index (second) finger will not reach this line in many cases. This short index finger is due to a gene which is dominant in the male and recessive in the female.

Horns of Sheep

We find an excellent illustration of this type of gene in the domestic sheep. The horned condition in sheep is due to a sex-influenced gene. Some breeds of sheep, such as the Dorset, have horns in both sexes; other breeds, such as the Suffolk, have horns in neither sex. The Dorset sheep are homozygous for the horned gene and the Suffolk sheep are homozygous for the hornless gene, so there is no distinction between the sexes in these pure breeds. When we cross pure-bred horned with hornless sheep, however, we get all horned males and all hornless females. These first generation offspring are all heterozygous, but the gene for horns acts as a dominant in the male and a recessive in the female. When two such heterozygous individuals are crossed, we get offspring in the ratio of three horned to one hornless among the males, but one horned to three hornless among the females. This is the expected ratio on the basis of a sex-influenced gene.

It is apparent from the survey which we have made in this chapter that sex plays an important part in the method of inheritance and expression of many genes. The possible influence of sex should be kept in mind as we continue our study of genetics.

PROBLEMS

Note — in all problems involving genes which might have any different expression in the two sexes indicate the results for each of the sexes separately.

1. Show the expected results from a cross between a red-eyed male and a white-eyed female *Drosophila*. Also, show the expected results if these first generation progeny are allowed to breed among themselves.

2. Suppose a young lady comes to you for advice in your capacity as a marriage counselor. She tells you that her brother has hemophilia, but both of her parents are normal. She wishes to marry a man who has no history of hemophilia in his family and wants to know the probability of her children having this disease. What would you tell her and how would you explain your conclusions?

3. A color-blind woman marries a man with normal vision. What kind of children would be expected from such a union?

4. A man sues his wife for divorce on the grounds of infidelity. Both man and wife have normal eyes, but there is a baby daughter who has *coloboma iridis*, a fissure in the iris of the eye. This character is known to be inherited as a sex-linked recessive. If you were the man's lawyer could you use this fact as evidence? If so, how would you explain the case to the jury?

5. A woman with normal vision marries a man with normal vision and they have a color-blind son. Her husband dies and she marries a color-blind man. Show the types of children that might be expected from this marriage and the proportion of each.

6. In the domestic fowl, silver plumage results from a dominant sex-linked gene, gold plumage from the recessive allele. Show the results of a cross between a rooster with gold plumage and a hen with silver plumage. If the first generation offspring are crossed themselves, what results would be expected in the F_2?

7. A barred rooster is crossed to a non-barred hen. One-half of the male chicks and

one-half of the female chicks are barred. The same rooster is crossed to a barred hen. Would you expect any non-barred chicks in the offspring? Explain.

8. In poultry it is highly desirable to know the sex of the young chicks, but this can be determined only by a technique which requires considerable skill and training. On the basis of the known method of inheritance of the barred feathers, outline a method of breeding which would make sex identification easy as soon as the feathers began to grow.

9. In cats there is a gene for coat color which is intermediate and sex-linked. There is one gene for yellow and one for black, but the heterozygote has a peculiar mixture of yellow and black which is called tortoise-shell. What type of kittens would be produced from a cross between a black female and a yellow male? If one of the female kittens, when matured of course, is crossed to the yellow male what type of kittens would be expected?

10. A man has hypertrichosis of the ears, a condition which appears to be due to a gene on the non-homologous portion of the *Y*-chromosome. He marries a normal woman. Show the types of children which they may expect to have.

11. In man *pseudohypertrophic muscular dystrophy* is a condition in which the muscles waste away during early life, resulting in death in the early teens. It is due to a recessive sex-linked gene. A certain couple have five children — three boys, ages 1, 3, and 10; and two girls, ages 5 and 7. The oldest boy shows the symptoms of this disease. You are their family physician and they come to you for advice. What would you tell them about the chances of their other children developing the disease?

12. A poultryman has a fine flock of Rhode Island Red hens which are excellent egg-layers. He wishes to raise more such hens so he buys a fine, healthy rooster and turns him loose in the chicken yard. When the pullets begin egg-laying, however, he finds that they never lay as well as the other hens did at the same age. You are the county agricultural agent and this man comes to you and asks an explanation for this. What would you tell him and what advice would you give him?

13. Show the type of offspring which would be expected from a cross between a heterozygous hen-feathered cock and a hen, homozygous for cock-feathering.

14. Assume that you are a consulting psychologist. A woman comes to you with a problem. Her mother is bald, but her father has a normal head of hair. Her older brother is rapidly losing his hair and will soon be bald. She is a circus acrobat who hangs by her hair as a part of her act. So that she can change her profession before it becomes too late if this should be necessary, she wants to know if she also might become bald. What would you tell her?

15. Some sheep have horns and some do not. Show the offspring which would be expected in a cross between a heterozygous horned male and a homozygous horned female.

16. In *Drosophila,* yellow body is due to a recessive sex-linked gene. A gynander is found with a yellow body on the male side and a gray body on the female side. Show how this condition could have arisen.

17. Let us assume that yellow body in the honeybee results from a recessive gene and black body from a dominant gene. A yellow queen, on her nuptial flight, mates with a black drone. She returns to the hive and begins her egg-laying task. These eggs hatch into workers and drones. What will be the body color of these bees?

18. A woman has a short index finger. Her husband is dead, but she has five sons who are now living away from home. What can you tell her about the length of index finger of these sons and how can you explain to her how you knew?

11

Lethal Genes

AS WE HAVE STUDIED THE DIFFERENT KINDS of genes in the preceding chapters, it has become apparent that genes differ in their effect on the viability of the organisms in which they are expressed. For instance, the gene which produces vestigial wings in *Drosophila* very definitely reduces viability, for individuals in which this character is expressed cannot fly at all. In a natural environment, consequently, their viability would be so low that few vestigial-winged flies could ever live to reproduce. This is understandable because of the clearly visible physical handicap which the gene produces.

Multiple Effects of Genes

There are many other genes, however, which have little effect on the appearance of the organism, and which yet may reduce viability to a great extent. For instance, some of the eye color mutations in *Drosophila* are accompanied by a reduction in viability. The color of the eyes seems to have nothing to do with the acuity of vision in the flies which show these mutations, and it is hard to understand how eye color could have any survival value. As a matter of fact, it does not, but the same gene which influences the production of pigment in the eyes may also have important functions of a physiological nature which are vitally concerned with the essential life processes of the organism. A mutation of this gene would have a phenotypic effect, but that might be the least part of its influence on the body. The mutation might also cause the gene to change its physiological effects and thus reduce the viability of the fly. Most genes are thus multiple in their influence and are involved in the physiological as well as the physical nature of the organism. In fact, the physical changes are manifestations of the physiological changes.

170

Detrimental Effect of Most Mutations

Most mutations which occur result in some slight detrimental effect on the organism without any detectable phenotypic expression at all. Such mutations can be detected only through tests on the viability and fertility of the organisms. Most genes, in any form of life, if they produce characteristics deviating from the normal, are going to be harmful to some extent, since the normal type represents a gene combination which has resulted from countless centuries of selection and is bound to be an efficient aggregation of genes or it would not otherwise have survived. It stands to reason, therefore, that any deviation from a phenotype which is already highly efficient is very likely to reduce the efficiency of the organism. Very few changes may happen to increase viability, but the great majority would be harmful, though often very slightly. Some of the deviations will be of such a nature as to produce only a slight reduction in viability; in other cases the effects will be more noticeable. Some will cause the death of the majority of the individuals expressing the genes, and some will be so extreme in their effects that all of the individuals in which the mutations are expressed will die. Genes which produce this last effect are called **lethal genes.**

NATURE OF LETHAL GENES

From the above discussion it will be understood that lethal genes are not of an entirely different nature from those we have been studying. Rather, they are merely genes which produce an effect which deviates so greatly from the normal that death results. For instance, there may be a recessive gene in man which causes internal adhesions of the lungs. A child homozygous for this gene might be able to survive embryonic development, but at birth, when he suddenly becomes dependent upon his lungs for his oxygen supply, he would die because his lungs could not expand properly. This would be a lethal gene. Being recessive, it could be carried by normal parents in the heterozygous state without any deleterious effects, but if two persons heterozygous for this gene were to marry, about one-fourth of their children would die at birth from this cause. Moreover, about two-thirds of the living children would be heterozygous for the defect. Thus even though persons expressing the character never lived to reproduce, it could be propagated indefinitely through the heterozygous individuals.

Lethal genes vary considerably in the time at which they exert their lethal effect. Through special techniques, some have been discovered which cause death of the zygote or embryo while it is still only a microscopic bit of protoplasm. Others have their effects in later stages. The time at which the influence makes itself felt depends upon the time when the organ whose abnormality causes death develops and functions. In man, defects of the kidneys, lungs, and digestive organs probably would not be lethal until birth, for these organs do not begin functioning until this time. On the other hand, genes which cause some types of heart defects would be lethal early in embryonic life, for the heart is a necessary part of the embryonic circulatory system and is one of the first organs to begin functioning.

FIG. 11–1 *Albinism in corn. These seedlings have sprouted from seed resulting from a cross between heterozygous plants. About one-fourth of the seedlings are albinos, as would be expected. They will die when they have exhausted the food which was stored in the seed for, without chlorophyll, they cannot manufacture food. Hence, this is a lethal gene.*

Lethal genes are quite common and widespread among most species which have been investigated sufficiently to detect them. Many have been found in plants. In corn, albinism is an interesting lethal. The albino condition is not lethal in the animal kingdom because pigment is not vital to existence, but green color is a necessary factor in the life processes of the green plants. The green-coloring material which is found in such plants is primarily chlorophyll, which is necessary for the manufacture of food. The plants which normally have chlorophyll have no other method of obtaining nourishment. Water and minerals can be absorbed from the soil and carbon dioxide obtained from the air, but there can be no food manufacture without chlorophyll. Plants which show the albino character fail to develop this important green material and hence are doomed to death early in their existence. A corn plant homozygous for this gene will sprout normally and grow a few inches in height, but, after it exhausts the food supply in the seed, it begins to shrivel and soon dies. Heterozygous plants continue to propagate the gene, and in many fields of corn one may see occasional albino plants from such heterozygous parents. Through artificial nourishment it is possible to keep albino corn plants alive. Glucose, the food which is normally manufactured by the chlorophyll, can be administered to the plant by immersing the leaf tips in glucose solution. When this is done properly, we see the phenomenon of a plant living, growing, and even reproducing when it is homozygous for a gene which is lethal under more normal growing conditions.

Frequency of Lethal Genes

Lethal genes are quite prevalent in most forms of life; in fact, most of us probably carry one or more lethals. We are protected from death only by the fact that we are heterozygous for these genes. In most marriages, both partners carry lethal genes, but the lethals are usually not homologous, so that the normal alleles of one parent prevent the lethals of the other from being expressed in the children. In marriages of closely related persons, however, there is a much greater likelihood that both parents will carry the same lethal genes and that there will be a correspondingly greater chance for their expression in the children.

A bit of simple arithmetic will illustrate this point. The average rate of spontaneous mutation seems to be on the order of about one in each 100,000 germ cells in a man for each gene when allowance is made for all the mutations with very slight effects. Now, if man has 20,000 genes in each germ cell, then one germ cell in about five will carry a new mutation at some one of these 20,000 loci. Since approximately one-fourth of all mutations seem to be lethal in nature, about one germ cell in twenty will carry a new lethal. The germ cells will also carry mutations which have occurred in past generations and which have been handed down to us from our ancestors. When these facts are considered, it is easy to understand why it is very likely that everyone carries one or more lethals. A group of leading geneticists has estimated that the average number of lethals carried by each person is two. Of course, there is some elimination of these genes, for when they become homozygous they die along with the person receiving them, but new mutations keep appearing. Thus, a sort of balance is established wherein the rate of elimination is about equal to the rate of appearance, and the number of lethals in the population remains about constant.

Chance of Lethal Expression

With so many lethals known to exist in the human population, the question naturally arises, "What is the chance that a lethal gene which you carry will be expressed in your children?" Let us assume that you carry a certain lethal, say a gene which causes a skeletal abnormality which results in death during birth as defective bones are crushed and injure vital organs. Then let us assume that statistics on birth records indicate that about 1 person out of each 5,000 carries this lethal gene. That means that there is only 1 chance in 5,000 that you will marry a person who will carry the same lethal, provided there is no close relationship between you. And, even if your marriage partner does carry it, there is only a chance of ¼ that you will have a child that is homozygous for it. This means the chance of an infant death from this gene is 1 in 20,000. On the other hand, if you marry your first cousin, the chances are greatly increased. If you carry the gene, then the chance that you received it from your father is ½. If he has it, then the chance that his sister (your aunt) also received it is ½; so the chance that both your father and your aunt carry it is ¼. Then, if your aunt has it, the chance that she transmitted it to her child (your cousin) is ½. Thus, the chance that your cousin carries this lethal gene is ⅛. If the two of you marry, then the chance that you will have a child that will be homozygous and die from its effects is 1 in 32. This is much greater than 1 chance in 20,000 when non-relatives marry and shows why not only lethals but other abnormalities due to recessive genes are far more likely to appear in the children of cousin marriages. It is not that they have any more harmful genes, but that they are just more likely to have the same ones.

Sex-Linked Lethals

The autosomal lethals which have their effects early in the embryonic development of the organism are difficult to detect, since they have no phenotypic expression in the offspring which are actually produced. However, some chicken

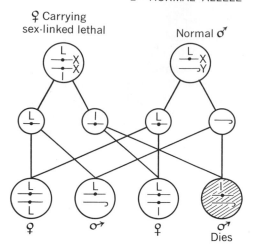

I – LETHAL
L – NORMAL ALLELE

♀ Carrying
sex-linked lethal

Normal ♂

FIG. 11–2 *Variation in the sex ratio caused by a sex-linked recessive lethal. In animals with the XY method of sex determination, the female may carry a recessive sex-linked lethal which causes one-half of her male offspring to die. Among the surviving offspring there will be twice as many females as males.*

Dies

eggs fail to hatch, some seeds fail to sprout, and some human pregnancies are terminated before a woman is even sure that she is pregnant. Some of these cases may be due to homozygous recessive lethals, but they are not usually considered in determining ratios among the offspring. Sex-linked lethals, however, are more likely to be detected because of the variation they cause in the sex ratio. If a female (with two *X*-chromosomes) is heterozygous for a recessive sex-linked lethal, then one-half of her male offspring will die since they receive their single *X*-chromosome from their mother and have no dominant allele to prevent the lethal effect. In small numbers of offspring such an effect might not be apparent, but in animals like *Drosophila* there will be a very noticeable effect which cannot be attributed to the laws of chance. A single female may produce a hundred offspring or more which will normally show a 1 : 1 ratio of the sexes. If one finds a case where there are twice as many females as males, then it is rather evident that some factor is operating to produce this disproportion.

The first positive evidence that has been uncovered to show that high energy radiation has a detrimental effect on human heredity has come from a study of the sex ratio among the descendants of those exposed to the bomb blast at Hiroshima. A reduction in the male offspring has indicated that the radiation caused recessive sex-linked lethals in the reproductive cells of the mothers. More details of this study are given in Chapter 21.

Intermediate Lethals

We learned in the early part of our study that many genes are either dominant or recessive. Some genes, however, prove to be neither dominant nor recessive, and express themselves in the heterozygous individuals but not as fully as in those which are homozygous. As we have seen, red and white four-o'clocks, when crossed together, yield pink offspring. Such genes were called intermediate genes. Since we have learned that lethal genes are no different in nature from other genes, it is not surprising to find that a similar intermediate condition exists in a comparatively few genes of this sort. An individual heterozygous for such a

FIG. 11–3 *A creeper hen. This hen is heterozygous for a lethal gene which has a phenotypic effect when heterozygous. The legs are shortened and the bird must creep along rather than walk normally. The wings also are shortened.* COURTESY CLYDE D. MUELLER.

gene would be viable, but would express some phenotypic character which results from the presence of the single lethal. An individual homozygous for this gene would not live. Sometimes such genes are called "semidominant lethals," but since they express themselves in the same manner as those which we have already studied as intermediate genes, there is no need to introduce a new terminology for them. An illustration from the domestic fowl will explain the results obtained in crosses including intermediate lethals.

In the domestic fowl there is a character which is called **"creeper."** Such chickens possess short, crooked legs and cannot walk normally; they creep about without raising their bodies off the ground. When two creeper chickens are crossed, they yield a peculiar ratio among the offspring. They produce chicks with normal legs and creeper chicks in the ratio of 1 : 2. Now this is different from any ratio we have yet studied. We have found 3 : 1, 1 : 1, and 1 : 2 : 1 resulting from monohybrid crosses, but so far no 1 : 2 ratio. We can find an explanation if we consider all the eggs which were incubated. Approximately

$$C^l - \text{LETHAL}$$
$$C^n - \text{NORMAL}$$
$$C^l C^n - \text{CREEPER}$$

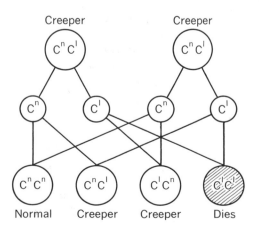

FIG. 11–4 *The results of a cross between two creeper chickens. One-third of the living offspring are normal and two-thirds are creepers. Those receiving the two lethal genes die in early embryogenesis. This is an example of a lethal gene with an intermediate effect when heterozygous.*

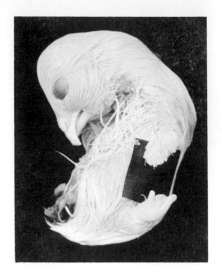

FIG. 11–5 *A homozygous creeper embryo. Note the legs, shortened to the point where only toes show, and the wings, represented by mere stumps. Most homozygous creepers die at about 72 hours of incubation, but this one lived until almost ready to hatch.* COURTESY WALTER LANDAUER.

one-fourth of these fail to hatch, and examination will reveal in each of these an embryo which has died at about the 4th day of incubation. If we include these embryonic chicks in our count, the ratio becomes 1 : 2 : 1, the typical ratio obtained from parents heterozygous for intermediate genes. It becomes evident from these results that the creeper condition in the chickens is a heterozygous expression of a gene which is lethal when homozygous.

Another classical case of intermediate lethals, in which the lethal effect is more obvious, is to be found in cattle. In England there are two breeds of cattle known as **Kerry** and **Dexter.** The Dexter has a somewhat heavier body, has shorter legs, and is a better beef-producer than the Kerry. There is a considerable demand for the Dexter cattle, but it is impossible to establish a pure-breeding herd of Dexters because these desirable characteristics of the breed result from the heterozygous expression of a lethal gene. When two Dexters are crossed, the offspring are produced in the ratio of 1 Kerry : 2 Dexter : 1 bulldog calf. The bulldog calf is usually still-born or dies a few days after birth. It has extremely short legs, a shortened muzzle which gives it a bulldog-like expression, and other deformities which cause death. This is the effect of the homozygous condition of a lethal gene. The Dexter characteristics result from the heterozygous condition; the Kerry is homozygous for the normal allele.

Breeders of cattle have learned that it is not a good policy to cross two Dexters to produce Dexter offspring, because only one-half of the offspring will be Dexters and one-fourth of the offspring will die. It is better to breed Dexters to Kerrys, for this cross yields one-half Dexters — the same number as the cross between two Dexters — and one-half Kerrys, with no bulldog calves. In the case of the roan cattle (see Chapter 5), we recall that 100 per cent of the roans can be produced if red is crossed with white. In the case of the Dexters, however, we cannot cross two individuals homozygous for the different genes because one of the genes if homozygous is lethal.

To state that it is impossible to produce 100 per cent Dexter cattle, however, would be to reckon without the ingenuity of man. We can say that it is not possible under normal conditions in nature, but a technique has recently been worked out whereby an individual carrying homozygous lethals in its gonads but not in the rest of its body can be created and used in breeding. Such an individual is created by gonad transplantation. This technique can be explained by a parallel case. In mice there is an intermediate lethal gene which produces a yellow body color when heterozygous and is completely lethal when homozygous, the embryo dying in an early stage of its development. Two yellow mice may be mated and, after allowing the embryos to start their development, the abdomen of the pregnant female can be opened and the embryonic mice removed. They will all be alive, but those which are homozygous for the lethal gene will show certain deformities that distinguish them from the others in the group. The ovaries may now be removed from the females among the homozygous embryos and transplanted into the bodies of newborn female mice from which the ovaries have been removed. These transplanted ovaries will grow, develop, and produce functional eggs, each of which will carry the lethal gene. Thus we can obtain offspring from a mouse that was never born and could never have been born alive because it was homozygous for a lethal gene which causes early embryonic death. A similar operation performed on embryonic calves would make it possible to produce 100 per cent Dexters if these animals were crossed with Kerrys.

In man several genes are suspected of being intermediate lethals. One of these, when present in the heterozygous state, produces persons with short fingers, a condition known as **brachyphalangy.** Such persons appear to have only two rather than the normal three joints to their fingers, as shown in Figure 11–6. X-ray studies indicate that this condition is due to the fact that the middle bone (phalanx) of the finger is greatly shortened and often fused with one of the other bones of the finger. Such a character is not common, and it is not often that two persons with this same gene would marry. But in one such marriage described by Mohr, one child was produced without any fingers or toes and was unable to survive. Two other children showed the characteristic short fingers and one was

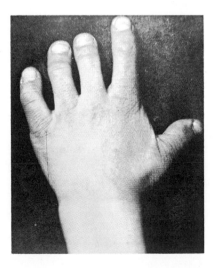

FIG. 11–6 *Brachyphalangy, an intermediate lethal in man. The hands of this man, who is heterozygous for this gene, appear to have only two joints to each finger. The middle bone is greatly shortened. Homozygosity of this gene seems to have a lethal effect.* COURTESY DAVID WHITNEY.

normal. This is the exact 1 : 2 : 1 ratio which would be expected from parents heterozygous for an intermediate lethal.

It is quite likely that many of the human genes which cause abnormalities when heterozygous, and which we have come to think of as dominants, are actually intermediate lethals similar to the one described above. Since they are rare, however, we have no cases of marriage between two afflicted individuals and no way to detect the effect of the homozygous gene.

DETECTION AND ELIMINATION OF LETHAL GENES

Intermediate lethal genes are much easier to detect than the more common recessive lethals because all individuals carrying an intermediate lethal will exhibit some phenotypic expression of the gene. As a consequence, it is much easier to rid a race of intermediate lethals — one must merely prevent reproduction of the easily recognizable heterozygous individuals. Recessive lethals, on the other hand, are more difficult to detect and eliminate. If the individual carrying the homozygous lethal lives to become large enough to be observed, the problem is simplified. As an example, in cattle there is a recessive gene, producing a condition described as **amputated,** which when homozygous results in calves without legs. Such a character is obviously lethal, but the calves may be born alive. This gene is very widespread in the Holstein breed of cattle because there was once a very famous bull, Gallus, who was crossed to many hundreds of cows during his lifetime on account of his fame in yielding high-producing milk-cows among his offspring. Unfortunately, it also happened that Gallus was heterozygous for the gene "amputated." In later generations this became apparent when descendants of Gallus were crossed with one another and "amputated" calves appeared in the offspring. By this time, however, the gene had become so widely distributed in the Holstein breed that it had given rise to a major problem for Holstein cattlemen. Such a gene could be eliminated by a testing program of the following sort.

Let us start with a cow and a bull who have produced an "amputated" calf. Both these parents must be heterozygous and can be used as test animals. The bull could be crossed repeatedly with other cows which are to be tested. If one cow produces seven or eight normal calves without any "amputated" offspring, it is most likely that this cow is homozygous for the normal gene. Such a cow could then be used as the female to establish a new herd. Various bulls could be tested in the same way by crossing them repeatedly to cows which had had "amputated" calves. After a bull had been found that yielded no "amputated" calves from such crosses, we could assume that he was homozygous for the normal gene and use him as the male to establish the new herd. Descendants from such tested individuals would, with high probability, be free of the gene. Of course, considering the comparatively small number of calves which can be produced from one cow during its lifetime, it might be possible for her to be heterozygous and yet fail to produce an "amputated" calf in the tests. Simultaneous testing of many cows would be better for this reason.

Recently perfected techniques of foster parenthood in cattle may greatly shorten the testing period required for cows. It is possible to remove the egg from a cow, fertilize it in a test tube, and implant it into the uterus of another cow where it will grow and develop into a calf with all the inherited characteristics of the cow that produced the egg. In this way it is hoped that fine cows may become the parents of large numbers of offspring, as is already possible with fine bulls. It would certainly give the breeder a chance to test cows for recessive characters in much less time and to get a sufficient number of calves to give a reliable ratio.

Recessive lethals which produce their effects in very early embryonic life are much more difficult to detect, but fortunately their effects are less serious in domestic animals. A cow carrying an embryo homozygous for a lethal that produces death within three weeks after fertilization would simply have an early abortion and could soon be bred again.

GENES LETHAL ONLY UNDER CERTAIN CONDITIONS

Before closing our discussion of lethal genes we should mention those that are not lethal under normal conditions but which under certain environments may cause the death of the organism. In other words, these are genes which produce a lethal effect only when a certain combination of hereditary and environmental factors occur together. An excellent illustration of this is provided by the dominant gene which produces the antigen responsible for the blood condition in man which is known as **Rh-positive.** We will study this in some detail in Chapter 12, but suffice it to say here that about 85 per cent of the people in the United States are Rh-positive and about 15 per cent are Rh-negative and do not carry this antigen. Now, no one in his right mind would say that the gene for the Rh factor is lethal when so many people carry it who live perfectly normal lives. Yet, under certain conditions it is lethal. When an Rh-negative woman bears an Rh-positive child she may become sensitized to the Rh factor and generate antibodies in her system which react violently with the Rh antigen. We commonly think of heredity as a one-way process — always from parent to child — but this is an unusual case in which the inheritance of the child influences the nature of the parent. When such a sensitized woman bears another Rh-positive child, this same reaction will occur in the body of the child, causing certain abnormalities known as *erythroblastosis foetalis.* Such babies are frequently born dead, or die shortly after birth unless special precautions are taken to save them. Thus in the uterus of a sensitized woman the dominant gene for the Rh factor may be a lethal.

The gene for albinism is always lethal in the plants which depend upon the presence of chlorophyll for the manufacture of their food, but in man this character does not seem to reduce viability in the protected environment of modern civilization. In a race which lacked the protection of clothing and housing, however, where all persons were exposed to intense sunlight, the gene could easily be lethal. Albinos lack the ability to produce the protective pigmentation in their skin and after extensive exposure to bright sunlight would become so badly burned that death would result.

Another illustration is provided by the gene for frizzled feathers in chickens. This gene causes the feathers to break off easily and results in birds which are almost naked. If such birds are kept in a reasonably warm environment they manage to survive, but if exposed to rigorous cold weather, without the insulation of a normal covering of feathers, death results. Hence the gene is lethal under one set of environmental conditions and not lethal under other conditions.

LETHAL GENES IN HAPLOID ORGANISMS

It is not possible to study lethal genes in haploid organisms as thoroughly as in the diploid forms because all genes in haploid organisms are expressed and a lethal will be eliminated in the first generation. Thus, every lethal which appears acts in the same manner as a dominant lethal in a diploid organism. It is possible, however, to study genes which are lethal only under certain conditions in haploid organisms. The wild-type mold, *Neurospora,* for instance, produces enzymes which can convert sugar and ammonia into the amino acid arginine. Hence, this mold can live on a medium which does not contain arginine. A mutant gene, however, fails to produce one of the enzymes necessary for the conversion and a strain of the mold carrying this gene cannot grow unless the medium contains arginine. Arginine will be present whenever there are proteins in most instances, for most proteins contain this amino acid. The mutant strain of *Neurospora* will not be handicapped, therefore, as long as the medium upon which it is growing contains proteins. If it is placed on a medium lacking these proteins, however, it cannot grow. Hence, the mutant gene acts as a lethal under these conditions.

PROBLEMS

1. Whenever sows (female hogs) are bred to closely related boars (male hogs), they tend to produce smaller litters of viable offspring than when they are outbred to non-related boars. Suggest an explanation for this on the basis of the facts which have been learned in this chapter.

2. On a certain college campus there is a maple tree. Seeds which fall from this tree sprout each spring, and a number of albino plants always grow from some of these seeds. Explain. Would it be possible to obtain seed from this tree through experimental means that would not produce any albino seedlings? If so, explain how it could be done.

3. Two *Drosophila* are crossed and yield eighty-two females and thirty-eight males. Such a great deviation from the expected 1 : 1 ratio could hardly be due to chance. Suggest an alternative explanation.

4. A poultryman buys a fine Rhode Island Red rooster and turns it out with the hens of his flock. He collects the eggs and incubates them in large numbers, but finds that about one-fourth of the eggs do not hatch. Later he finds that about two-thirds of the chicks which hatch are males. Suggest an explanation for this.

5. In *Drosophila*, star-eye is the heterozygous expression of a lethal gene. Two star-eyed flies are mated and produce 180 living offspring. About how many of these would be star-eyed?

6. Show the type of offspring which would be expected from a cross between a Dexter bull and a Kerry cow.

7. Suppose the ovary is removed from a newly hatched chick (there is but one functional ovary in birds) and the ovary from an embryonic, homozygous creeper is transplanted. When the bird matures it is mated with a normal rooster. Show the types and proportions of offspring which will be expected.

8. Multiple telangiectasia is a genic trait in man which is characterized by an enlargement of some of the finer blood vessels of the nose, tongue, lips, face, and fingers. Heterozygous persons are viable, but in homozygotes there is an extreme enlargement of these blood vessels which then rupture and cause the death of the afflicted person within a few months after birth. Two persons with the minor form of this trait marry. Show their possible children.

9. Brachyphalangy is also an intermediate type of lethal in man. A man with the typical, short fingers of the heterozygote marries a woman with a multiple telangiectasia. Show the type of children which they may expect to have.

10. Mexican hairless dogs have very little hair and few teeth. When crossed with dogs with a normal amount of hair, they ordinarily produce one-half with hair and one-half without hair. But when two Mexican hairless dogs are crossed, about two-thirds of the puppies are hairless, about one-third have hair, and in addition, there are some deformed puppies that are born dead. Show how these results may be explained.

11. In dogs, wire hair is dominant over straight hair. A pure-bred wire-haired terrier is crossed with a Mexican hairless dog. (These hairless dogs are homozygous for straight hair, but the gene is not expressed, for there is no hair to be straight; this is epistasis.) Show the kind of hair which will be expected in the living puppies that result from this cross.

12. As a farmer examines his new-born pigs he notes that in many cases two or three in a litter are born with hydrocephalus. This condition is characterized by a swollen head caused by excessive internal pressure on the brain cavities. Such pigs die at birth or shortly thereafter. He reads in a book on genetics that this condition is caused by a recessive lethal gene. Outline a breeding program by means of which he can eliminate this gene from his hogs.

13. On a fox ranch in Wisconsin a new type of coat color appeared as a mutation among silver foxes. This was a silvery blue coat with symmetrical white spots on it. It was called a platinum coat. It appeared to be inherited as a dominant, but, when two platinum foxes were crossed, about one-fourth of the pups were pure white and died before weaning, one-half were platinum, and one-fourth silver. The platinum is in great demand and brings a premium price. How can the ranch produce the maximum number of platinum foxes without getting any of the lethal whites?

14. Statistical studies indicate that there is a slightly larger proportion of males among the still-births and neonatal (newborn) deaths in the hospitals of the United States. In the light of the facts learned in this chapter and in our study of the method of sex determination in man, what could be a possible explanation of this? Could this also be a factor in male death rate in later life?

15. Lethal genes are much less common in haploid organisms than in diploid organisms. Explain.

12

Multiple Alleles and Blood Group Inheritance in Man

IT IS AN INTERESTING FACT THAT SCIENTISTS quite frequently discover the existence and nature of certain parts or products of living systems by reason of their absence or deficiency. For instance, the effects of insulin on the human body were first determined by studying diabetics whose pancreas does not secrete a sufficient amount of this hormone to maintain certain normal body functions. As long as a person secretes sufficient amounts of this hormone he is simply normal — there is no indication of the presence of the hormone, no clue that it exists. The relation of the testes to the sexual characteristics of a male vertebrate animal becomes most evident by contrast after the removal of these glands — another instance where the effect of the male hormone is most discernible by its absence. The great majority of the vitamins have been discovered by the syndromes which develop in cases of their deficiency. As long as we get our minimal requirement of the vitamins, there is no indication that they are present in our food — it is only through their deficiency that their function becomes evident.

In the same way, the presence of genes for normal body development becomes apparent only when we find alleles which produce characteristics that deviate from the normal. The number of known genes, therefore, has been limited to those cases in which we find at least two varieties of genes at the same locus. We could not detect the existence of a gene for straight wings in *Drosophila,* nor could we locate it on the chromosomes, unless we had found a gene for curly wings which is an allele of the gene for straight wings. Because of this fact we can neither locate nor study the majority of genes because we have not been able to find alleles which produce some deviation from the normal expression of these genes. Even in such an extensively investigated species as *Drosophila melanogaster,* in which we have found and located over five hundred genes, there still remain many times this number of genes for which we have found no alleles that produce a detectable phenotypic difference. There can be little doubt, however, that many will be discovered in future investigations.

ORIGIN OF ALLELES AND MULTIPLE ALLELES

We may now raise the question of how alleles, that is, genes responsible for phenotypic variants in the same character, may arise. Why do some persons have free lobes at the base of their ears while others have attached lobes? We know that the conditions of the ear lobes is inherited, and studies on many family histories indicate that people have attached ear lobes only when they are homozygous for a gene controlling this condition. We can say, therefore, that the gene for attached ear lobes is a recessive allele of a gene for free ear lobes. This means that the gene for attached ear lobes occupies the same locus on the chromosome as the gene for free ear lobes, but as the ear lobes develop it causes the skin at the base of the lobes to be attached to the head. If we assume that free ear lobes was the original condition in the human species, then we can assume that the gene for attached ear lobes arose through mutation of the gene for free ear lobes. Of course, we know that a number of genes are involved in the production of ear lobes, but a change in a particular one of these genes causes an alteration of the pattern of development to produce the attached condition. As this mutation became distributed throughout the population, it gradually began to express itself as individuals homozygous for this gene were produced.

The genes which we have used as examples in the previous chapters have been presented as each possessing a single allele — albino-normal pigmentation, colorblind-normal vision, purple flowers-white flowers. Some, we found, have three phenotypic expressions in the offspring — e.g., red-roan-white shorthorn cattle — but this condition has resulted from the interaction of a single pair of alleles. In this chapter we wish to emphasize the point that there is no reason why there should be only two varieties of a gene. If a gene can mutate to produce a variant allele, then there is no reason to suppose that it cannot mutate at some other time to produce another allele different from the first. This would give us three different genes which occupy the same locus on the chromosome, although we must keep in mind that only one of the three could be on any one chromosome and no more than two of the alleles could be present in a single normal diploid organism. Also, it is possible that one of the mutant alleles might mutate, at some later time, so as to increase the total to four, and so on. We use the term **multiple alleles** to refer to all such genes that occupy the same locus on a chromosome. As genetic studies progress we find that more and more genes have multiple alleles.

THE WHITE-EYE SERIES IN DROSOPHILA

In 1910, T. H. Morgan, working at Columbia University, discovered a fruit fly with white eyes in a vial of flies with the normal red eyes. The gene for white eyes arose as a mutation of a gene which is on the X-chromosome and which is involved in the production of the eye pigment. By testing with other genes, it was later found that this gene was located 1.5 units from the "left" end of this chromosome. It was not long, however, before another gene was discovered which produces flies with an eye color somewhere between red and white. This gene was called "eosin" because the color of the eyes it produces is similar to that of the eosin dye which is widely used in biology laboratories. It was also found to be located at 1.5 on the X-chromosome, and it proved to be an allele of both red and white. This was the first case of multiple alleles to be found.

Since that time other alleles at this same locus have been found. Some of these alleles have arisen as mutations from red, while others have arisen as mutations of one of the alleles which in turn had arisen as a mutation of red.

When one begins to work out crosses involving multiple alleles, the question arises as to what symbols to use. It is customary to use the first letter of the characteristic which deviates from the normal (wild type) characteristic as the symbol for all of the genes at one locus. For instance, we used a to represent the gene for albinism and A to represent the gene for normal pigmentation. This also agrees with our custom of using the small letter for the recessive gene and a capital for the dominant. With multiple alleles, however, we run into complications. It was all right when there were only two known alleles at a locus; we used w for white and W for red eyes. When eosin was discovered, it was found to be recessive to red, but intermediate when crossed with white. What letter should be used to represent eosin? Since we already have selected the letter w to represent two genes at this locus, it would not be well to use an entirely different letter for eosin. A solution to the problem was provided by using w for the basic letter and using the first letter (or letters) of the new alleles after the w in the position of an exponent. For instance, eosin may be represented as w^e. The symbols which have been selected for the entire series of white-eye alleles are represented as follows: W, red; w^w, wine; w^{co}, coral; w^{bl}, blood; w^c, cherry; w^a, apricot; w^e, eosin; w^b, buff; w^t, tinged; w^h, honey; w^{ec}, ecru; w^p, pearl; w^i, ivory; and w, white.

These are listed approximately in the order of the intensity of the color. The wild type, red, is dominant to each of the other alleles, in the series, but these others show intermediate inheritance when brought together. For instance, a fly heterozygous for the genes coral and white will have an eye color quite similar to apricot as a result of the intermediate effect. Figure 12–1 shows diagrammatically how some of the multiple alleles for eye color in *Drosophila* arise.

Multiple alleles have been found in many other forms of life. Some of these are mentioned in the problems at the end of this chapter. The study of the blood groups in man illustrates one of the most interesting cases of multiple alleles, and also throws important light on the nature of genes and their reactions. Since there is a very practical importance to the inheritance of blood groups in man, we will consider the subject at some length.

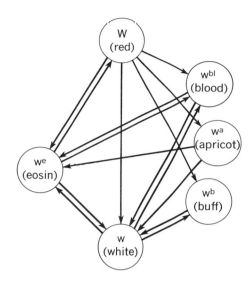

FIG. 12-1 *The direction of mutations at the white-eye locus of Drosophila, as tabulated by Gulick. This shows that multiple alleles may arise as direct mutations of the original wild-type gene, or by mutation of mutations which have arisen from this gene. Also, it will be noted that many of the mutations have been observed in two directions, thus indicating that the mutations are reversible.*

THE BLOOD GROUPS IN MAN

The transfusion of blood from one person to another was tried as early as the eighteenth century in France and later in England. Some cases were successful, but in others the blood of the donor proved incompatible with that of the recipient, and the recipient died a short time after the transfusion was given. When blood was mixed outside of the body, it was found to mix smoothly in some cases, whereas in others there would be clumping (agglutination) of the blood cells. The reason for this was not understood until about 1900, when Landsteiner removed blood cells from the plasma and on some occasions recombined the two elements of the blood. He noted that a smooth combination always resulted when one's cells were recombined with one's own plasma, but when plasma from one person was mixed with cells from another person, the mixture would sometimes be smooth but in other cases would result in clumping of the cells. This led to the discovery of the blood types. It was found that all persons could be classified according to four blood groups. At first these were designated by Roman numerals, but today the letter system is used almost universally. By this system the four blood types are called **O, A, B,** and **AB.**

Antigens and Antibodies

Before we can continue with our study of blood types we must digress for a time to gain some understanding of **antigens** and **antibodies,** for the blood is typed according to the presence or absence of these substances, as shown in Figure 12–2. Antigens are substances which form a part of almost all proteins and are capable of stimulating the production of specific antibodies. Antibodies are substances produced by animals in response to contact with foreign antigens and react specifically to particular antigens. We specify foreign antigen in this definition because an animal will normally not produce antibodies in response

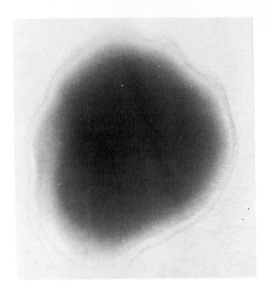 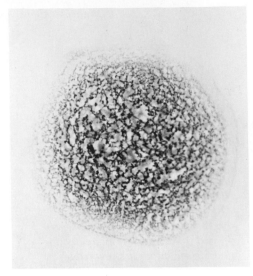

FIG. 12–2 *Blood typing. On the left is a drop of serum from type A blood and on the right one from type B blood. To each has been added a drop of blood from a person being tested. The serum on the right, containing anti-A antibodies, clumps the blood, but the serum on the left, containing anti-B antibodies, does not. Hence, the person being tested is type A.*

to the presence of antigens which are a part of its own body. This will be seen much more clearly by means of an illustration.

Let us suppose that we inject some of the albumen from a hen's egg into the body of a guinea pig. There is nothing poisonous about the white of an egg, so the guinea pig shows no unfavorable reaction to this treatment. Ten days later, let us inject the same guinea pig with more albumen from a hen's egg. In this case the guinea pig will probably go into spasms within a few seconds and in all probability will die. What has happened in the body of the guinea pig between injections to give such a vastly different reaction? Well, egg albumen contains antigens which are foreign to the guinea pig. These antigens introduced within its body have stimulated the guinea pig to produce antibodies. Upon the second injection, there occurs a reaction between these two kinds of substances which is so severe that it causes tissue damage and leads to the death of the guinea pig. Had albumen from a turkey's egg been used for the second injection, it is very likely that no reaction would have occurred, for each antibody is specific for the antigen which produced it and would not react to a different antigen.

Antibody production is very valuable in the establishment of resistance and immunity to disease. If you have typhoid fever, or receive "shots" of killed typhoid germs, you will produce antibodies specific for the antigens of the typhoid germs. As long as you have a good concentration of these antibodies in your system, you remain immune to typhoid. Should any typhoid germs gain entrance to your body, they will be destroyed through the antigen-antibody reaction and you do not develop typhoid fever. On the other hand, you may suffer because you develop antibodies to an antigen in ragweed pollen and get the antigen-antibody reaction in the nasal cavity whenever you inhale this pollen. This causes the sneezing, running nose, and irritation which is commonly called "hay fever." All of the various allergic diseases — and there are many — result from this sort of reaction to some antigen.

Blood Types

Now let us see how all this is related to blood types. Landsteiner found that the red blood cells of man may contain two distinct antigens, and the letters **A** and **B** were chosen to represent them. Persons with type **O** blood possess neither of these antigens, type **A** persons have the **A** antigen, type **B** persons have the **B** antigen, and type **AB** persons have both. These blood types have very important clinical significance, because blood plasma may contain antibodies which cause clumping (agglutination) of red blood cells containing any one of the corresponding antigens. Of course, no person could carry an antibody which would affect the antigens in his own red blood cells, but the blood of other persons may contain such antibodies. This is important in blood transfusions, for no blood should be given a person whose plasma contains antibodies that will clump the red cells as fast as they enter his body.

Figure 12–3 shows the antigen-antibody composition of the four types of blood. It will be noted that every type of blood contains all of the anti-**A** and anti-**B** antibodies it can without agglutinating its own cells. One of the unexplained mysteries of serology concerns the presence of these natural antibodies in the blood. If a person has antibodies for the typhoid fever bacillus in the

FIG. 12–3 *The relation of the A and B antigens and antibodies to the four major blood types. Note that the name of the blood type indicates the antigen composition of the red blood cells.*

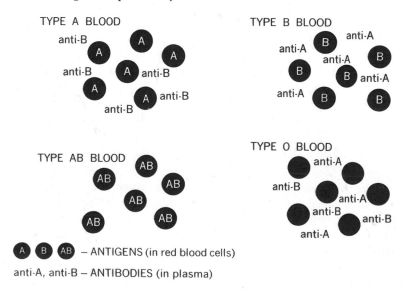

plasma of his blood, that is a sure sign that he has had the bacillus in his body at some recent time. When plasma containing such antibodies is mixed with typhoid bacteria, the bacteria will agglutinate, much like the red blood cells when incompatible blood is mixed. No case is known where antibodies to disease organisms occur naturally in the blood. Then why should a person with type **A** antigens have anti-**B** antibodies in his plasma which will agglutinate the cells from a person with type **B** blood? This question remains to be answered by future research. So far as is known, these are the only two antibodies which may occur naturally in the blood — all others which may be present have been produced by contact with the antigen to which they are specific.

Before leaving this discussion we should make it clear that not all antigen-antibody reactions result in cell agglutination. Some antibodies (precipitins) cause a precipitation of the protein containing the antigen; others (antitoxins) cause a neutralization of toxins; still others (lysins) cause the cells containing the antigen to be dissolved. Those antibodies which are related to blood typing are sometimes called **agglutinins** because of the agglutination reaction which they bring about; and the blood type antigens upon which they have their effect are called **agglutinogens.** Many other antigens and antibodies, such as those of typhoid, are of this type.

Inheritance of Blood Types

With this background we are ready to return to our study of inheritance of blood types. We know that the **A** antigen is produced by an autosomal gene. No person can carry this gene without having the **A** antigen, and, except for the very rare event of mutation, no child can have the **A** antigen unless at least one of his parents had it. In this respect the gene appears to act as a dominant, but it is not dominant in the same sense that the gene for normal pigmentation is dominant over the gene for albinism. When a person carries the gene for **A** antigen and the gene for **B** antigen, his blood will contain both antigens. Hence neither gene is dominant over the other, a condition similar to intermediate inheritance, but which is more accurately described as a case of lack of dominance. Each of the blood group alleles seems to act independently and is not suppressed by the presence of any of the others. Because of this rather unusual method of inheritance, it has become customary to use the letter **I** (Isohemagglutinogen) as a basic symbol for the genes at this locus, with a second letter as an exponent to indicate which variation of the allele is represented. For instance, I^A represents the gene which produces the **A** antigen, and I^B represents the gene which produces the **B** antigen. Persons heterozygous for these two genes will have both antigens in their blood, and it will therefore be classed as type **AB.** The third gene in the series is I^O. This gene is usually considered recessive to the other two, since a person heterozygous for the **A** and **O** genes is type **A**, and a person heterozygous for **B** and **O** is type **B**. The type **O** individuals must therefore be homozygous, $I^O I^O$. Table 12–1 shows the various genotypes which produce the four blood types.

TABLE 12-1

The Genotypes which Produce the Four Blood Types

Genotype	Blood type
$I^O I^O$	O
$I^A I^A$ & $I^A I^O$	A
$I^B I^B$ & $I^B I^O$	B
$I^A I^B$	AB

Medico-Legal Applications of Blood Group Inheritance

The method of inheritance of blood types is so definite and well understood that it is often of great value in legal cases of disputed parentage. King Solomon might never have had to make use of his profound wisdom in deciding on the custody of the child which was claimed by two women had he known about blood types. A modern case of this nature came to light recently in a court trial. A young couple had taken what they supposed to be their baby home from the hospital, only to discover a tag on the child bearing the name of another woman who had borne a child at about the same time. This woman had also taken a baby home from the hospital on the same day, but she felt that she had her own child and would not consent to an exchange when told what had happened. Luckily, however, it was possible for blood typing to settle the dispute. The first baby was type **O**, but the couple who took this baby home from the hospital were **O** and **AB**. The second couple had a baby with type **A** blood, but both parents were **O**. Thus it was possible to say definitely that the babies had gone to the wrong parents, since neither couple could have been the parents of the child which they had taken home. The exchange was made and everyone was satisfied.

Another case was that of a young woman in Hollywood who sued a famous actor for the support of her child, claiming that he was its father. By typing the blood of the three persons involved, it was possible to say that the man could not have been the father of the child. The child was **B**, the mother was **A**, and the accused man was **O**. The true father must have been either **B** or **AB**. It is an interesting commentary on our American courts that the actor was ordered to support the child anyway. This was possible because in California the laws do not recognize blood-typing evidence as binding. Table 12–2 shows the possible types of children which may result from parents with the different blood types. Another kind of legal case which may be unsnarled by blood typing is that in which a man sues his wife for divorce on the ground that she has borne a child by another man.

It can easily be seen from the table that when evidence is based only upon the four blood types discussed so far, there will be many cases which are inconclusive. For instance, in a paternity suit it might be found that the true father of a child must be either type **A** or **AB**. The accused man, let us say, is type **A**. This does not by any means prove that he is the father, for there are many other men with type **A** blood. Should he have been of type **O**, we could say definitely

Blood type of parents	Possible blood types of children
O × O	O
O × A	O, A
O × B	O, B
O × AB	A, B
A × A	A, O
A × B	O, A, B, AB
A × AB	A, B, AB
B × B	B, O
B × AB	A, B, AB
AB × AB	A, B, AB

TABLE 12–2

Possible Blood Types of Children from Parents of Various Blood Types

that he could not have been the father, but since he is of type **A,** we can only say that he might have been the father. Some variations of antigen **A** have been discovered which narrow down the indeterminable cases considerably. Three different grades of **A** antigen have been discovered which can be distinguished by clinical tests. These have no separate significance in blood transfusion, but they extend the possible number of **ABO** blood groups to eight — three varieties of type **A,** three varieties of type **AB,** and one variety each of **O** and **B.**

Another discovery makes possible a still further reduction in the number of inconclusive cases. An extract from the blood of the eel, as well as an extract of a plant (*Vlex europus*), will clump cells of persons carrying the gene I^O. These extracts cause some agglutination when mixed with type **O** blood. Of greater importance, however, is the fact that they will cause agglutination of the cells from type **A** and **B** persons if they are heterozygous. Thus, we can now distinguish the blood of heterozygous persons from that of homozygous persons. This reduces the number of possible blood types in the offspring in some cases. For instance, a heterozygous type **A** man could have children with type **O** blood, but a homozygous type **A** man could not have such children.

The Secretor Trait

As investigators studied the blood groups from various angles it was discovered that some persons have **A** or **B** antigens in the body secretions (from eyes, nose, salivary glands, and mammary glands) as well as on the red blood cells. Those with this characteristic are known as secretors and can be identified even by a bit of dried saliva which would remain on an envelope that had been licked or by a cigarette butt. It is distinguished by mixing the secretion with serum which contains antibodies for **A** or **B** antigens. If the antigen is present in the secretion it will react with and neutralize the antibodies and the serum will lose its ability to agglutinate cells bearing **A** or **B** antigens. A chemical explanation has been found for the secretor trait. Persons who are secretors have water-soluble antigens and, therefore, some of these pass out into the body secretions.

The non-secretors, on the other hand, have antigens which are only alcohol-soluble and cannot be dissolved out in the secretions which are aqueous. Thus, the secretors can be identified by tests on the blood as well as on the body secretions. The ability to secrete these antigens is due to a dominant gene, *S,* with the recessive allele, *s,* representing the non-secretor condition. About 77 per cent of the people in the United States are secretors.

Other more recent studies indicate that the antigens of the **ABO** series may also be found in cells of the body other than blood cells. For instance, Coombs demonstrated their presence in the epidermal cells of the skin by an ingenious technique. He placed serum containing anti-**A** on the skin of a person with type **A** blood. Then he added red cells of type **A** and found that these cells clumped and adhered tightly to the surface of the skin like little pancakes. When the same serum and cells were placed on the skin of a person with type **B** or type **O** blood there was no such adherence. Others have found an agglutination reaction for a variety of isolated cells, leading us to feel that all of the body tissues probably carry the antigens of the **ABO** blood groups.

The Rh Blood Antigens

Another most interesting series of multiple alleles relating to blood antigens was discovered by Wiener, Levine, and others through studies of the blood of monkeys. These, known as the **Rh blood antigens,** were mentioned briefly in Chapter 11 but will be discussed in greater detail here. When blood from a rhesus monkey was injected into guinea pigs, it was found that the guinea pigs produced antibodies which will agglutinate the red cells of all rhesus monkeys. This indicates that the red cells of all monkeys of this species contain a particular antigen. It was designated **Rh.** When human blood was tested by this guinea pig serum, it was found that the cells of some persons clump, whereas the blood of other persons is not affected. It was therefore concluded that some persons have the same antigen (**Rh**) as that found in the blood of rhesus monkeys, while others do not have it. Those with the antigen were designated **Rh**-positive; those without it, **Rh**-negative. No persons have been found who contain the natural anti-**Rh** antibodies, but it has been found that an **Rh**-negative person can develop these antibodies if exposed to the **Rh** antigen. This has important significance in transfusion. Let us assume that an **Rh**-negative person has been injured in an automobile accident and is in serious need of immediate transfusion. A donor of a suitable **ABO** blood group is found and the transfusion is given. The donor, however, happens to be **Rh**-positive. There will be no reaction, if the person has never had a positive transfusion before, since there are no natural antibodies to **Rh** antigens, but the presence of the **Rh** antigen on the cells received will stimulate the recipient of the transfusion to generate **Rh** antibodies. Now let us suppose that this same person needs another transfusion, let us say, 1 year later. Perhaps the same donor is selected and the transfusion is given. This time the cells of the donor will clump as they react with the antibodies now present, and the recipient may die. As a result of this discovery, it has now become common procedure to type blood for the **Rh** antigens as well as for the **A** and **B** antigens. No **Rh**-negative person should ever be given a transfusion from an **Rh**-positive person, even though it may be the first transfusion.

The **Rh** factor also has great significance in childbirth. It has been found that an **Rh**-negative woman who has been sensitized by **Rh**-positive blood may give birth to abnormal children, if she is married to an **Rh**-positive man, as shown in Figure 12–4. Her children may inherit the **Rh** antigen from their father. If so, the antibodies of the mother may pass through the placenta and cause damage to the red cells of the child during the last few months of pregnancy. This causes the disease known as *erythroblastosis foetalis,* which consists of an anemia due to the hemolysis (breakdown) of the red blood cells in the foetus and a consequent jaundice as the blood vessels in the liver become clogged with the broken cells and bile is absorbed by the blood. The disease takes its name from the fact that the blood contains immature, nucleated red blood cells which normally are found only in the bone marrow where these cells are formed. These cells are not efficient carriers of oxygen. The disease may be so severe as to cause death before birth (stillbirth) or neonatal death (death within a few days after birth). The disease is also known as hemolytic disease because of the hemolysis of the red blood cells.

An **Rh**-negative woman may be sensitized by a transfusion of positive blood and, as a result, her first child could have this disease. It is unfortunate that she may also become sensitized simply by carrying a positive child within her body. Some of the cells from the embryo may seep into her own blood stream and cause this. The antibodies do not reach sufficient strength before the birth of this child to cause any great damage as a rule, but subsequent positive children may be afflicted. In many cases, however, a negative woman can bear a positive child without becoming sensitized sufficiently to cause serious damage to subsequent positive children. Hence, it is not a foregone conclusion that the second positive child from a negative woman will have *erythroblastosis.* Physicians can

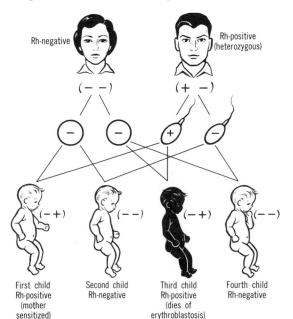

Rh-negative

Rh-positive (heterozygous)

(− −)   (+ −)

(− +)   (− −)   (+ +)   (− −)

First child Rh-positive (mother sensitized)

Second child Rh-negative

Third child Rh-positive (dies of erythroblastosis)

Fourth child Rh-negative

FIG. 12–4 *The types of children which may result from a marriage between an Rh-negative woman and an Rh-positive man (in this case heterozygous). Of course, the children will not necessarily appear in this sequence, but the mother must be sensitized to the Rh factor before a child will be produced with erythroblastosis foetalis.*

test a woman's blood for the level of antibodies (titre) and determine in advance whether any difficulty is to be anticipated. Thus forewarned, they can take steps to prevent damage whenever the tests indicate that it is advisable. The physicians can be ready with blood for transfusion and may give the newborn infant a complete change of blood to replace the damaged cells of its own body. If the baby can survive the first few days, it will have no more difficulty because the source of the antibodies is gone when the attachment with the mother is severed.

A study in Manchester, England, showed that about one child in each twenty-three born to **Rh**-negative women married to **Rh**-positive husbands have this disease. When we consider that some of these children will be **Rh**-negative and that the first positive child will not normally be afflicted, this figure shows a rather high probability of affliction for those positive children following the first positive child.

In the light of this discussion the question may have arisen in your mind as to why we do not have *erythroblastosis* arising from other blood antigen-antibody incompatibility. For instance, what about the type **O** woman bearing a type **A** child. She has antibodies which will react with her child's, why don't they? There are some cases where they do. A study by Waterhouse and Hogben showed by statistical study that marriages of type **O** men with type **A** women gave more children with type **A** blood than the reciprocal type of marriage where the men were **A** and the women **O**. It appears that the difference is due to foetal incompatibility. Such incompatibility occurs very early in pregnancy, during the first 7 or 8 weeks, instead of very late as in **Rh** incompatibility. Also, a few other blood antigens have on rare occasions been known to result in *erythroblastosis*.

The **Rh** antigen is produced by a dominant gene *R,* and the absence of the antigen (**Rh**-negative) results when a person is homozygous for the recessive allele *r*. Thus it can be understood that negative parents can have only negative children, but positive parents, in accordance with the laws of simple Mendelian inheritance, can have both types of children. The dangers of pregnancy outlined above result only from those marriages between an **Rh**-positive man and an **Rh**-negative woman. Since about 85 per cent of the white population of Europe and America are **Rh**-positive, this leaves the **Rh**-negative woman in rather a dilemma when she chooses a husband — that is, if she knows about the problem. Lawrence Snyder relates an amusing case of a young technician who did all the blood typing at a large hospital. She herself was **Rh**-negative and had learned the full significance of the **Rh** factor. Whenever she typed a man's blood and found it to be negative, she took a good look at him and made it a point to become acquainted with the patient if he turned out to be an eligible male. In the fullness of time she married one of the men she had met in this way.

Complexities of the Rh Factor Inheritance

It was not long after the discovery of the **Rh** factor that evidence began accumulating that here again was a case of multiple alleles and it has turned out to be one of the most complex of all the multiple allele systems known in any form of life. Just as subdivisions of the **A** antigen were found, it is possible to find three major subdivisions of the **Rh** antigen. An American, A. S. Wiener, proposed that there were eight main alleles at the locus for this characteristic on the chromosome. This was possible because some of the alleles produced two or three of the positive antigens. By this system the following method of designation can be used:

Gene symbol	Positive antigens produced
r	**none**
R_0	$\mathbf{R_0}$
R'	$\mathbf{R'}$
R''	$\mathbf{R''}$
R_1	$\mathbf{R_0}$ and $\mathbf{R'}$
R_2	$\mathbf{R_0}$ and $\mathbf{R''}$
R_x	$\mathbf{R_0}$, $\mathbf{R'}$, and $\mathbf{R''}$
R_y	$\mathbf{R'}$ and $\mathbf{R''}$

This discovery had no bearing on the previous findings from a clinical standpoint except that it was found that nearly all of the difficulties of embryonic hemolysis came about as a result of the presence of the $\mathbf{R_0}$ antigen. For some reason this antigen seemed to be stronger in stimulating antibody production than the others. Unfortunately, however, this is by far the most common positive antigen.

The English geneticist, R. A. Fisher, proposed a theory of pseudoallelism to explain the relationship. This hypothesis holds that there are actually three genes involved in the **Rh** antigens and that these genes are so close together on the chromosome that they ordinarily move and act as one gene. Such genes are known as **pseudoalleles.** He proposed letter symbols for the three dominant positives, *CDE,* and the recessive alleles, *cde.* Thus an **Rh**-negative person would have the genotype *cde/cde.* If any one of the dominant genes is present the person would be considered **Rh**-positive. The gene *D* would produce the antigen designated as $\mathbf{R_0}$ by Wiener; *C* is equivalent to *R'*; and *E* to *R''*. Thus, instead of assuming that one gene, R_1, produced two antigens ($\mathbf{R_0}$ and $\mathbf{R'}$), we would say that a person had the two genes, *D* and *C,* in his genotype. His complete genotype might be any one of the following: *CDe/cde, CDe/cDe, CDe/Cde,* or *CDe/CDe.* In Table 12–3 are shown the frequencies in the English population of the commoner and some of the rarer genotypes according to both systems of designation.

There is no great point of divergence between these two theories; one merely assumes that there are multiple variations of one gene, the other that there are three separate genes lying side by side and very close together. The Fisher theory seems to be more in accord with recent discoveries in biochemical genetics.

TABLE 12-3

Genotypes and Frequency of Some of the Rh-Positive Blood Groups (From Sorsby)

Common genotypes		Approximate frequency in population in percentage
Wiener	Fisher	
R_1/r	CDe/cde	35.00
R_1/R^1	CDe/CDe	20.00
R_2/r	cDE/cde	12.00
R_2/R^2	cDE/cDE	2.00
R_1/R^2	cDE/CDe	13.00
R_0/r	cDe/cde	2.00
Rarer genotypes		
R'/r	Cde/cde	0.75
R''/r	cdE/cde	0.85
R_0/R^0	cDe/cDe	
R'/R'	Cde/Cde	
R_x/r	CDE/cde	
R_y/r	CdE/cde	

Other Blood Antigens

As the work with blood antigens progressed, it occurred to workers in this field that there might be some antigens in the human blood for which the human body never produces antibodies. Landsteiner and Levine injected various samples of human blood into guinea pigs and found two different antigens which call forth antibodies in the serum of the guinea pigs. They called these **M** and **N** antigens, and they are produced by the genes M and M^n. So far, no allele for the absence of these antigens has been found. These antigens have no significance in transfusion since human beings do not produce antibodies which react with them, but they are of value in increasing our knowledge of antigens and their method of inheritance, as well as in medico-legal cases. People may be typed as **M, N,** or **MN.**

This work paved the way for many other discoveries on the complex antigenic nature of the human blood. For instance, a woman named Mrs. Kidd gave birth to a child who showed some degree of antigen-antibody hemolysis, yet there was no indication of any **Rh** incompatibility. It was found that she had become sensitized to an antigen in her husband's blood cells which she did not carry. It was called the Kidd factor in her honor, and tests have shown that about 77 per cent of the people in the United States are Kidd-positive and 23 per cent Kidd-negative. A Mrs. Kell demonstrated a similar antibody production, but this antigen was different from that of Mr. Kidd, so this was called the Kell factor. Then Mrs. Lewis, Mrs. Duffy, Mrs. Lutheran, and Mrs. Cellano all had their names preserved for posterity because of antibody production against different antigens in the red blood cells of their husbands. The Cellano antigen turned out to be the one present in Kell-negative people.

New antigens have been and will be discovered to extend this list. For example, the Diego factor, so far found only in American Indians. Altogether thirteen different blood antigen systems have been found, the latest being a sex-linked antigen known as **Xg.** Only very rarely do any of them cause any maternal complications to compare with the **Rh** factor, so their primary significance lies in their use to obtain a better understanding of gene action and as a means of identification in cases of disputed parentage or criminal investigation. When all of the various combinations of the blood groups are compounded, we find that the possibilities run high into the millions. Thus, a drop of blood left at the scene of a crime might be just as positive in identifying a criminal as his fingerprints. It should be brought out, however, that some of the antigenic differences can be distinguished only by experts, and such experts must do the tests if they are to be reliable in legal proceedings.

Thus we see that the human blood is indeed a strange and complex substance and an understanding of its genetics will do much to help us understand many of its physiological reactions which have been a mystery to man down through the centuries.

PROBLEMS

1. In *Drosophila* the wild-type red eye color is dominant over a number of different eye colors, including cinnabar (*cn*) and ⌐epia (*se*). When sepia-eyed flies are mated to cinnabar-eyed flies the offspring have the wild-type red eyes. From these results would you conclude that sepia and cinnabar are alleles? Diagram this cross to support your answer.

2. In corn there is a gene for sun red which causes the kernels to turn red when exposed to sunlight. There is an allele for weak sun red which causes a weaker red color to appear, and another allele for dilute sun red which causes a still weaker color. Emerson obtained the following results from various crosses: sun red × weak sun red gave all sun red in the F_1 and a 66 : 16 ratio in the F_2; sun red × dilute sun red gave all sun red in the F_1 and 998 : 314 in the F_2; weak sun red × dilute sun red gave all weak sun red in the F_1 and 1,300 : 429 in the F_2. Show the relationships between these three alleles with respect to dominance.

3. A heterozygous weak sun red plant is crossed with a dilute sun red plant. Show the expected ratio in the first generation.

4. In rabbits, full color, *C*, is dominant; Himalayan albinism, c^h, produces white rabbits with colored legs, ears, and noses; albinism, c^a, produces all white rabbits. Himalayan is dominant to albinism, but recessive to full color. What types of offspring would be expected from a colored male, heterozygous for Himalayan, crossed with an albino female?

5. Two Himalayan rabbits are crossed. They yield seven Himalayan and five albino rabbits. Could this be an example of a 3 : 1 ratio? (Use principles learned in Chapter 7 to determine this.) If so, what is the genotype of the parents?

6. Apricot, white, and coral are eye colors in *Drosophila* which are sex-linked alleles. Heterozygous coral and white give an eye color very near apricot. Show the results from a cross between two flies with apricot colored eyes although the

female is heterozygous for white and coral. (Heterozygous coral and apricot yields an eye color similar to cherry.)

7. A man has type **A** blood and his wife has type **B** blood. A physician types the blood of their four children and is amazed to find one of each of the four blood types among them. He is not familiar with genetics and calls upon you to explain how such a thing could happen. What would you tell him?

8. A couple preparing for marriage have their blood typed along with the other required blood tests. Both are **AB**. They ask you what types of blood their children may have. What would you tell them and how would you explain your conclusions?

9. A woman sues a man for the support of her child. She has type **A** blood, her child type **O**, and the man type **B**. Could the man be the father? Explain your answer.

10. Further tests of the persons in Problem 9 show that both the man and the woman are **Rh**-negative while the child is **Rh**-positive. Would this information be of any value in the case? Explain.

11. A wealthy, elderly couple die together in an accident. Soon a man shows up to claim their fortune, contending that he is their only son who ran away from home when a boy. Other relatives dispute this claim. Hospital records show that the deceased couple were blood types **AB** and **O**, respectively. The claimant to the fortune was type **O**. Do you think that the claimant was an impostor? Explain.

12. A woman bears a child with *erythroblastosis foetalis* at her second delivery. She has never had a blood transfusion. On the basis of this information, classify the woman, her husband, and both children as to **Rh** blood type. (Assume that all cases of *erythroblastosis foetalis* are caused by **Rh** incompatibility.)

13. A woman is **Rh**-positive and both of her parents are **Rh**-positive. She marries a man who is **Rh**-negative. Is there any possibility that they may have any **Rh**-negative children? (Ignore mutation in your considerations.)

14. Police records in a large city show that a woman left her baby in its carriage outside a supermarket. When she returned from shopping it was gone. She later found a baby which had been left at a foundling home shortly thereafter and claimed that it was her baby. Blood typing was done with results given below. Evaluate her claims on the basis of this evidence.

Mother	A	MN	**Rh**-negative
Father	O	N	**Rh**-positive
Foundling	A	M	**Rh**-positive

15. All Chinese are **Rh**-positive and about 85 per cent of the white race are **Rh**-positive. Suppose you find 10,000 cases where Chinese men have married white women in the United States. In how many of these marriages would there be a possibility of **Rh**-induced *erythroblastosis?* In 10,000 cases where the men are white and the women Chinese, how many would run this danger?

16. A young lady about to be married learns that she was born with *erythroblastosis* and is greatly perturbed lest some of her children have the same condition. Her fiancé refuses to have their blood tested, saying that it is all a lot of foolishness, that women have been having babies for centuries without all this "new-fangled **Rh**-business" and that they can go on without it in the future. What could you tell her that might ease her anxiety?

17. Another young lady learns that her fiancé was the fifth child in a family in which the second and fourth children were born with *erythroblastosis*. She asks you to

tell her what her chances are that she could have any children with *erythroblas-tosis*. What could you tell her?

18. A merchant comes to work one morning and finds that his place of business has been burglarized during the night. Detectives find that the thief cut himself on the sharp edge of the money box and left a little puddle of blood behind. This blood is analyzed for antigens and found to be type **A, Rh**-negative, **M,** and secretor. The frequencies of these in the general population in percentages are: 40, 15, 30, and 77. Later a suspect is arrested and his blood is found to match that found in the store. He claims that this is mere coincidence. What are the chances that he is telling the truth?

13

Multiple Gene Inheritance and Quantitative Characteristics

MOST OF THE INHERITED CHARACTERISTICS which we have studied so far have been sharply classified into distinct groups. We have studied sheep with horns and sheep without horns, guinea pigs with short hair or long, black hair or white. There is no doubt about the presence or absence of these characters or of many of the others we have considered. Cattle may be red, roan, or white. Peas are either round or wrinkled, green or yellow. People either have hemophilia or they do not, and they are of blood type **O, A, B,** or **AB.** All of these may be called **discontinuous traits,** since there is a sharp line of distinction between one group and another. There are many characteristics, however, which are not so clearly defined, but show quantitative variation from one extreme to another. These may be called **continuous traits.** Since it is not possible to explain such variations on the basis of two or more alleles at a certain chromosome locus, other explanations for them must be sought.

INHERITANCE OF CONTINUOUS TRAITS

Height in Corn

Some of these variations can be explained on the basis of **environmental** factors alone, many are due to a combination of **environmental and hereditary** factors, and a few may be due to **hereditary** factors only. The height of the stalk in corn is inherited, but we all know that soil conditions, the amount of fertilizer used, insect and fungus parasites, weather, and water play an important part in the height which is attained. We could obtain a homozygous variety of corn through repeated self-fertilization, but there would be variations in the height of the plants which grew from the seed of this variety because of environmental factors alone, even though the height would be much more uniform than that of a field of corn which had a mixed heredity. On the other hand, if we raised a mixed population in a greenhouse with as nearly the same environment as humanly possible, we would still find quantitative variations in height due to genes. These variations would not fall neatly into just two classes, short and tall, as would be expected from dominant-recessive inheritance. Neither would they be distinctly short, medium, or tall, as would be expected in intermediate inheritance of a single variant pair of alleles. The height of the corn stalk is inherited through a number of different genes located at different loci on the chromosomes. Such a character is said to exhibit multiple gene inheritance. In such cases it is difficult to pick out the effects of single genes and to say whether they are dominant, recessive, or intermediate. We can see the over-all effect of the interaction of the genes, however, and study them quantitatively.

Stature in Man

Body height in man is another good illustration of this principle. Of course, here again environment plays an important role — diet, disease, and other environmental factors all have their effects in the attainment of hereditary potentialities of stature, as illustrated in Figure 13–1. Even identical twins, who have the same heredity, show some variations in height, especially when raised apart.

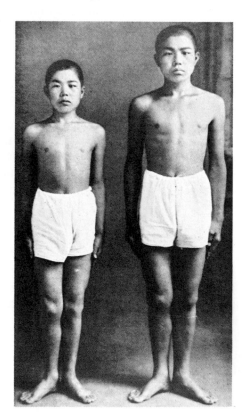

FIG. 13–1 *Environmentally-induced variation in stature. These 15-year-old twins were about the same physical size until 5 years of age. At that time, the one on the left had a disease which caused some deterioration of the pituitary gland. He developed diabetes insipidus and now weighs thirty-one pounds less than his twin brother who was not so affected.* FROM TAKU KOMAI AND GORO FUKUOKA, JOURNAL OF HEREDITY.

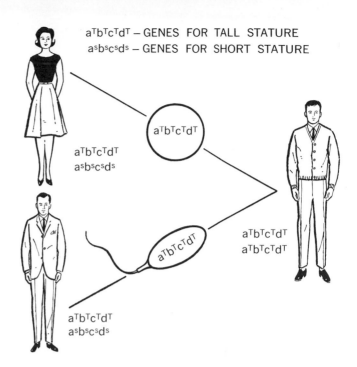

$a^Tb^Tc^Td^T$ – GENES FOR TALL STATURE
$a^sb^sc^sd^s$ – GENES FOR SHORT STATURE

$a^Tb^Tc^Td^T$
$a^sb^sc^sd^s$

$a^Tb^Tc^Td^T$

$a^Tb^Tc^Td^T$
$a^sb^sc^sd^s$

$a^Tb^Tc^Td^T$

$a^Tb^Tc^Td^T$
$a^Tb^Tc^Td^T$

FIG. 13–2 *How medium-sized parents may transmit genes for tall stature to a child. For this illustration we have assumed that there are only four genes involved, all of which are intermediate. Actually, there are more genes than this involved, and some are dominant or recessive, but the principle of transmission is the same. Also, this diagram does not indicate the influence of environment.*

And non-identical twins, who differ in some of their genes, show a greater variation in height because of the differences in both genes and environment. A group of non-related children of the same age who have been raised in an orphanage show a considerable range of variation. Since the environment of such persons is very similar, we can conclude that height in man is influenced by heredity, and because there are so many gradations, by a number of genes at different loci.

Such multiple gene inheritance is certainly nothing unusual. Indeed, it is to be expected that such a characteristic as body stature in man would depend on the reactions of many genes which are related to the general growth of the body. Various mutations of the individual genes involved would give variant alleles at a number of different loci. This tends to give the quantitative effect. So far as heredity is concerned, one's stature depends upon the interactions of these numerous genes. Thus it is easy to see how two medium sized parents can produce a tall child, for the parents can be heterozygous for a number of genes for tallness which may become homozygous in the child (Figure 13–2). There are a few genes which may be epistatic to an entire series of genes which influence a quantitative character. For instance, there is a gene which produces dwarfism in man because of its effect on the pituitary gland. It causes an undersecretion of the growth hormone, and the afflicted individual is a dwarf even though he may carry a preponderance of genes for tallness at the other loci of genes which influence body stature. This is known as the hypopituitary type of dwarf. Such a person could have a child who would be above average in stature.

Skin Color Inheritance in Man

Skin color is a body characteristic which is dependent upon quite a number of genes. As with body stature, we see many gradations in the color of the skin from very dark to extremely fair, even among the members of the Caucasian race. Again, environment plays its part, and to minimize the effects of suntan, we should conduct our observations in midwinter if we wish to determine the true effects of heredity. The color of the skin is dependent upon the amount of a pigment, **melanin,** which is deposited in the skin. The total amount deposited is dependent upon genes and on the amount of sunlight the skin receives. The Negro race, typically, has a rather heavy deposit of this material regardless of exposure to sunlight. According to detailed studies by Curt Stern, the difference in skin color which exists between the Caucasians and the Negroes appears to be determined by genes at from four to seven different loci on the chromosomes.

For purposes of illustration let us choose the estimate of four. Each of these gene pairs will segregate independently, so we will have a tetrahybrid type of gene assortment in the offspring of mixed marriages. The genes involved all appear to be intermediate in their effects, so the skin pigmentation of the offspring of a Negro-Caucasian cross is the intermediate pigmentation known as mulatto. When two mulattos marry they may have children with nine degrees of pigmentation including white and Negro. A child with white skin, in all probability, would show many characteristics of the Negro race, since there are other genes which distinguish the two races besides those involved in skin color. A white child thus produced might have typical negroid facial features or hair characteristics. The reverse, of course, would be true of a child with Negro skin coloring.

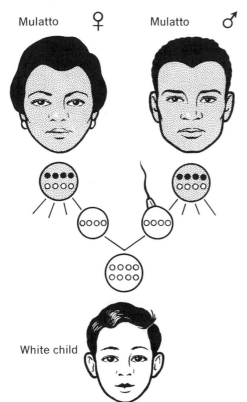

Mulatto ♀   Mulatto ♂

White child

FIG. 13–3   *Mulatto parents can produce a child with skin as fair as the Caucasians through gene segregation. Here we assume four gene pair differences between the Negro and Caucasian. When each of the parents contributes all four genes for the fair skin, a child is produced as indicated.*

FIG. 13–4 *Quantitative variation in skin color of Caucasian and Negro races in the United States. There is continuous variation in both races and some overlapping because of racial amalgamation. Due to gene segregation, there are a few classified as Negroes with a skin as fair as the fairest Caucasian. The continuous variation of a considerable degree among Caucasians indicates that there are many genes involved, but the differences between the two races appear to be due to four to seven of these genes.* DIAGRAM FROM CURT STERN.

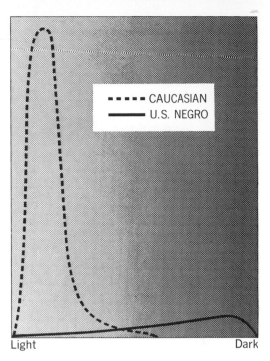

On the basis of four gene differences for skin color, a child with typical Caucasian skin color would be expected in 1 birth out of every 256 and the same probability would hold for Negro skin color. For five gene loci the probability would fall to 1 in 1024 and for six it would be only 1 in 4096. The number of degrees of skin color in between the two extremes would also increase as the number of gene loci increases.

It is sometimes possible to determine the number of gene loci by the proportion of the two extremes in second generation offspring. Thus, $\frac{1}{16}$ of each of the two extremes would indicate two loci, $\frac{1}{64}$ would indicate three, $\frac{1}{256}$ would indicate four, and so on. As the number of loci increases, however, there is very likely to be increasing difficulty in distinguishing the different classes of offspring. This is the difficulty encountered in trying to establish the exact number of loci involved in racial skin color differences. It becomes difficult to distinguish the pure white and the pure negroid skin colors when other classes are very similar to it.

In attempts to analyze the genes for skin color which distinguish the two races, it has been suggested that the Negro race carries four pairs of genes which cause the extra deposits of melanin. The alleles of these in the Caucasian race neither add to nor substract from the amount of melanin. The first type is called a **contributing gene,** while the second is known as a **neutral gene.** The pure Negro, having eight contributing genes, has the greatest pigmentation of all; the mulatto, with four contributing genes, is in between; and those with all eight neutral genes show the fairest skin. Both races, no doubt, possess other contributing and neutral genes which cause the skin color variation within the races, but here we are considering only the estimated four gene loci involved in the skin color differences between the two races. These terms, contributing and neutral, might well be applied to other genes which we have studied. For instance, the genes for antigens in the red blood cells are contributing genes, while the genes for the absence of antigens are neutral in their effect.

MULTIPLE GENES IN OTHER FORMS OF LIFE

Color in Wheat

Nilsson-Ehle made a very interesting study on the color of grain in **wheat,** which showed that the factors operate in a way quite similar to those in our illustration of skin color in man. He found that, when he crossed a variety of wheat bearing dark red kernels with a variety having white kernels, he obtained an intermediate shade of red in the first generation. When these were crossed among themselves, he obtained colors which ranged from dark red to white, about 1/16 of the grains having each of these latter colors. Further examination revealed that the grains could be separated into five distinct groups with a ratio of about 1 : 4 : 6 : 4 : 1. By application of the laws of probability we can see how this would lead to the conclusion that two segregating genes are responsible for the color differences in the grains of wheat. Those with red grains contain four contributing genes for red (two pairs of alleles); the white-grained wheat, on the other hand, contains four neutral genes. The first generation contains two of the contributing genes for red, which produces the intermediate color. The color in the second generation depends upon the segregation of the genes for red, and the five groups are produced according to whether they possess 4, 3, 2, 1, 0 genes for color, respectively.

At a later date, however, Nilsson-Ehle crossed another strain of red-grained wheat with white and found a different ratio. Only 1/64 of the second generation grains were red and 1/64 white. Also, there were seven varieties of coloration instead of five. The ratio between these varieties, ranging from red to white, was approximately 1 : 6 : 15 : 20 : 15 : 6 : 1. This is explained by the fact that this particular strain of wheat carried six contributing genes for the red color (three pairs of alleles).

Size in Chickens

Unfortunately, however, not all problems in the inheritance of quantitative characteristics are so simple to understand. Thus when Punnett crossed the small Sebright Bantam chickens with the larger Golden Hamburg chickens, the first generation offspring were intermediate in size. The second generation showed variation, but to his surprise included some fowls which were larger than either parent and some which were smaller than either. How can we explain this on the basis of a segregation of genes affecting body size from the original parents? Punnett explained it by assuming that there are four pairs of genes which affect body size in these birds. Let the letters *ABCD* represent the genes for large size, and the letters *abcd* represent those for small size. If we assume that the Golden Hamburgs have the genes *AABBCCdd,* then they would be distinctly larger than the Bantams, because they have six genes for large size and only two for small size. The Bantams could be *aabbccDD,* which would make them small because they had six genes for small size and only two for large size. The first generation offspring would all be heterozygous, *AaBbCcDd,* which would result in intermediate weight if we assume that these genes are intermediate in their effects. If a large number of F_2 offspring are produced, there will be some homozygous for all the genes for large size, *AABBCCDD.* These will therefore be larger than the Hamburg parents, who carried two genes for small size. An equal number of the offspring will receive all eight genes for small size, and these will be even smaller than the Bantam parents.

This principle makes clear how new and desirable phenotypes may be produced in domestic animals and plants through outbreeding to other races and varieties with obviously undesirable characteristics and then selecting for the desired characters. For instance, we can increase the size of a chicken that is already large by breeding it to the tiny Bantam; and, conversely, we can develop chickens smaller than the Bantam by outbreeding the Bantam to the large Hamburg. In a similar way, it is possible to increase the yield of a variety of corn by outcrossing it with a rather poor, non-productive variety. In other words, even the poor varieties usually contain some genes which are desirable.

Coat Color in Cattle

It would be possible to cite many other crosses of this kind from many other forms of life, all of which seem to follow the principles which underlie the cases presented above. Some of these are included in the problems at the end of this chapter. Here, however, it will be more useful to present one more instance which illustrates a slightly different type of multiple gene inheritance. Certain breeds of cattle have spotted coats, while others have coats of solid color. It has been found that this distinction is due to a single pair of alleles. The gene for spotting is recessive, *s,* while the gene for solid color is dominant, *S.* This character is inherited as a simple Mendelian factor. The degree of spotting is quite variable, however, as can be seen in Figure 13–5. In some, the spotting is

FIG. 13–5 *Multiple gene effects in Holstein cattle. These three cows are all homozygous for the recessive gene for spotting, yet they show great variation in the degree of spotting. This is due to the influence of multiple, modifying genes.*

localized around the head and neck, in others there are scattered spots of color on the body, while in still others colored spots cover the major portion of the body. The degree of spotting is quantitative, and we would expect it to be due to multiple genes. And this seems to be a correct assumption, for through selection it is possible to establish a herd of cows with a large amount of spotting or with a small amount. If a cow happens to carry the dominant solid gene, *S,* it will reveal no spotting of any kind, no matter what aggregation of genes for degree of spotting are present. This is a condition called epistasis (see Chapter 6). The gene for solid color is epistatic over the group of genes which regulate the degree of spotting.

Multiple Genes for Eye Color in Drosophila

One of the best illustrations of the interaction of genes is the case of the multiple genes for eye color in *Drosophila*. We learned in the previous chapter that there is a series of multiple alleles which affects the color of eyes in the fruit fly. There are other genes, at different loci, however, which also affect the color of the eyes. The white-eye series is located at 1.5 on the *X*-chromosome, but another mutation has been discovered, at 44.0 on the third chromosome, which produces a scarlet eye. Still another gene, on the second chromosome at 54.5, produces a purple eye. This shows us that genes at these loci are involved in the production of color in the eyes of *Drosophila* as well as the gene at 1.5 on the *X*-chromosome. The following list includes a number of the different gene loci which may be included in this series: carnation, *car;* garnet, *g;* pink, *p;* purple, *pr;* ruby, *rb;* scarlet, *st;* vermilion, *v;* and white, *w*. A fly with red eyes contains at least one dominant allele for each of the above recessive genes which produce eye colors different from red. Red eyes are found in practically all fruit flies collected in the wild state. If we were to give the complete gene formula for eye color for a fly with carnation eyes it would be: *car car G G P P Pr Pr Rb Rb St St V V W W*. In common practice, however, we assume that all the genes not specified are of the normal dominant type, and we would write it only *car car*.

This once more brings out the very important fact, which we cannot emphasize too often, that as a rule an inherited characteristic is not the product of a single pair of genes but results from the interaction of genes at many different loci. When one of these genes mutates the phenotype of the characteristic may be altered, but that does not mean that only the genes at this locus produce the characteristic. When the gene for white eyes in *Drosophila* was discovered, it was thought that the gene for eye color was at this location and that the normal allele of white produces red eyes. We know now that the normal allele of white is only one of a large group of genes which contributes to the formation of the red pigment in the eyes. The dominant alleles must be present at every one of the eight loci mentioned in order to have flies with the wild-type red eyes. The same is true, no doubt, of many other characters for which we have as yet observed only one type of mutation from the normal or wild type. In all forms of life, as new mutations at different loci are discovered, the number of characters known which exhibit multiple gene inheritance will increase.

The concept of multiple gene inheritance can be better understood it we consider the method of gene action briefly at this point. A single gene seems to control the formation of a single polypeptide. Some enzymes and structural proteins are only single polypeptides and, therefore, are formed by the action of single genes. The **A** and **B** blood antigens, for instance, are produced by the action of single genes. More often, however, the enzymes and structural proteins are polymeres, combinations of two or four separate chains of two or more kinds of polypeptides. Hence, two or more gene loci are involved in the production of these substances. The adult human hemoglobin, for instance, is a protein made up of four polypeptide chains; two, known as alpha chains, are determined by one gene locus, and the other two, known as beta chains, are determined by another gene locus. Thus, variations in the nature of human hemoglobin can be brought about by mutation in either of two genes located at separate loci on the chromosomes. An alteration of just one of the nucleotide pairs which makes up a rung on the DNA ladder of either one of the two genes will be sufficient to result in an altered form of hemoglobin which can have far-reaching effects on the individual involved. Such cases are discussed in Chapter 17.

The Standard Deviation

When we study quantitative variations of characteristics influenced by multiple genes it is often quite helpful to make a statistical analysis in order to determine the part played by the genes on an objective, analytical basis. One of the statistical tools used for this purpose is known as the standard deviation, which is represented by the Greek letter, a small sigma (σ). The standard deviation shows the degree of spread of a normal curve of distribution. If we are studying some characteristic such as body stature in a large group of men, say in an army camp, the mean, $\bar{m}$, will give us the average, but the standard deviation will tell us whether the men are somewhat uniform in height or whether there is considerable variation from one extreme to the other. If we plotted the heights on a curve, it would follow a pattern known as the curve of normal distribution. Such a curve, however, may be steep or widespreading, according to the degree of variation which is found. When we calculate the standard deviation, we find that about 68 per cent of the men deviate from the mean by no more than the standard deviation. If we assume that the standard deviation is 3 inches, then about two-thirds of the men would be within 3 inches of the average. In other words, if the mean is 5 feet, 8 inches, then two-thirds of the men would lie between 5 feet, 5 inches and 5 feet, 11 inches. Ninety-five per cent of the men will deviate no more than twice the standard deviation, in this case 6 inches from the mean, and only 0.03 per cent will lie outside the area marked by three times the standard deviation.

Another useful measurement is the standard error of a mean. This measures the reliability of the mean of the sample which has been chosen to estimate the variability of a population. This takes into account the size of the sample and the degree of variation. When there is little variation in a population, a

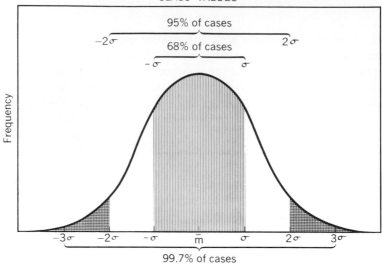

FIG. 13–6 *The curve of normal variation from a mean. Whenever the frequency of occurrence, such as the number of people in a large group, is plotted against the class value of some variable factor, such as body height, we tend to get a bell-shaped curve. When we mark off the standard deviation on either side of the mean, we find that about 68 per cent of the cases fall within this region and 95 per cent fall within the area covered by twice the standard deviation. Only 0.3 per cent will fall outside three times the standard deviation.*

smaller sample will be needed than when the variation is great. The formula for obtaining the standard error is simple once the standard deviation is known. It is as follows:

$$s_{\overline{m}} = \frac{\sigma}{\sqrt{n}}$$

The mean which has been obtained from the particular sample used in the study will be expected to lie within one standard error of the true mean for the entire population about two-thirds of the time. Expressed in another way, if means were obtained from many random samples of a population, about two-thirds of the means would be within one standard error, either above or below, the true mean of the entire population. We usually express the standard error by placing it after the mean with a plus sign over a minus sign between the two figures. For the example of height in man, we could give the mean obtained from one sample in inches as 68 ± 0.78 inches. Thus, the worker has some indication of the reliability of the mean he has obtained from a particular sample.

You can surely understand all this much better with an illustration of the use of these statistical methods on a particular problem. Let us assume that a producer of frozen, frying chickens wants to raise chickens which will be of a uniform size so that he can market a standardized product. He finds that cockerels of breed A and breed B both reach an average size of about 3 pounds after 8 weeks of special, uniform feeding, but there appears to be a greater

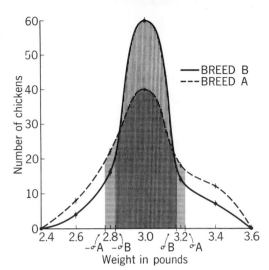

FIG. 13–7 *Chart to show the distribution and standard deviation of two breeds of chickens. It can be seen that breed B is more uniform in size.*

variation in breed A. Before deciding definitely to use breed B, however, he makes tabulations of 100 cockerels of each breed of the same age and uses the standard of deviation to obtain an objective evaluation. Breed A is analyzed in Table 13–1. The first column of figures gives the weight range in pounds divided into classes of equal range, and the second column shows the class value or mid-point of each class. The third column (f) shows the number of cockerels at 8 weeks of age that fall into each weight group. The fourth column shows the product of the frequency times the class value. Then we obtain the mean by simply dividing our total number of chickens, n, into the sum of the frequency times the class value, $\Sigma\,(fv)$. We find this mean to be almost exactly 3 pounds.

TABLE 13–1

Size Distribution of Chickens in Breed A

Wt. range in pounds	Class value	Number of chickens in each group		Deviation of class value (v) from mean		
	v	f	fv	d	d^2	fd^2
2.5–2.7	2.6	8	20.8	−0.4	0.16	1.28
2.7–2.9	2.8	22	61.6	−0.2	0.04	0.88
2.9–3.1	3.0	40	120.0	0.0	0.00	0.00
3.1–3.3	3.2	18	57.6	0.2	0.04	0.72
3.3–3.5	3.4	12	40.8	0.4	0.16	1.92
Σ		100	300.8			4.80

$$\overline{m}\ (\text{mean}) = \frac{\Sigma fv}{n} = \frac{300.8}{100} = 3.008 \pm 0.022$$

$$\sigma\ (\text{stand. dev.}) = \sqrt{\frac{\Sigma fd^2}{n}} = \sqrt{\frac{4.80}{100}} = 0.219$$

$$s_{\overline{m}} = \frac{\sigma}{\sqrt{n}} = \frac{0.219}{\sqrt{100}} = 0.022$$

The fifth column shows the deviation of the class value from the mean, the sixth shows the squares of these deviations, and the seventh shows the product of each frequency times its deviation squared. The standard deviation is then obtained by taking the square root of the sum of the last column divided by *n*. Our answer (0.219) tells us that about two-thirds of the cockerels raised will deviate less than 0.219 pounds from the three pound mean and, of course, about one-third will show a deviation greater than this.

This deviation is greater than the breeder would like to have, however, so 100 cockerels from breed B are analyzed by the same technique, as shown in Table 13–2. Here we see that indeed the weights are grouped more closely around the mean. The standard deviation in this instance is only 0.1672, so the breeder can expect a more uniform product for his business. About two-thirds of the cockerels of this breed will vary no more than about one-sixth of a pound from the mean. We can accordingly assume that breed B has a much greater homozygosity with respect to genes which govern body weight.

The standard errors of the two groups of chickens also have value in showing the chicken breeder how much the mean weight of his samples might vary from the mean weight of all chickens of these particular breeds. Since his samples were large, 100 chickens in both cases, the standard errors are small and show that these samples are reliable as a measure of the weight variability of the two breeds.

TABLE 13–2

Size Distribution of Chickens in Breed B

Wt. range in pounds	Class value v	Number of chickens in each group f	fv	Deviation of class value (v) from mean d	d^2	fd^2
2.5–2.7	2.6	4	10.4	−0.4	0.16	0.64
2.7–2.9	2.8	16	44.8	−0.2	0.04	0.64
2.9–3.1	3.0	60	180.0	0.0	0.00	0.00
3.1–3.3	3.2	14	44.8	0.2	0.04	0.56
3.3–3.5	3.4	6	20.4	0.4	0.16	0.96
Σ		100	300.4			2.80

$$\overline{m} = \frac{\Sigma fv}{n} = \frac{300.4}{100} = 3.004 \pm 0.017$$

$$\sigma = \sqrt{\frac{\Sigma fd^2}{n}} = \sqrt{\frac{2.80}{100}} = 0.1672$$

$$s_{\overline{m}} = \frac{\sigma}{\sqrt{n}} = \frac{0.1672}{\sqrt{100}} = 0.017$$

STATISTICAL MEASUREMENT OF QUANTITATIVE VARIATION

Coefficient of Correlation

Whenever we wish to study the effect of one variable factor on another variable factor, we can employ the statistical tool known as the coefficient of correlation (r), which is somewhat similar to the standard deviation in its method of determination. In order to be of greatest value, all factors should be as nearly constant as possible except the two variables being evaluated. As an example let us take a case from agriculture.

Suppose we wanted to determine whether the yield of corn in a field was closely correlated with the amount of nitrate of soda applied. We know that heredity plays a part in corn yield and, in order to minimize this, we must choose seed corn of a uniform hereditary background and mix it thoroughly before planting. Then we must select an area for planting where the soil is level and of uniform texture and slope so as to minimize the influence of other environmental conditions. After the corn is planted we add the nitrate of soda in one plot at the rate of 50 pounds per acre; in other plots we apply it at the rate of 60, 70, 80, 90, and 100 pounds per acre, respectively. When the corn is harvested we carefully measure the yield of each plot in terms of bushels per acre. Table 13–3 shows what the coefficient of correlation would be if the yield was increased exactly 2 pounds for each 10 pounds of fertilizer applied. This is a perfect positive correlation.

TABLE 13–3

Perfect Positive Correlation

X lbs. fertilizer	Y yield of corn	dx deviation of X from Av.	dy deviation of Y from Av.	$(dx)^2$	$(dy)^2$	$dx \cdot dy$
50	38	−25	−5	625	25	125
60	40	−15	−3	225	9	45
70	42	− 5	−1	25	1	5
80	44	5	1	25	1	5
90	46	15	3	225	9	45
100	48	25	5	625	25	125
Σ 450	258	00	00	1750	70	350
Av. 75	43					

$$r \text{ (coefficient of correlation)} = \frac{\Sigma dx \cdot dy}{\sqrt{\Sigma (dx)^2 \cdot \Sigma (dy)^2}}$$

$$= \frac{350}{\sqrt{1750 \cdot 70}} = \frac{350}{350}$$

$$= 1.0 \text{ or } 100\%$$

TABLE 13–4

Perfect Negative Correlation

X r units of X-rays in 1,000s	Y viable eggs per 100	dx	dy	(dx)²	(dy)²	dx·dy
1	90	−2	40	4	1600	−80
2	70	−1	20	1	400	−20
3	50	0	0	0	0	0
4	30	1	−20	1	400	−20
5	10	2	−40	4	1600	−80
Σ 15	250	00	00	10	4000	−200
Av. 3	50					

$$r = \frac{\Sigma dx \cdot dy}{\sqrt{\Sigma(dx)^2 \cdot \Sigma(dy)^2}} = \frac{-200}{\sqrt{10 \cdot 4000}}$$

$$= -1.0 \text{ or } -100\%$$

Of course, if this experiment were to be actually done, we would expect some variation because of the uncontrollable factors in the heredity and environment, and we might find that the coefficient of correlation was 0.86 (86 per cent) as an example. This is still a very high positive correlation and is highly significant. We commonly consider any coefficient above 0.25 as significant and above 0.50 as highly significant.

When the coefficient comes out 0.0 we have no correlation at all. As the figure moves to the minus side of zero, then the correlation begins to be negative, and so on, until the coefficient of −1.0 would be perfect negative correlation. As an example, let us assume that we are carrying out an experiment to determine the correlation, if any, between irradiation with X-rays and the viability of the eggs of *Drosophila*. We express the amount of irradiation in terms of roentgen units (r) and the number of eggs that hatch as a percentage. Let us assume that our hypothetical results are as follows: 1000 r, 90; 2000 r, 70; 3000 r, 50; 4000 r, 30; 5000 r, 10. It is quite evident from these results that the viability of the eggs is in an inverse ratio to the amount of irradiation. Table 13–4 shows how this may be worked out to give a coefficient of correlation of −1.0, a perfect negative correlation. Our actual results would certainly not be so perfect, but we would expect a very high negative correlation.

Human Stature

Now that we have some understanding of the method for working out the coefficient of correlation, we may apply it to our problem of stature in man. We must begin by considering the effect of sex. Women as a group are not as tall as men. There is a definite overlapping of heights, but the average female height runs about 5 inches less than that of the average male. In other words, simply being a female cuts about 5 inches off a woman's height. If we are going to study a mixed group, therefore, we must find some way to convert both sexes to the same relative terms. Extensive studies have shown that if a woman will multiply her height by 1.08 she will obtain a figure that will represent her height had she been a man. A woman who is 5 feet, 4 inches tall (64 inches) would have been about 5 feet, 9.12 inches tall (69.12 inches) had she been a man. Since it is customary to make studies of stature in terms of male height, we must multiply all female heights by 1.08 before beginning our study on the inheritance of this character. All the figures used in our illustrations have thus been converted.

Table 13–5 shows the results of a comparison between the heights of parents and their children which was made in a group of fifty-two students in a class in genetics at Stetson University. In order to avoid unnecessary multiplication, the heights are expressed in terms of inches above 5 feet, and all are in terms of male height. Those who had an average parent height falling between 5 feet, 5 inches and 5 feet, 7 inches were given a class value of 6 (66 inches), and were found to have an average child height of 8 (68 inches). The chart shows how this was done for each of the other five classes. The coefficient of correlation was found to be .894, which is a very high positive correlation.

Although our sample is much too small to allow us to draw conclusions of great significance from these results, it certainly does indicate that, in this group at least, the height of the children is on the average similar to that of the parents. The means of the parents and children, however, show that the children tend to exceed their parents. This is an observed fact which can be correlated with

TABLE 13–5

Correlation in Height of Parents and Children

X Av. parent height to nearest inch above 5'	Y Av. height of children in inches above 5'	dx	dy	$(dx)^2$	$(dy)^2$	$dx \cdot dy$
7	8.0	−2.5	−1.95	6.25	3.80	4.87
8	8.5	−1.5	−1.45	2.25	2.10	2.17
9	10.5	−0.5	0.55	0.25	.30	−0.27
10	9.5	0.5	−0.45	0.25	.20	−0.23
11	11.8	1.5	1.85	2.25	3.42	2.77
12	11.4	2.5	1.45	6.25	2.10	3.62
Σ 57	59.7	0.0	0.00	17.50	11.92	12.93
Av. 9.5	9.95					

$$r = \frac{\Sigma dx \cdot dy}{\sqrt{\Sigma (dx)^2 \cdot \Sigma (dy)^2}} = \frac{12.93}{\sqrt{17.50 \cdot 11.92}}$$

$$= 0.894 \text{ or } 89.4\%$$

other statistical studies. It is generally concluded that our improving conditions of nutrition and health care have been reflected in an increasing degree of expression of the potentialities of the genes which influence stature.

Also, another important genetic principle is brought out by a study of the degree of variation in the parents and children. The children of parents at the two extremes tend to fall nearer the average. In other words, the children of very short parents tend to exceed the parents' height, while the children of very tall parents tend to average less than the parents' height. We can measure this by obtaining the standard deviation for both groups. This is easy to do now because we have already done most of the calculating necessary. Thus, the standard deviation for the parents is:

$$\sigma = \sqrt{\frac{\Sigma(dx)^2}{n}} = \sqrt{\frac{17.50}{6}} = 1.7 \text{ inches}$$

For the children it would be:

$$\sigma = \sqrt{\frac{\Sigma(dy)^2}{n}} = \sqrt{\frac{11.92}{6}} = 1.4 \text{ inches}$$

The smaller standard deviation for the children illustrates the principle which was first noticed by Francis Galton, cousin of Charles Darwin, and he called it "the law of filial regression." It appears that, on the average, the offspring of parents which lie at an extreme of a quantitative variation do not receive the same combination of genes and environmental factors which caused their parents to deviate so greatly from the population mean.

Thus we see that there is much that we can learn about quantitative inheritance through the use of comparatively simple statistical measurements. Such methods are especially valuable in human populations where we cannot have the control over mating that is possible in many other forms of life.

Assuming a constant environment, if all of the genes involved in quantitative traits were of an intermediate nature, there would be no regression. All of the offspring would be expected to average the same as the parents. The fact that there is regression in most quantitative characteristics indicates that some of the genes must be dominant or partially dominant. A person with extreme height, for instance, may carry some recessive genes for smaller stature in a heterozygous state. A marriage between two persons of this nature would produce children who would be of a shorter stature, on the average, than the parents because some of the recessive genes for shorter stature would become homozygous. This would account for the regression toward the mean. This does not mean that the population would tend to become more uniform in height. Some of the parents nearer the mean will produce children with just the right combination for either extremely tall or extremely short children, so the population distribution with respect to height would tend to stay about the same.

The principle of regression is also clearly evident when there is a selection for one extreme in experimental organisms. Suppose you wish to establish a race of extremely large rats. From a group of rats raised in the same environmental conditions, you select the largest 10 per cent as parents. If all traits affecting size were intermediate, the offspring would average the same size as the parents. Actually, however, you would find that the offspring average considerably less than the parents because of regression, but the mean of the offspring will be greater than the mean of the original population. Continuing selection of the largest 10 per cent for several generations would probably result in a race which had a mean near the largest of the original population. This would come about because of a gradual elimination of the recessive genes for smaller size. We shall learn more about the application of statistical methods in quantitative human inheritance in Chapter 25.

PROBLEMS

1. Which of the following inherited human traits would you think are induced by variations in genes at a single locus and which by variations in multiple genes: *aniridia,* absence of the iris of the eye; intelligence; cleft palate, a fissure in the roof of the mouth; *vitiligo,* areas of the skin which become depigmented; size of the mouth, musical ability? Explain why you decide as you do in each case.

2. A certain man is rather tall and his wife is of small stature although neither of them represents extreme deviation from the average. Out of four children they have a boy and a girl of about average height, one daughter about the size of the mother, and a son considerably taller than his father. Assuming that environmental factors were about the same for all, how would you explain these results?

3. Two circus midgets marry and have three children of normal stature. How can you explain this?

4. A Holstein bull is solid white and is mated a number of times with a cow which is white except for a small amount of black spotting around the head. About half the calves are solid white and half have black spotting over a large part of the body. Explain.

5. One breed of rabbits has ears about 4 inches long and another breed has ears about 2 inches long. When the two types are crossed, the offspring have ears about 3 inches long. In the second generation of offspring, however, the ears show a great range of variation from 2 to 4 inches. Out of 1,000 rabbits, three have 2-inch ears and five have 4-inch ears. How many pairs of genes seem to be involved in ear length? Explain.

6. In certain plants, such as wheat, the number of seeds produced by a plant is more variable than the weight of the individual seeds. Ignore possible environmental influences and explain this condition on a genetic basis.

7. The yield of oats per plant was found to range from 4 grams to 10 grams in a mixed population. Seeds from one plant were selected and subjected to several generations of self-fertilization. When seeds from this stock were planted in the same field, the yield ranged from 6 to 9 grams per plant. From these results would you conclude that the yield of oats per plant was due to heredity, environment, or a combination of the two? Give reasons for your answer.

8. Let us assume that in squash there are three pairs of contributing genes which influence the weight of the fruit. A plant with genes *aa bb cc* bears fruit weighing about 3 pounds. A plant with genes *AA BB CC* bears fruit weighing about 6 pounds. Thus each of the contributing genes adds about ½ pound to the weight

of the fruit. (We assume that all the genes have equal value.) Cross a 3-pound plant with a 6-pound plant and show the approximate weight of the first generation. Cross one of these first-generation plants with a 6-pound plant and show the approximate weight of the fruit of the plants which would be produced.

9. In Holstein cattle the degree of spotting depends upon a series of multiple genes. If we assume that five pairs are involved, how many different degrees of spotting could be produced by crosses between two cattle showing the medium grade of spotting, heterozygous for all five genes?

10. A pair of red-eyed *Drosophila* are crossed and yield 120 offspring. Among these some have scarlet eyes, some pink, some ruby, some red, and some flies have eye colors that are not like any in the pure stocks. Explain these results, showing the genotype of the parent flies.

11. On one of the Caribbean Islands there has been unrestrained intermarriage of the Negro and Caucasian races for many years. In one of the public schools of the island a study of the skin color was made of the pupils enrolled and an attempt was made to classify them into nine groups corresponding to the number of contributing genes for negroid pigmentation from 0 (white) to 8 (negroid). Results are given in the table below. Determine the mean number of contributing genes carried by these pupils and the standard deviation from this mean.

Contributing genes	No. of persons in each group
0	3
1	8
2	20
3	34
4	44
5	40
6	33
7	12
8	6

12. If we assume that the island was populated originally by an equal number of Negroes and Caucasians, what was the average number of contributing genes possessed by the people during the early history of the island? Have differential reproduction, immigration, and other factors altered this figure if we accept the data obtained in Problem 11 as a typical sample of today's population? What percentage of increase or decrease in contributing genes has there been?

13. In the offspring of first generation mulattos the mathematical expectation of distribution according to contributing genes from 0 to 8 is: 1 : 8 : 28 : 56 : 70 : 56 : 28 : 8 : 1. Use these figures instead of the numbers given in Problem 11 and obtain the standard deviation of this ratio.

14. Compare the results of Problem 11 with those of Problem 13 and tell whether the deviation in the observed group is greater or less than the deviation of the mathematical ratio. List any possible explanations which you might think of which can explain any variation from the ratio.

15. A Mormon living in Utah during the last century had twelve children by a wife named Esther. The adult heights of these children are given below in male values (female heights have been multiplied by 1.08). Male height in inches above 5 feet: 6.5, 7.1, 7.7, 8.1, 8.3, 8.4, 8.7, 8.9, 9.1, 9.5, 10.1, 11.2. Tabulate these heights in a vertical column, consider the frequence of each as one, and determine the mean height of these children and the standard deviation.

16. The same man had a second wife, Ruth, and also had 12 children by this wife.

The male heights above 5 feet for these were: 6.9, 7.7, 8.8, 8.9, 9.0, 9.1, 9.3, 9.4, 9.6, 9.7, 10.5, 10.7. Determine the mean height of these children and the standard deviation.

17. From the results obtained in the previous problems which of the two wives would you say appeared to be the most homozygous with respect to genes influencing body stature? Explain.

18. Ruth claimed that she came from a stock of tall ancestry and couldn't understand why it was that Esther bore the tallest child of the group. From the results and conclusions you have obtained, what could you tell her to explain this?

19. Use graph paper or make a simple graph and plot the distribution of the children of Esther as given in Problem 15. On top of this plot the distribution of the children of Ruth as given in Problem 16. For each curve mark off the upper and lower limits of the standard deviation. You can use a technique similar to that in Fig. 13-7 if you wish.

20. A scientific farmer finds that in a certain variety of corn the average weight of the dried kernels taken from the center of the ear is .38 gram, but he notices that there is considerable variation in their size. He wonders if heredity could have anything to do with size and if it would be worthwhile to select the largest seed for planting in order to get the largest kernels in the offspring. He selects five sizes of seed and plants them in five plots of ground. He carries out self-pollination to avoid bringing in other genes. He then weighs the dried kernels of corn which are produced and determines the average weight per kernel for each plot. The results are given below. Determine the coefficient of correlation and tell him if you think selection of seed size is of any value in producing large seed in the offspring.

Wt. range of parents in grams	Mid-class value in hundredths of grams above .30	Average wt. of kernels in hundredths of grams above .30
.33–.35	4	4.8
.35–.37	6	5.8
.37–.39	8	9.2
.39–.41	10	8.3
.41–.43	12	10.3

21. Determine the standard deviations for the parents and for the offspring in Problem 20. Does the "law of filial regression" seem to be operating here? Explain your answer.

22. It is sometimes said that natural selection tends to result in animals which are larger in the colder latitudes than the animals of the same family in warmer latitudes. It is known that the larger an animal is, the less body surface is exposed in proportion to his body weight. Heat is lost through radiation from the body surface, so it is obvious that a larger size would be of advantage in a cold climate and vice versa for a warm climate. A group of deer is measured for each of several latitudes, and the average heights at the shoulders are given below. From these figures determine the coefficient of correlation between latitude and size and state your conclusions.

Latitude North	Average height of male deer at shoulder (in feet)
20°	2.2
30°	3.4
40°	3.6
50°	4.6

14

Linked Genes, Crossing Over, and Chromosome Mapping

IN OUR DISCUSSION OF DIHYBRIDS IN CHAPTER 6 we purposely omitted any consideration of two pairs of genes which are located on the same pair of chromosomes; we are now in a position to consider what happens to genes which are so situated. If any two genes are picked at random, it is most likely that they will not be on the same chromosome pair. Since each chromosome bears a large number of genes, however, it stands to reason that if a number of different genes from one species are studied, sooner or later one will find two genes on the same pair of chromosomes. It is quite evident that two such genes would not show the same independent assortment which characterizes genes located on different pairs of chromosomes, but rather that they would be linked together in heredity by virtue of their common chromosome. Thus all the genes on one chromosome are said to be **linked genes** and tend to move *en bloc* during meiosis when the chromosomes are segregated into two haploid groups. It happened that Mendel did not study any dihybrid crosses involving linked genes. Had he done so, he would have had to modify his law of independent assortment, for linked genes are not assorted independently.

LINKED GENES

The existence of linkage was predicted before it was discovered. In the laboratory at Columbia University, where so many of the early genetic discoveries were made, Sutton as early as 1903 outlined the theory that chromosomes are the bearers of the units of heredity. He reasoned that, since the number of hereditary units is much larger than the number of pairs of chromosomes, then each chromosome pair must contain a number of pairs of genes. Furthermore, since the chromosomes move as units during meiosis, all the genes on one chromosome will be linked together. As a result, each species of animal or plant will have a specific number of linkage groups of genes which will correspond with the number of chromosome pairs (or the haploid number of chromosomes) found in that species. Three years later the first case of linkage was demonstrated by two English geneticists, Bateson and Punnett, in the sweet pea. They found that a pair of genes which influences the color of flowers was linked to another pair which influences the shape of the pollen grains. Plants with red flowers and spherical pollen grains were crossed with plants having purple flowers and cylindrical pollen grains. A test cross was made of the first generation plants, and instead of the independent segregation of these two qualities which was expected according to Mendel's law, the flower color and pollen shape tended to remain together as in the parental types. The actual ratio obtained was about 7 : 1 : 1 : 7 instead of 1 : 1 : 1 : 1, which would be expected if the genes had recombined freely. This result led Bateson and Punnett to suspect that these two pairs of genes were on the same chromosome and that this fact prevented free recombination. The fact that there were any recombinations at all indicates that linkage is not absolute — there is some recombination, but it is much less frequent than was found for non-linked genes. Later in this chapter we shall learn how these recombinations of linked genes occur.

Continued studies supported Sutton's theory and showed that the number of linkage groups which could be determined by genetic crosses corresponds to the number of pairs of chromosomes which could be demonstrated cytologically. The sweet pea had seven linkage groups of genes and seven pairs of chromosomes. *Drosophila melanogaster* had four linkage groups and four pairs of chromosomes. It thus became possible to determine the number of chromosomes in the cells of a plant or animal by studying the gene linkage without ever seeing a chromosome. Of course, a large number of genes must be studied before one can feel reasonably sure that genes from each chromosome have been included. In practice, both genetic crosses and cytological investigations are usually carried on at the same time.

CROSSING OVER

Linkage introduces a phenomenon which is of great importance from the standpoint of evolution. The continued existence of any natural, wild population of plants or animals depends upon continual natural selection. Certain genes result in characteristics which are beneficial in the particular environment in which the organisms are living, while other genes have effects which are detrimental. Natural selection results in the maintenance and propagation of the beneficial genes, and a corresponding reduction of the harmful ones. But what about those cases where both beneficial and harmful genes are found on the same chromosome? Does selection occur for the over-all effect of the entire group of genes on the chromosome? The answer is "No." A mechanism exists by which a group of genes on one chromosome sometimes changes place with a similar group of genes on the homologous chromosome. This process is known as **crossing over,** and it explains the occurrence of the two recombination classes in the second generation of the sweet pea cross made by Bateson and Punnett.

Mechanics of Crossing Over

Crossing over occurs during the early prophase of the first division of meiosis while the chromosomes are closely paired. You will recall that at this stage each chromosome is double, so that each pair of chromosomes consists of four strands or chromatids. The chromatids become twisted about one another, and often there occurs a breakage at identical levels and a reattachment of portions of the chromatids from homologous chromosomes. Figures 14–1 and 14–2 illustrate this process better than words can describe it. During the later

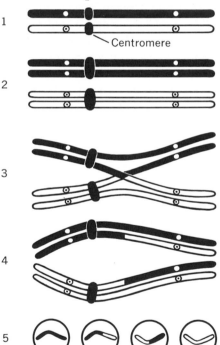

FIG. 14–1 *Diagrammatic representation of crossing over. 1. The two homologous chromosomes before meiosis. 2. Each chromosome becomes double as it forms two chromatids. 3. One chromatid from each chromosome exchanges a portion with a homologous chromatid from the other chromosome; this is crossing over. 4. The two chromosomes pull apart in the anaphase of the first division of meiosis. 5. These represent four spermatids (or one ootid and three polar bodies). Note that two of them carry crossover chromosomes and two do not.*

FIG. 14–2 *Crossing over in the chromosomes of the grasshopper. This highly magnified photograph shows a pair of chromosomes in the prophase of the first meiosis in a living cell taken from the testes. The four chromatids can be seen and four chiasmata between them indicate regions where crossing over has taken place between the opposite chromatids.*

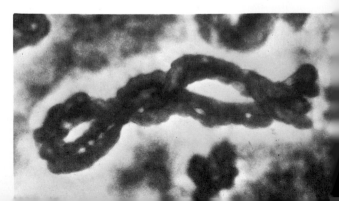

part of the prophase, when the chromatids separate from one another slightly, a distinct cross, known as a **chiasma,** is visible. The evidence seems to indicate that the chiasmata (the plural form of the word) represent regions where crossing over has already taken place. As the centromeres pull the two chromosomes that have undergone crossing over to the poles, we find that one of the chromatids of each chromosome will bear a portion of a chromatid from the homologous chromosome. After the second division of meiosis, two of the four resulting cells will contain chromosomes with the recombination of genes brought about by the crossing over of the chromatids. In this way new combinations of linked genes occur.

Effects on Genetic Crosses

Let us now illustrate the effect of this cytological phenomenon on the ratio of offspring from a genetic cross. Suppose we have two mutant genes in *Drosophila* and we wish to determine whether they are on the same or different chromosomes and, if they are on the same one, how frequently crossing over will occur between them. One of these genes produces a black body rather than the more common gray one, and the other produces tiny, vestigial wings rather than the normal long wings. Suppose we cross a male homozygous for both these recessive genes (black and vestigial) with a wild-type female which is homozygous for the dominant genes for gray body and long wings. The members of the F_1 generation are all of the wild type, for both mutant genes are recessive. Now let us obtain virgin females from these offspring and mate them with males which are homozygous for the double recessive. You may remember that this is what we call a test cross. If these two genes are on different chromosomes, we will expect a ratio of approximately 1 gray, long: 1 black, vestigial: 1 black, long: 1 gray, vestigial on the basis of independent assortment. Our actual results, however, are as follows:

Wild-type (gray, long)	965	} Parental types
Black, vestigial	944	
Black, long	206	} Recombinations
Gray, vestigial	185	
Total	2300	

Total crossovers, 391
Percentage of crossovers, 17

Determining Distance Between Genes

As we look at these figures it is immediately evident that we do not have a 1 : 1 : 1 : 1 ratio. We can therefore conclude that these two mutant genes are on the same chromosome; that is, they are linked. The ratio is about 1 : 1 for the parental types, with a lower number of recombinations which have resulted from crossing over. In this case, these recombinations represent 17 per cent of the total number of flies obtained. The percentage of crossing over which is obtained between different linked genes varies according to the distance between the genes on the chromosome. It is evident that crossing over will not produce new combinations of genes unless it occurs between the linked genes. Hence those genes which are closest together on the chromosome will have the least amount of crossing over between them, and a correspondingly greater amount of crossing over will be found between genes that are farther apart. The percentage of crossing over therefore indicates the relative distance between the linked genes being studied. It is customary to express this distance in terms of units, each unit representing one per cent of crossing over. Thus we can say that the genes for black and vestigial are 17 units apart on the chromosome. We will modify this procedure somewhat after further study.

An important question might logically be raised at this point. That is, do these crossing over units represent definite real distances on the chromosomes? Detailed cytological studies have been made in an effort to answer this question. From them we may conclude that units of crossing over only approximate the actual distances on the chromosome, for there is some variation in the ease with which chiasmata are formed in different portions of the chromosome. In *Drosophila* there is less crossing over in proportion to chromosome length near the centromere of the two V-shaped chromosomes and close to the ends than there is in the middle of each arm. The genes near the centromeres, therefore, are actually farther apart than the crossover percentages would indicate when we compare them with the distances between the genes in the central regions. Similar variability occurs in other species. A cytological and a genetic chromosome map are compared in Figure 14–3.

We might compare the genetic method of determining distances between genes on a chromosome to a railroad timetable as a means for determining

FIG. 14–3 *Comparison of cytological and genetic chromosome map of the second chromosome of Drosophila melanogaster, after the work of Dobzhansky. The chromosome as mapped by genetic crosses is shown at the bottom with some of the genes which have been located by crosses. The lines running to the chromosome at the top show the location of portions of this genetic chromosome on the actual chromosome as determined by cytological studies. It can be seen that the correspondence is only approximate since there is less crossing over in proportion to length in the center and at the ends of the chromosome.*

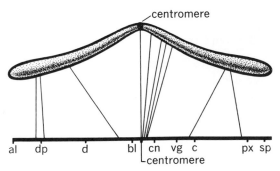

distances between stations. If a train requires 20 minutes to travel between stations *A* and *B* and 40 minutes to travel between stations *B* and *C,* we might assume that *C* is twice as far from *B* as *A* is from *B*. Such calculations are based on the assumption that the train travels at the same rate of speed between the stations. It is possible, however, that there may be mountains between *B* and *C,* which are not present between *A* and *B,* and will slow down the train's speed. If so, the actual distance between *B* and *C* would be less than the elapsed travel time indicates. In a similar way, various factors which influence the facility of crossing over in different regions of the chromatids will cause variation in the percentage of crossing over obtained.

To carry the analogy a step further, it is possible to determine the sequence of stations on a railroad line by a comparison of elapsed time between stations. The train requires 20 minutes to travel between *A* and *B*, 40 minutes to travel between *B* and *C,* and 60 minutes to travel between *A* and *C*. It is obvious from these time intervals that *A* and *C* must be the farthest apart and *B* must be in between. In a similar manner, if we are studying three genes on one chromosome, the two genes which show the greatest amount of crossing over between them must be the farthest apart, and the third gene must be in the middle. This principle is used in preparing **chromosome maps** indicating the sequence and relative distances between genes on the chromosome. The matter is discussed in greater detail later in this chapter.

If we had reversed the sexes of our *Drosophila* test cross and used heterozygous males, crossing them with the homozygous double-recessive females, we would have noted an interesting thing: there would have been no crossing over in the gametes of the male. All the offspring would have been of the two parental types in a ratio of about 1 : 1. This unusual result arises from the fact that there is no crossing over in the males of dipteran insects or the females of the silkworm. For the great majority of living organisms crossing over occurs in both sexes, but it may be more frequent in one sex.

Advantages of the Test Cross

The test cross is almost invariably used to determine linkage and crossover values. This type of cross brings out all the recessive genes which are present in the chromosomes of the hybrids. If we crossed the F_1 among themselves, the recessive characteristics would show in approximately one-fourth of the offspring, rather than in one-half as is the case with the test cross. Thus, a greater number of offspring would be necessary in order to feel reasonably certain of a representative sample. Also, it would be much more difficult to calculate crossover values between genes in an F_2 progeny because allowance would have to be made for the larger number of dominant characteristics.

Double Crossing Over

The linkage studies on the genes for black body and vestigial wings in *Drosophila* showed about 17 per cent crossing over. Before we accept this as a final figure for a chromosome map, however, let us make other crosses with an-

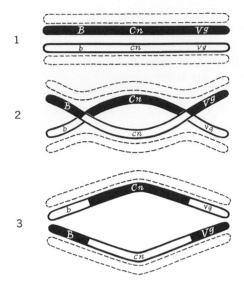

FIG. 14–4 *Double crossing over: how homologous chromatids may undergo crossing over at two points at the same time. To show how double crossing over affects linkage, three genes are shown on the chromosome (from Drosophila). These are black, cinnabar, and vestigial on one chromatid, and the three wild-type alleles on the other. The chromatids which are not involved in the crossing over are shown by the broken lines.*

other gene. Let us introduce a recessive gene for the bright red "cinnabar" eye color. When we cross a black cinnabar fly with one of the wild type of the opposite sex, and test cross the offspring, we find 9 per cent crossing over between these two pairs of genes. When we cross a cinnabar-vestigial fly with a wild-type fly and test cross the offspring, we obtain 9.5 per cent crossing over. From these results we conclude that cinnabar lies about half-way between the genes for black body color and vestigial wings, but the total amount of crossing over between the latter genes is 18.5 per cent rather than the 17 per cent expected on the basis of our previous cross. Why this discrepancy? The answer lies in the occurrence of double crossing over, that is, of two crossovers occurring simultaneously in the same cell between these two loci. At times, two or more chiasmata form between paired chromosomes. When a double crossover occurs, the genes lying outside the crossed regions will retain their original associations and leave no evidence of crossing over in the offspring. This is illustrated in Figure 14–4.

If the genes studied are close together on the chromosome, however, double crossing over cannot occur between them. If crossing over occurs at one point, a second crossover will not occur within a certain distance of it. This phenomenon is known as **interference.** In *Drosophila* it has been shown that interference prevents double crossing over almost altogether between genes ten units apart or less. The degree of interference diminishes as the genes considered are farther apart, and it is practically nonexistent between genes forty or more units apart.

We may conclude from these considerations that 18.5 represents the distance between black and vestigial more accurately than 17.0 does, since the value 18.5 was obtained by adding crossover values less than ten. As a final check on these results, it would be well to make a **trihybrid** or **three-point cross** using all three pairs of genes at once. If we cross black, cinnabar, vestigial flies with wild-type mates and test cross the offspring, our results might be as shown below (data from an actual cross).

Wild-type (gray body, red eye, long wings)	332 ⎫	Parental types
Black, cinnabar, vestigial	326 ⎭	
Black, cinnabar, long	36 ⎫	Crossovers between *cn* and *vg*
Gray, red, vestigial	34 ⎭	
Black, red, long	35 ⎫	Crossovers between *b* and *cn*
Gray, cinnabar, vestigial	31 ⎭	
Black, red, vestigial	2 ⎫	Double crossovers —
Gray, cinnabar, long	4 ⎭	*cn* and *vg; b* and *cn*
Total parental types	658	82.25 per cent
Total crossovers between *cn* and *vg* (includes double crossovers)	76	9.5 per cent
Total crossovers between *b* and *cn* (includes double crossovers)	72	9.0 per cent
Total crossovers between *b* and *vg* (double crossovers counted twice)	148	18.5 per cent
Crossovers between *b* and *vg* if double (crossovers are not included)	136	17.0 per cent

These results illustrate how the double crossovers may be included in considerations of linkage so as to obtain more accurate relative distances between genes as a basis for chromosome mapping. We may obtain the distance between black and vestigial by adding the distance between black and cinnabar to the distance between cinnabar and vestigial. This gives us 18.5. Or we may add all the crossovers and determine their percentage of the total. In doing this, however, we must remember to count the double crossovers twice, for each double crossover represents two exchanges of segments between the chromatids. These results also show us how we obtained the original figure of 17 per cent when there was no gene between black and vestigial to indicate whether double crossovers were occurring or not.

Locating Genes on the Chromosome

With these crossover distances worked out we now know the distances between these three genes in terms of crossover units. (One unit represents 1 per cent of crossing over.) We can now locate two of these genes on the chromosome if we know the location of the third. To illustrate, let us assume that we know from previous studies that black is located at 48.5. We can add 9 onto this and get a tentative location of cinnabar at 57.5. Vestigial would then be at 67.0 by the same line of reasoning. We have neglected, however, to take one important point into account. With the information we have obtained the two genes might just as easily lie on the other side of the gene for black. This would place them at 39.5 and 30.0 respectively. To determine which is correct we would have to make another crossover test using another known gene on the chromosome. Suppose we choose curved wing, which is known to be located at 72.5, and we find a crossover percentage of 8.5 between this gene and vestigial. This gives us our needed information and we can place the gene for vestigial at 67.0 and cinnabar at 57.5. If we are working with a species where we do not already have these convenient genes of known chromosome location we must be content with the information on the distance between genes until we can build up sufficient data to establish a chromosome map.

Determining the Sequence of Genes

If we are studying the crossovers between three or more gene loci at one time without benefit of previous information, we need to determine the **sequence** of the genes on the chromosomes as well as the distances between them. Since *Drosophila* has been studied so extensively in this connection, we again choose three pairs of genes from this insect to illustrate the method. Suppose we select flies with three mutant genes producing curled wings, glass eyes, and hairless bodies, respectively. Our first tests show that the three mutant genes are linked and that curled and glass are recessive, but hairless is dominant. Our next problem is to determine their sequence on the chromosome and the crossover distances between them. The test cross results turn out as follows:

1. Wild-type (straight wings, faceted eyes, haired body)	410
2. Curled, glass, hairless	405
3. Curled, faceted eyes, haired	57
4. Straight, glass, hairless	63
5. Curled, glass, haired	26
6. Straight, faceted eyes, hairless	29
7. Straight, glass, haired	4
8. Curled, faceted eyes, hairless	6
Total	1000

A preliminary examination of these figures shows that the first two classes represent the 1 : 1 ratio of the parental types. The other groups represent recombinations formed by crossing over. Since there can be only four classes of single crossovers, it is evident that the two smallest classes must represent double crossovers. Our problem is to determine the sequence of the genes on the chromosome. Is it curled, glass, hairless; or is it curled, hairless, glass; or does curled come between the other two? To determine this, we might recall by our analogy with the train schedule that the genes which show the greatest amount of recombination will be on either end and the third gene, obviously, will be in the middle. When we total all of the classes where curled and glass are not found together (classes 3, 4, 7, 8), we obtain the figure 130. Likewise, when we total all classes representing recombinations between curled and hairless (3, 4, 5, 6), we obtain a higher figure, 175. Recombinations between glass and hairless total 65. Since 175 is the highest of the three figures, we may assume that curled and hairless are on the ends, with glass in between.

It is also possible to determine sequence by examination of the double crossovers. The two smallest classes probably represent double crossovers, in this case classes 7 and 8. In our original cross the genes for curled, glass, and hairless were on the same chromosome. Double crossover recombinations will remove the middle gene from this group and leave the two end genes together. In the double crossover classes, glass is the gene which is removed from the other two. We can therefore assume that glass is in the middle with curled and hairless on either side. This agrees with our deductions by the first method.

Now that we know the sequence of genes, we can determine the **crossover percentage** and the **relative distances** between these three genes. To obtain all the crossovers between curled and glass, we add classes 3, 4, 7, and 8. This gives us 130, which is 13 per cent of the total number of flies. We can now say that curled and glass are approximately 13 units apart. By adding classes 5, 6, 7, and 8, we find that glass and hairless are about 6.5 units apart. We can add these two and find that curled and hairless are about 19.5 units apart.

Crossing Over Between Sex-Linked Genes

Sex-linked genes are all on the same chromosome in the great majority of organisms, though a very few have multiple sex chromosomes. We use the terms **incompletely sex-linked genes** and **Y-linked genes** to refer to those which may be present on the Y-chromosome. Thus, in most studies, genes which are sex-linked are also linked by virtue of a common chromosome. The crossover values and linkage relationships are somewhat easier to study for such genes because one sex normally carries only one set of these genes and both dominant and recessive sex-linked genes are expressed. For instance, in animals with the XY type of sex determination, crossing over of sex-linked genes occurs only in the females, and it is not even necessary to test cross these females in order to determine the crossover values. (Incompletely sex-linked genes and Y-linked genes are not considered as sex-linked genes in our discussion.) All sons of these females will carry one of the mother's X-chromosomes. Since the males are simplex (carry only one set of sex-linked genes), they will show all the genes on this chromosome. Thus it is possible to allow the F_1 flies to breed among themselves and examine the flies of the F_2 to determine the percentages of crossing over. Or the F_1 females may be bred to any available males — it makes no difference — and the crossing over can still be determined by examination of their male offspring. This generally means less work and attention than is necessary for a test cross.

Linkage Studies in Man

Even in human beings, it has been possible to work out some percentages of crossing over for linked genes through studies of pedigrees where linked characteristics are passed down through several generations. As would be expected, the knowledge of sex-linked genes is greater than that for autosomal genes. Haldane studied data from various sources and found three cases of crossing over between the genes for deutan (the common green) color blindness

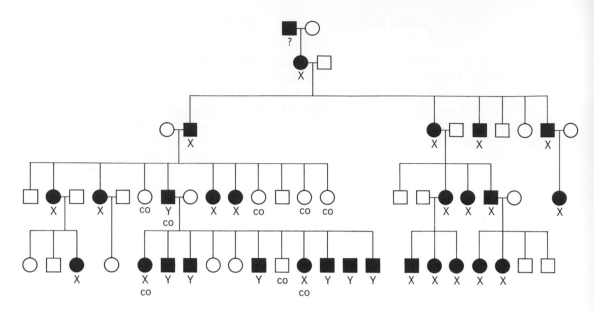

FIG. 14–5 *Crossing over in man. A pedigree of retinitis pigmentosa, a condition of pigmentation of the retina resulting in partial or complete blindness. It is caused by a dominant, incompletely sex-linked gene. It may be on the homologous portions of the X- or the Y-chromosomes and crossing over may take place between genes on this portion of these chromosomes. The large letter under each afflicted individual indicates whether this person carries the gene on the X- or Y-chromosome. The small letters (co) indicate those individuals who represent crossovers. For instance, the man on the left in the third generation must carry the gene on his X-chromosome, since he received the gene through his mother. He has a son with the affliction, however, which indicates that the gene has crossed over to the Y-chromosome. Also, he has four unafflicted daughters who represent crossovers; they would all be afflicted if they received the entire X-chromosome of their father.*

and hemophilia in thirty-four individuals. These results give a crossover value of 8.9 per cent. As further studies were made on other pedigrees, the recombination value was established more precisely at 11.4 per cent. Later the gene for the absence of the enzyme, glucose-6-phosphate dehydrogenase (G6PD), in red blood cells was found to be 6.5 crossover units from deutan color blindness and 17.9 units from the gene for hemophilia. Then the sex-linked gene for the blood antigen, **Xg,** was found to be 31.9 units from the gene for G6PD. This gives us a sequence of genes and the distances between the genes on the human X-chromosome.

Since crossover data for linked autosomal genes are much more difficult to obtain, our information is rather limited. There have been a few cases reported, however. The gene for the **ABO** blood groups is linked with the gene for the nail-patella syndrome with a 10 per cent crossover value. The gene complex for the **Rh** blood antigens and the gene for elliptocytosis are linked with a crossover value of 3 per cent. The gene for the Duffy blood antigen and the gene for cataract of the eye (dominant form) are linked with, crossover values not definitely established. The genes for hemoglobin **B** and hemoglobin **S** are also linked, but no crossovers at all have yet been observed between them.

Linkage Studies in Bacteria

Since bacteria are haploid and do not have anything like meiosis, it is not possible to study linkage distances by means of percentages of crossing over. The microbial geneticists, not to be deterred by such a stumbling block, have worked out methods of establishing the sequence of genes and distances between genes in bacteria. You will remember that *Escherichia coli,* the common colon bacillus, has a circular chromosome and, in conjugation, the chromosome of the donor cell opens out and a portion of the chromosome is transferred to the recipient cell (see Chapter 8). When a male strain (*Hfr*) which contains known genes is mixed with a female strain (*F⁻*) without these genes, a portion of the male chromosome containing some of the known genes will pass from the male to the female. The two strains are mixed with about 20 times as many females as males, to be sure that each male finds a conjugant partner. Thus, it is possible to test the cells remaining after conjugation and be sure that the results are from female cells alone and not from left-over male cells which did not conjugate.

The number of genes which are transferred in conjugation depends upon the time the two strains remain together. Normally, the conjugating partners separate before the entire chromosome has been injected, but they can be separated at any time by placing them in a kitchen blender. After separation, the recipient cells can be tested by placing them on various media to see which genes have been transferred in a specific time. For instance, if we have male bacteria with certain known genes and leave them in conjugation with females for 5 minutes before separation, we will find that only the gene *arg D* (a gene which makes the enzyme ornithine transcarbamylase) has been transferred. If we leave the bacteria in conjugation for 10.3 minutes, however, we also get *pho* (a gene which makes the enzyme, alkaline phosphatase). When separated 16 minutes after conjugation began, we add a third gene *gal A* (producing the enzyme galactokinase). A 24-minute conjugation will allow a fourth gene *try C* (producing the enzyme, indole 3-glycerol phosphate synthetase) to pass to the female cell. On the basis of these results we can determine the sequence of these four genes and the relative distance between them by using time rather than frequency of crossing over as a criterion. One minute of time conventionally represents one unit of distance on the bacterial chromosome. This is shown in Figure 14–6.

Two important questions remain to be cleared up. First, how does the inserted chromosome segment from the donor become incorporated into the chromosome of the recipient and how is the excess portion of a chromosome taken out of the cell? One theory holds that the chromosome segment injected by the male pairs up with the circular chromosome of the female so that homol-

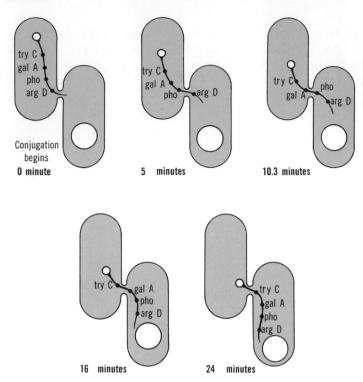

FIG. 14-6 *How distances between genes of the chromosome are determined according to the time elapsed in conjugation of Escherichia coli.*

ogous genes lie opposite one another. Then, there is double crossing over which incorporates some of the injected genes into the recipient chromosome. This would occur at the time the DNA strands are duplicating preceding mitosis. Another theory which has held favor with microbial geneticists is known as **copy-choice.** It is assumed to work in the following manner. Before replication the two strands of DNA uncoil and form single strands. Each single strand then replicates its complementary strand and two double DNA strands are present, each a duplicate of the original. If another strand is nearby, however, the newly-developing strand could be formed for a time on one, then form on the adjoining strand for a time, and finally go back and finish its formation on the original strand. This would have the same end result as crossing over, a chromosome incorporating genes from the donor bacterium. It might be possible that this could also be the mechanism for chiasma formation and recombination of chromosome portions in higher organisms, rather than an actual crossing over of chromatids. The haploid state in the bacterium is restored during cell divisions following the conjugation. Whatever extra chromosome portions are present, be they from the donor or recipient, will be thrown off, and the cell will again have a single chromosome and haploid genes.

There is one objection to the copy-choice mechanism. In meiosis of diploid organisms it could explain crossing over only between two of the chromatids in a tetrad of four, yet cytological evidence shows that all four chromatids may be involved in some instances. If the chromosomes become duplicated first and then the exchange takes place, any two of the opposing chromatids could exchange places. Some geneticists today feel that perhaps both methods take place.

If you have been thinking carefully during this discussion, another question will surely have occurred to you. If the donor chromosome is always first injected at the same place, and rarely is the entire chromosome injected during conjugation, then it would seem that there would be little chance for recombination of the genes near the episome. These genes are furthest removed from the end of the chromosome which is injected first and only rarely would sufficient chromosome be injected to include these genes. On the other hand, those genes lying at the end opposite from the episome would appear to have a relatively high rate of recombination, since they always enter the cell first. This unbalance of gene exchange does not take place in large populations which consist of mixed strains because the point where the episome becomes attached to the chromosome varies in different strains. Hence, the genes which enter a recipient cell first will vary and each part of the chromosome will be in a position to go first in some strains.

Linkage Studies in Neurospora

Although *Neurospora* has haploid somatic cells as a whole, it also has a diploid zygote and there is a chance for crossing over to take place in the first division of meiosis of this zygote. The gamete nuclei, each with seven chromosomes, unite within a cell which then elongates to form a sac, the ascus. The diploid nucleus, formed by the union of the haploid nuclei, then undergoes meiosis to form four haploid nuclei. These four are arranged in a definite sequence, the top two come from one nucleus of the first division of meiosis and

FIG. 14–7 *Copy-choice theory of gene exchange in bacteria. As a new chromatid is formed it may replicate for a time from the chromosome portion which has been injected by the donor bacterium. This process accomplishes the same sort of gene recombination as would be accomplished by crossing over of chromatids.*

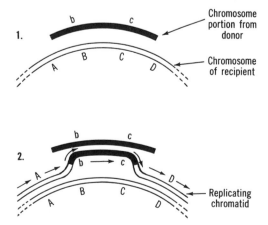

the bottom two come from the other nucleus. Next, there is a mitotic division of each of the nuclei to give eight haploid nuclei. Each of these becomes encased in a hard outer coating and is a spore, which can grow into a new mold. Microbial geneticists can remove each spore individually and grow them on separate media, so it is possible to observe segregation. For instance, if an albino strain is crossed with the wild-type pink strain, the spores will usually be segregated into a 4 : 4 arrangement. The four at the top may grow into albino molds and the four at the bottom may grow into pink molds, or vice versa.

In some cases, however, a different sequence of the spores is found. The top two may produce albino molds, the second two may produce pink molds, the third two may produce albino, and the fourth or bottom two may produce pink. A sequence of this nature is an indication that crossing over has taken

FIG. 14–8 *How crossing over can take place in Neurospora. When no crossing over takes place the spores are arranged in sequence with four from each parent in order. When crossing over takes place, however, the spores are in various combinations of pairs.*

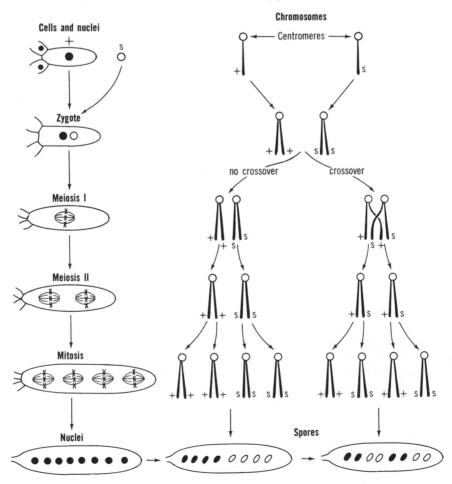

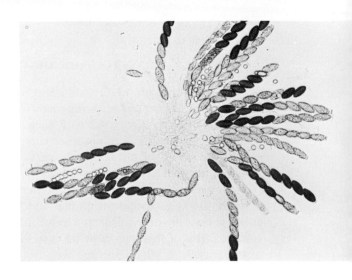

FIG. 14–9 *Photographic proof of genetic recombination in Neurospora. These ascospores were produced by a cross between a strain with slowly maturing spores and one with normally maturing spores. The slowly maturing spores appear light. Most of the asci contain the 4–4 non-crossover distribution of spores, but some show the results of recombination since the spores are in pairs of two.* COURTESY DAVID R. STADLER.

place. The chromosomes of *Neurospora* have terminal centromeres and a crossover which takes place between the centromere and a specific gene, such as the gene for albinism, will result in a rearrangement of the sequence of the genes in the spores. This is shown better than words can describe in Figure 14–8. You can actually see the results of crossing over in the photograph in Figure 14–9. This is a cross between a strain with normally maturing spores and a strain with late maturing spores. Those with a delayed maturation are light in color because the protective outer spore coat has not yet been formed, while the spores with a normal maturation time are dark because the coat has already been formed. By studying the frequency of crossing over between the centromeres and different genes on the chromosomes, it is possible to learn the sequence of genes and the relative distances between them.

FIG. 14–10 *Some of the better known genes in mice arranged according to established linkage groups and some independent genes for which linkage has not been established.* COURTESY MARGARET C. GREEN.

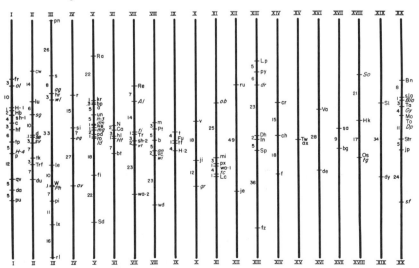

CONSTRUCTION OF CHROMOSOME MAPS

In any organism, when a sufficient number of genes have been located, it is possible to construct chromosome maps which show the number of chromosomes, the linkage groups associated with each pair of chromosomes, the sequence of genes on each of the chromosome pairs, and the relative distance between the genes on each of the chromosome pairs. All this can be done by means of the techniques which we have learned.

As an illustration, let us suppose that we wished to construct a chromosome map for a certain species of animal which had not been previously investigated either genetically or cytologically. Our first problem would be to collect as many variant forms of this animal as we could find and to obtain pure-breeding stocks of those variants which are due to genes. Next we would breed these together and set up linkage groups which should correspond to the number of pairs of chromosomes. Then we would study each linkage group separately and thoroughly to determine the crossover percentage between each two pairs of genes in each linkage group. By careful analysis of these results we could arrange the genes in proper sequence on the chromosomes. Let us suppose that there are six linkage groups which we designate with roman numerals. In group V, let us suppose that we have located thirteen genes, which we will designate as *a* through *m*. Crossover data show that *a* and *m* are the genes farthest apart. Let us select *a* as our starting point and call it 0.0. If *b* shows 5.3 per cent of crossing over with *a*, we locate *b* at 5.3 on the chromosome. Then if *c* shows

FIG. 14–11 *A genetic map of the genes on the circular chromosome of Escherichia coli based upon time of transference of genes during conjugation. Each unit represents one minute of time.* REDRAWN FROM TAYLOR AND THOMAN.

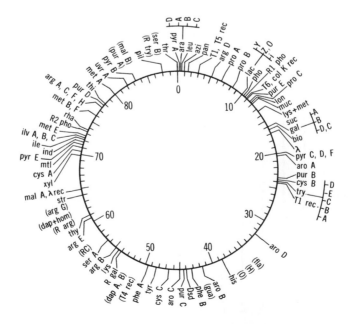

TABLE 14-1

Location of Some Better Known Genes of Drosophila melanogaster as Determined by Chromosome Mapping

X-chromosome I	Chromosome II	Chromosome III	Chromosome IV
0.0 y—yellow body	0.0 net—net veins	0.0 ru—roughoid eyes	0.0 sv—shaven bristles
0.0 sc—scute bris.	1.3 S—Star eyes	0.2 ve—veinlet wing	0.0 ci—cubitus inte-ruptus venation
0.6 br—broad wing	11.0 ed—echinoid eyes	19.2 jv—javelin bristles	0.0 gvl—grooveless scutellum
0.8 pn—prune eyes	12.0 ft—fat body	26.0 se—sepia eye	0.2 ey—eyeless
1.5 w—white eyes	13.0 dp—dumpy wing	26.5 h—hairy body	
3.0 fa—facet eyes	16.5 cl—clot eyes	41.4 Gl—Glued eye	
5.5 ec—echinus eyes	41.0 J—Jammed wing	43.2 th—thread arista	
6.9 bi—bifid wings	48.5 b—black body	44.0 st—scarlet eyes	
7.5 rb—ruby eyes	51.0 rd—reduced bristles	45.3 cp—clipped wing	
13.7 cv—cross-veinless wings	54.5 pr—purple eye	46.0 W—Wrinkled wing	Y-chromosome
18.9 cm—carmine eyes	55.0 lt—light eye	47.0 in—inturned bristles	
20.0 ct—cut wing	55.9 ti—tarsi fused	48.0 p—pink eye	male fertility
21.0 sn—singed bristles	57.5 cn—cinnabar eye	48.7 by—blistery wing	long bristles
27.7 lz—lozenge eyes	67.0 vg—vestigial wings	50.0 cu—curled wing	male fertility
32.8 ras—raspberry eyes	72.0 L—Lobe eye	58.2 Sb—Stubble bristles	(no locations worked out)
33.0 v—vermilion eyes	75.5 c—curved wing	58.5 ss—spineless bristles	
36.1 m—miniature wings	100.5 px—plexus veins	59.0 Rf—Roof wing	
43.0 s—sable body	104.5 bw—brown eyes	62.0 sr—stripe thorax	
44.4 g—garnet eyes	107.0 sp—speck wing	63.1 gl—glass eye	
51.5 sd—scalloped wings		66.2 Dl—Delta veins	
56.7 f—forked bristles		69.5 H—Hairless bristles	
57.0 B—Bar eyes		70.7 eb—ebony body	
59.5 fu—fused veins		90.0 Pr—Prickly bristles	
62.5 car—carnation eyes		91.1 ro—rough eyes	
66.0 bb—bobbed bristles		93.8 Bd—Beaded wing	
		100.7 ca—claret eye	
		104.3 bv—brevis bristles	

Data from Bridges and Brehme.

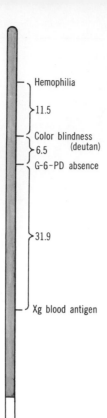

3.4 per cent crossing over with *b* and about 8.7 per cent with *a,* we locate *c* at 8.7 on the chromosome. We would locate the other ten genes in a similar manner, probably speeding the procedure by combining our studies to include several genes at each test. We would now have a provisional chromosome map of one chromosome. We say "provisional" because further study might reveal more accurate crossover data which would cause us to readjust our unit distances on the map. Also, we might find a new gene which falls outside the range of our provisional map. For instance, we might find a new gene which shows 2.2 per cent crossing over with *a* and 7.5 per cent crossing over with *b*. Such a gene could not be between *a* and *b*, nor could it be to the "right" of *b*. It must, therefore, be to the "left" of *a*. We would then put the new gene at 0.0 on our map and advance all the other genes by 2.2 units. We would follow this procedure for each of the six linkage groups in order to have a complete map of all the chromosomes. Also, this work would probably be correlated with cytological investigations which corroborated our genetic findings before we could feel that our map possessed any high degree of accuracy.

The above description is of course an over-simplification of the task. The amount of work involved is so great that it is doubtful whether any one person could prepare an accurate chromosome map during his entire lifetime. The maps which have been made usually represent the combined efforts of many individual research workers. Extensive chromosome maps have been worked out for quite a number of organisms.

PROBLEMS

1. A large number of genes in corn have been investigated. These all fall into ten distinct linkage groups. On the basis of this information, how many chromosomes would you say are present in the somatic cells of a corn plant? Explain your answer.

2. *Drosophila melanogaster* has four pairs of chromosomes and quite a number of genes have been located on each of the chromosomes. Suppose you discover a new gene. Tell how you would determine on which of the chromosomes this new gene is located.

3. In the somatic cells of the sweet pea there are fourteen chromosomes. If any two genes are chosen at random, what is the probability that both will be in the same linkage group? Assume that the chromosomes are all of about the same length. Explain your answer.

4. In the sweet pea the genes for red flowers and spherical pollen grains belong to the same linkage group. A new gene is discovered which produces variegated leaves. What are the chances that it will be linked to the first two genes?

5. In rabbits black and short hair are characters resulting from two dominant genes. The recessive alleles of these genes produce brown and long hair. When we mate homozygous black, short-haired rabbits with brown, long-haired rabbits and test cross the offspring we obtain the following results:

black short	29
brown long	33
black long	35
brown short	27

From these results, would you conclude that these two genes are located on the same chromosome? Why? If your answer is yes, what is the percentage of crossing over?

6. In corn there is a dominant gene for colored seed and another dominant gene for full seed. The recessive alleles of these genes produce colorless seed and shrunken seed. Plants homozygous for colored full seeds are crossed with colorless shrunken, and a test cross of the F_1 yields the following results:

colored full	190
colorless shrunken	198
colored shrunken	7
colorless full	5

Would you say that these two genes are linked? If so what is the percentage of crossing over?

7. In rabbits two recessive genes produce a solid body color and long hair respectively, in contrast to a spotted body color and short hair which result from the dominant alleles. The results from a cross between the heterozygous spotted short-haired rabbits and solid long-haired rabbits are as follows:

spotted short	48
spotted long	5
solid short	7
solid long	40

In terms of crossover units, how far apart are these two genes on the chromosome?

8. Let us assume that some of the heterozygous F_1 rabbits from problem 7 are mated with homozygous spotted long-haired rabbits, and 100 offspring are obtained. What types of coat characteristics would be found and about how many of each would be expected.

9. In rabbits, black is dominant over brown coat color, but neither of these colors can show when the animal is homozygous for the recessive gene for albinism. Albinism is said to be epistatic to the other two genes. The gene for albinism is not an allele of the genes for black or brown. Albino rabbits which are homozygous for brown (*aabb*) are crossed with black rabbits (*AABB*). The F$_1$ are all black (*AaBb*). A test cross of these yields the following results:

black	66
albino	100
brown	34

Examine these results carefully and determine if these genes are linked. If they are linked, what is the percentage of crossing over?

10. Let us assume that we have three recessive genes (*abc*) in a hypothetical animal. When individuals showing these three recessive characters are mated with the homozygous dominant and a test cross of their offspring is made, we obtain the following results:

Normal (*ABC*)	61
abc	56
aBC	14
AbC	8
ABc	2
Abc	12
aBc	10
abC	3

What is the sequence of these genes on the chromosome? Show how you obtain your results. How many units apart are these genes?

11. If previous studies of the case in Problem 10 show that *a* is located at 5.2 on the chromosome, where would you place *b* and *c* on the chromosome?

12. The genes for garnet eye and forked bristles are both sex-linked characters in *Drosophila*. They are 10 units apart on the *X*-chromosome. Wild-type males (*G F*) are crossed with garnet, forked females (*gg ff*) and the F$_1$ are crossed among themselves. What percentage of each of the four possible types of offspring would you expect among the males? Among the females?

13. If the sexes of the parents in Problem 12 were reversed and garnet, forked males were crossed with homozygous wild-type females, what results would be expected in the F$_2$?

14. Studies of tomatoes show a number of linked genes. Let us cross plants with pubescent epidermis (*p*), ovate fruit shape (*o*), and compound inflorescence (*c*) with plants having the dominant alleles — smooth epidermis (*P*), oblate fruit (*O*), and simple inflorescence (*C*). The test cross of the F$_1$ yields 13.9 per cent crossing over between *p* and *o*, 12.8 per cent between *o* and *c*, and 23.2 per cent between *p* and *c*. If we assume that *p* is at the "left" end of the chromosome, what is the location of *o* and *c*? Why is the crossover value between *p* and *c* less than the sum of the crossovers between *p* and *o*, and *o* and *c*?

15. A fourth gene, for dwarf plant size (*d*), is found to be linked with the above three genes. When crossed with *p* and *o*, it shows 17.2 per cent crossing over with *o* and 3.5 per cent crossing over with *p*. Reconstruct your chromosome map on the basis of this new information and show the location of all four of the genes.

16. At least two mutant genes are needed in studies to determine gene position on the chromosome for higher organisms, but in *Neurospora* only one mutant gene is necessary. Explain.

17. In one donor strain of *E. coli* the gene RC (regulation of RNA synthesis) is transmitted into a recipient cell after 8 minutes of conjugation. In another strain the same gene is not transmitted until after 21 minutes of conjugation. How would these two donor strains differ in order to have such a difference in time of transmission of the same gene.

18. Explain how copy-choice differs from chromatid exchange as an explanation of gene exchange in *E. coli*.

15

Chromosomal Aberrations

IN THE NORMAL COURSE OF MEIOSIS chromosomes synapse and subsequently segregate in an orderly manner, and gametes are regularly produced with the haploid chromosome number. The gene loci retain the same sequence they have had for many generations past. In genetic research, however, countless numbers of individuals are studied, and in the course of time certain rare cases come to light which show variations in the structure, number, or arrangement of the chromosomes. Such variations may produce phenotypic changes, changes in the expected genetic ratios, or changes in the linkage relationships of certain genes. Just as variations in the expression of the genes enable us to learn a good deal about the nature of the genes, these chromosomal variations are of inestimable value in studies of the nature of the chromosomes and their relation to the genes.

CHROMOSOME BREAKAGE

We have learned that chromosome strands normally undergo breakage and reattachment to homologous strands during the prophase of the first division of meiosis in the process known as crossing over. This is very common, of course, and no abnormal chromosome or gene arrangement results. It is possible, however, for breakage to occur without reattachment to the homologous strand, and when this occurs, various sorts of abnormal chromosomes may result. Some of these are important in evolutionary changes of species, or because they produce phenotypic changes like those due to simple mutation.

240

Deficiency or Deletion

A chromosome without a centromere is like an automobile without a driver — it has no power to move properly during cell division, and hence will very often not become properly included in the nucleus of a new cell which is formed. That is, should a portion of a chromosome become detached from the remaining portion of it which bears the centromere, it will usually be left behind in the cytoplasm as the chromosomes follow their centromeres to the poles and the new nuclear membranes are formed. Such a block of genes, excluded from their normal surroundings in the nucleus, soon disintegrate. This leaves a chromosome within the nucleus which lacks those genes that were present on the deleted portion. A germ cell may be formed carrying such a chromosome deficient in the deleted genes. When this germ cell unites with another germ cell carrying a normal chromosome, a zygote is produced that carries the particular group of genes in a single dose.

As an illustration, we may observe the effects of chromosome deletion in mice. There is a recessive gene in mice which produces an odd mode of behavior called **waltzing.** Mice homozygous for this gene cannot walk in a straight line but continually turn from side to side or go around in circles, rather like waltzers on a dance floor. This is due to a defect of the semicircular canals in the inner ear which makes it impossible for the afflicted mice to maintain normal equilibrium, so that they act as if they were in a constant state of intoxication. The gene is autosomal, and a cross between homozygous normal and waltzing mice

FIG. 15–1 *How a deletion of a portion of a chromosome in a mouse can produce a waltzing offspring from a cross between a homozygous normal mouse and a homozygous waltzer. In the absence of the dominant allele, the recessive gene for waltzing expresses itself.*

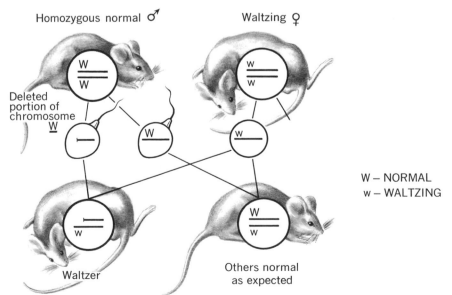

Homozygous normal ♂   Waltzing ♀

Deleted portion of chromosome

W – NORMAL
w – WALTZING

Waltzer   Others normal as expected

normally results only in normal offspring, all of which are heterozygous. In one such cross, however, a waltzing female appeared among the offspring. The chromosomes of this female and a number of her offspring were later examined cytologically and it was found that a section of one of the chromosomes was missing. A deletion must have occurred in the germinal tissue of the normal parent of this female. When a germ cell with this abnormal chromosome united with a germ cell which carried the gene for waltzing, the zygote carried no normal dominant allele of this gene. Hence, the gene for waltzing expressed itself, like any gene present in a single dose whether recessive or dominant. Deletions of this nature have enabled geneticists to locate the gene on the particular portion of the chromosome on which it is carried.

Later breeding tests showed that, when zygotes are formed which are homozygous for the chromosome with the deletion, they do not survive. The deletion has a lethal effect when it is homozygous. This is to be expected, since the deleted portion of the chromosome no doubt carries many genes which are necessary for the formation of a viable mouse. This is true of most deficiencies: the chromosomes with a deleted portion can be maintained only in the heterozygous state, for an individual cannot live when there are no genes at all for certain vital body characteristics. There are a few cases, however, in which it has been shown that individuals with chromosomes that have certain very small deletions can survive even when homozygous for them. Such a case has been found on the *X*-chromosome of *Drosophila*. Of course, any large deletion which occurs on the *X*-chromosome will have a lethal effect, and all males which receive the chromosome with that portion missing will die, since males carry no alleles for the missing genes. A very small deficiency has been found, however, which includes at least two genes near the "left" end of the *X*-chromosome and which may exist in viable males and homozygous females. This indicates that genes included in the deficiency are not absolutely vital to the existence of the individual. A number of similar cases have been found in various animals, but the portion of the chromosome involved has always been very small. In some plants, however, an entire chromosome may be missing without fatal results.

Two types of deletions are possible as illustrated in Figure 15–2. A **terminal deletion** is one in which a portion is lost from the end of the chromosome. An **intercalary deletion** is one in which a portion of the chromosome is lost from within, leaving the ends of the chromosomes intact. This type of deletion may occur when a chromosome breaks in two places and the end pieces become attached to each other. The two ends of the portion deleted from the center of the chromosome may become attached to form a ring-shaped structure. Of course, it is possible that the ring-shaped portion will include the centromere, in which case the terminal portions will be lost and the ring retained. Such ring-chromosomes have been found, but most are not stable and cannot be maintained because of abnormal meiosis. The intercalary deletions are by far the more common of the two types.

Chromosomes with deletions form interesting figures when they undergo **synaptic pairing** with normal chromosomes in meiosis. The attraction holds

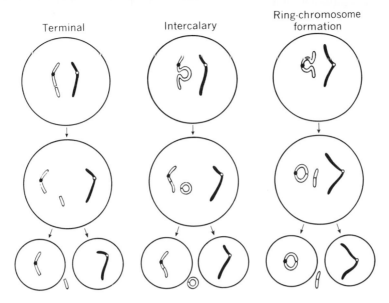

FIG. 15–2 *The method of origin of different types of deletions. A terminal deletion results when a portion is lost from the end of a chromosome. An intercalary deletion results from the loss of a portion within the chromosome. This usually results from a ring formation. When the ring includes the centromere, this portion remains within the nucleus and the two terminal portions are lost. These drawings are highly diagrammatic and are not intended to show the actual method by which deletions occur.*

only for the homologous parts of the paired chromosomes, and hence the portion of the normal chromosome which corresponds to the deleted portion of its mate buckles, as shown in Figure 15–3.

Inversion

Occasionally a portion of a chromonema will break out of the center to produce a deletion, and the broken portion will become reattached in its original position but with the two ends reversed. This is most likely to happen when the chromosome forms something of a loop, since such twisting makes it easy for the broken ends of the deleted portion to become reattached in the new position (see Figure 15–4). Inversion is most likely to occur in meiosis, and will result in a germ cell carrying a chromosome with genes in reverse order on the inverted portion. Let us suppose that the original chromosome carried genes *abcdefghi*

FIG. 15–3 *Synapsis of a normal chromosome (above) with a chromosome having an intercalary deletion (below). The portion of the normal chromosome corresponding to the deleted portion of its mate bulges upward.*

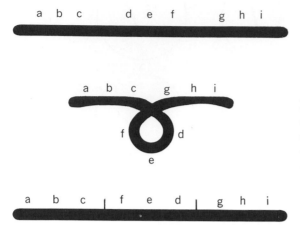

FIG. 15–4 *Inversion. The sequence of genes is reversed when a portion of the chromosome changes position, as shown in this diagram.*

in that particular sequence, and an inversion of the central portion of the chromosome occurred. The new sequence might be *abcfedghi*. The location of the three central genes would no longer conform to their positions as previously determined by chromosome mapping. When a germ cell carrying such a chromosome unites with a normal germ cell, there is usually no detectable effect of the inversion on the phenotype of the offspring, for all the genes are present in normal numbers. When synapsis occurs in subsequent generations, however, the cytological figures show abnormal configurations as the normal chromatids become eliminated during meiosis, and the gametes which are formed normally do not contain any chromosomes with crossovers in the region represented by the inversion. Quite often a geneticist may wish to maintain a series of genes on a chromosome in a heterozygous state without any crossing over. This is possible provided there is an inversion on one of the two paired chromosomes. An inverted *X*-chromosome, known as *ClB,* is very valuable in mutation studies. The method of its use is presented in Chapter 19.

Inversions probably play an important part in evolution, for, if two groups of organisms are separated for a long period of time, they may separately accumulate so many inversions of different kinds that normal synapsis and segregation cannot take place between them and the offspring of such crosses cannot produce viable gametes. When the hybrid becomes sterile, we may justifiably refer to the two groups from which the parents came as separate species. More about this is presented in Chapter 24.

Duplication

Another type of chromosomal aberration, which may occur simultaneously with and as a result of a deletion, is known as **duplication.** It is possible for the deleted portion of a chromosome to become attached to a break in another chromosome. The deleted portion of one chromosome may become attached to a chromosome with a centromere, so that the former does not become lost in the cytoplasm but reproduces itself during mitosis in a normal manner. A germ cell receiving such a chromosome may carry two doses of the genes which lie on this transferred or duplicated portion. And when it unites with another germ cell it will produce a zygote having three doses of these genes. This results in interesting ratios among the offspring, and we have learned much about the nature of gene action through the study of such results. Duplication has no doubt played an important part in evolution, since in this way additional genes may be added to the complement which is already present. Through mutations these may then become different from the homologous genes at the old location, and eventually the chromosome carrying the transferred piece may carry a segment of entirely new sorts of genes. Mutations which previously have been lethal may now arise and be tolerated, inasmuch as the physiology of development is protected by having at least one locus of the two to carry on in the event that the other mutates. Such genes may actually produce effects which are beneficial to the race. Thus the previous limits set to mutation can be transcended, and genes with new effects may come into existence.

Duplication may also arise as a result of unequal crossing over. This will be discussed in Chapter 16 in connection with the origin of the gene for bar-eye in *Drosophila*.

Translocation

Not only may the sequence of linked genes be upset by various chromosomal aberrations, but it is possible for a block of genes to shift from one linkage group to a different one. This is done by translocation. Probably the only way in which chromosome segments may be transferred to non-homologous chromosomes is by **reciprocal translocation,** which occurs when portions break from two non-homologous chromosomes and exchange places. The process is illustrated in Figure 15–5. A cell heterozygous for a translocation produces very interesting synaptic figures, since the genes tend to pair with their alleles even

FIG. 15–5 *Translocations in Drosophila. These two drawings represent actual examples of translocations which have been found and studied both genetically and cytologically. On the left a portion of the third chromosome has become attached to the end of a second chromosome. At the same time, apparently a very small piece of the end of the second chromosome has moved to the broken end of the third. On the right is another reciprocal, but there has been unequal translocation between the second chromosome and the X-chromosome.*

though they are located in a portion which has become translocated to a non-homologous chromosome. Such figures stand out very clearly in the chromosomes of the salivary glands of *Drosophila*. Figure 16–6, page 268, shows pairing between translocated portions of the second and third chromosomes in a cell heterozygous for the translocations. In the process of reciprocal translocation, it is possible for one chromosome to get both centromeres while the other is left with none. The chromosome without a centromere will be lost, of course, but the other may be pulled apart to make two chromosomes when the two centromeres move to opposite poles.

Translocations can often be predicted before they are seen. H. J. Muller found one strain of *Drosophila* in which a group of genes including scarlet, which normally is on the third chromosome, showed linkage to the genes of the second chromosome. Cytological examination showed that the third chromosome was much shorter than usual, while the second chromosome was longer than usual. Thus cytology confirms the results of genetic breeding experiments.

Translocations have, no doubt, played an important part in evolution. Changes in the chromosomes, additions and subtractions of genes, changes in the linkage groups — all these are vital parts of the process of constant change and adaptation which we call evolution.

Position Effect

One of the great genetic discoveries resulting from studies of chromosome breakage has been what are called **position effects.** We have already mentioned the fact that new characteristics resulting from transfers of portions of chromosomes often result from changes in quantity, so that an individual with three genes of a certain kind, for instance, may differ from an individual with two. A number of chromosomal aberrations have been found in *Drosophila,* however, where there was no change in the quantity of genic material and yet there was a new phenotypic effect. These cases have shown that some genes may exert a different effect from the normal one when their relationship with other genes is altered. The genes which lie near a point of breakage will find themselves associated with a different set of neighbors no matter whether the breakage results in an inversion, a deletion, a duplication, or a translocation. Most genes go right on in their new positions just as if no change had been made, but a number of cases have been discovered in *Drosophila, Oenothera* (the evening primrose), and in maize, in which the gene exerts a different effect in its new position. An excellent example of this involves the production of the bar-eye in *Drosophila*. This is discussed in Chapter 16, since it is best explained by studies of the salivary gland chromosomes.

ABERRATIONS DUE TO BREAKAGE OF HUMAN CHROMOSOMES

Since the discovery of greatly improved methods of studying human chromosomes, it has been possible to find many examples of chromosome breakage in man with resulting aberrations. As an example, an extensive study of patients with chronic granulocytic leukemia, a fatal affliction, have revealed a consistently short chromosome 21. A part of the long arm of this chromosome is missing. It is evident that a terminal deletion has occurred and the absence of the group of genes from the deleted portion upsets the genic balance and results in the leukemia. Since this shortened chromosome was discovered in Philadelphia, it is known as the **Philadelphia chromosome.** Apparently the deletion can occur in cells of the blood-forming tissues at any time of life, since the disease can make its appearance at any age from childhood to old age. Cultures made from leucocytes taken from the blood will show the Philadelphia chromosome, but cultures made from the skin or other body cells do not show the aberrant chromosome. A deletion which occurs in a single cell of the bone marrow which produces blood cells may replicate repeatedly until a large clone of cells with the deletion are present and the leukemia develops. Careful analysis of the DNA content of the Philadelphia chromosome, in comparison with that of the normal chromosome 21, shows that it has lost about 39 per cent of its gene volume. This lost DNA represents about 0.5 per cent of the total DNA content of a normal diploid human cell.

Many translocations have also been found in human chromosomes in recent years. Examples have been found in each of the groups of chromosomes of the six distinguishable sizes. Most of these translocations produce no sign of abnormality in the carriers where the deletion of one chromosome is balanced by the presence of the deleted portion on another chromosome. About one-half of the offspring of the carriers of such translocations, however, will be grossly abnormal because of either a triploid or a haploid condition of a group of genes. This is shown in Figure 15–6.

ANEUPLOIDY AND POLYPLOIDY

In addition to the various sorts of chromosomal aberrations which may result from breakage and reattachment of chromosome segments, there are cases in which entire chromosomes may be lost or duplicated. This, of course, upsets the ratio between the genes and often results in pronounced phenotypic effects. The changes may involve a single chromosome or perhaps several chromosomes; such cases are known as **aneuploids.** Other changes involve entire genomes (haploid sets of chromosomes); such cases are known as **polyploids.**

Aneuploids

One of the first cases of aneuploid chromosome distribution was discovered by Bridges during his early work on *Drosophila.* After his discovery that the gene for white eyes was on the X-chromosome, he noted that in examinations of the offspring of crosses between white-eyed females and red-eyed males he occasionally found offspring which did not follow the normal pattern of sex-linked inheritance. The normal offspring of such a cross will be red-eyed females and white-eyed males, but by examination of thousands of flies he found a few white-eyed females and a few red-eyed males. Bridges reasoned that these cases

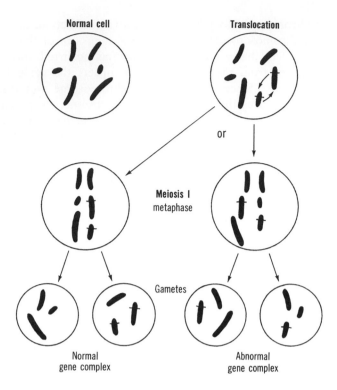

FIG. 15–6 *Diagram to show how a translocation will result in gametes, one-half of which will have a normal gene complex and one-half of which will be abnormal. The abnormal gametes will either have some genes in double quantity or there will be a deficiency of some genes. Such abnormal gametes give zygotes which are either triploid or haploid with respect to certain genes.*

perhaps resulted when the two X-chromosomes failed to separate during oogenesis, for this failure would result in some eggs having two X-chromosomes and others having none. Those carrying the two X-chromosomes would yield normal individuals if fertilized by Y-carrying sperms, for this would result in a female with the two X-chromosomes of the mother, which would produce white eyes. The extra Y-chromosome seems to have no effect on the appearance or fertility of such females. If such an egg is fertilized by an X-carrying sperm, however, the offspring will have three X-chromosomes, which makes it a "superfemale," as we learned in Chapter 8. Such flies are completely sterile and have a very low viability, for they seldom live long enough to be seen. The type of egg without any X-chromosome at all might be fertilized by an X-carrying sperm, which would result in a male showing the red eyes of its father. Such males will lack a Y-chromosome and consequently will be sterile, since this chromosome carries genes for male fertility; but phenotypically, males without a Y-chromosome are indistinguishable from normal males. Should a Y-carrying sperm fertilize an egg without an X-chromosome, the offspring dies almost at

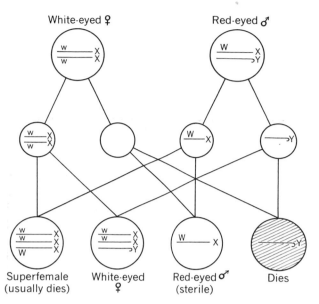

W – RED EYES
w – WHITE EYES

White-eyed ♀   Red-eyed ♂

FIG. 15–7 *Non-disjunction of the X-chromosome in Drosophila. In rare cases, the two X-chromosomes adhere in meiosis and both go to one pole of the spindle. This produces some eggs with two X-chromosomes and some with none. When these eggs are fertilized by normal sperms, exceptional types of offspring will be produced as indicated.*

Superfemale (usually dies)   White-eyed ♀   Red-eyed ♂ (sterile)   Dies

once, since it completely lacks the important group of genes on the *X*-chromosome. These four possibilities are shown in Figure 15–7.

Cytological observations later confirmed Bridges' suggestions. For some reason, during the first stage of meiosis in oogenesis the two *X*-chromosomes occasionally fail to be pulled apart; both chromosomes move to one pole and become incorporated into the nucleus of one of the daughter cells, leaving the other cell without an *X*-chromosome. This comparatively rare occurrence is known as **non-disjunction.**

One exceptional stock of *Drosophila* was found in which non-disjunction regularly occurred in all divisions of the primary oocyte. Females from this stock invariably produce eggs which carry either two *X*-chromosomes or none. This unusual occurrence results from permanent **attachment of the X-chromosomes.** The pair of attached *X*-chromosomes form a large V-shaped figure in dividing cells. This stock of flies is extremely useful to the geneticist, for it is

X   X   Y

FIG. 15–8 *The chromosomes of a female Drosophila with attached X-chromosomes. This condition results in non-disjunction at every meiosis. Both X-chromosomes carry the gene for yellow body color, which is a convenience in distinguishing these females. A Y-chromosome from the male parent is also present.*

possible to cross any type of male with females of this strain and obtain male offspring which will be like the male parent in their sex-linked characters. The original attached *X*-chromosomes both had the gene for yellow body color, with the result that the attached *X*-females are yellow phenotypically. Now, many other sorts of attached *X*-females exist.

Non-disjunction may occur in other chromosome pairs as well as in the sex chromosomes. It has occurred in the fourth chromosome of *Drosophila* to give some offspring with three of these chromosomes and some with only one. These two types are called triplo-IV and haplo-IV, respectively. The haplo-IV flies do not contain two complete genomes, and as a result are not as robust or healthy as normal flies. They can be recognized phenotypically by their shortened bristles and a roughened condition of the eyes. The triplo-IV flies are somewhat more difficult to recognize, yet they can be identified easily by cytological examination.

The term **trisomic** is used to refer to the condition where one chromosome is present in triplicate. In some cases this triple structure may not be formed from normal complete chromosomes, but rather from broken parts of chromosomes which contain a centromere. The term **monosomic** is used to refer to those cases where there is only one chromosome rather than the normal two in the cell. Bridges has found that in *Drosophila melanogaster* either the monosomic or the trisomic condition of one of the large V-shaped chromosomes is lethal. Apparently the genic balance is too greatly upset when all the genes on one of these long chromosomes are present in a triple or a single dose rather than the normal double dose. The haplo-IV fly is rather low in viability, in spite of the fact that the fourth chromosome contains comparatively few genes. If the alteration of the normal dosage of genes in such a tiny chromosome can have such a pronounced effect on viability, it is easy to understand how the alteration of dosage of the large number of genes on the longer chromosomes can be so extreme in its effect as to be lethal. This is not true, however, of the comparatively long *X*-chromosome, since one-half of the normal flies which are produced are haplo-*X*, namely, the males.

Non-disjunction is known to occur in other species as well as in *Drosophila*. One of the most complete studies has been made in the Jimson weed, *Datura stramonium,* by the American geneticist A. F. Blakeslee. This plant normally has twelve pairs of chromosomes. When non-disjunction of one of the pairs takes place, a plant is produced with either twenty-five chromosomes (eleven pairs and a set of three homologous chromosomes) or with twenty-three chromosomes (eleven pairs and one unpaired chromosome). There is a recognizable phenotypic effect when one chromosome is present in triplicate. In the normal diploid the genes are in a balance which produces the normal phenotype, but with an extra chromosome present that balance is upset. According to this concept, a different phenotype would be expected for each chromosome which could be present in triplicate. This was found to be true; Blakeslee found twelve different phenotypes which deviate from the normal, each of which was due to the presence of a different chromosome in triplicate. Cytological studies confirm this as the correct interpretation. Thus, we see that trisomy can have different effects according to which chromosome has the trisomic condition. Corresponding monosomic patterns were not found, although non-disjunction should form just as many monosomics as trisomics. They were not found because the monosomic condition is lethal.

Non-disjunction of Human Chromosomes

In the studies of human chromosomes, cases of non-disjunction have also been discovered. These cases have provided us with an explanation of many phenotypic abnormalities which have been known for many years, but the cause of which has previously eluded our best efforts at detection. In Chapter 8 we learned how non-disjunction of sex chromosomes can have very profound effects on the development of sexual characteristics, as well as other characteristics, of individuals which may be produced from abnormal gametes.

Non-disjunction of autosomes may also take place and give rise to gametes which are diploid with respect to one of the autosomes. When such a gamete is fertilized by union with a normal gamete from the opposite sex, a zygote is produced which is trisomic with respect to the autosome involved. The best known example of trisomy of an autosome in man involves one of the smaller chromosomes, the pair designated as number 21. Human chromosomes are numbered by pairs according to size, the largest autosome pair being designated as number 1 and the smallest autosome pair being designated as number 22. The sex chromosomes are numbered 23, even though the X-chromosome is one of

FIG. 15–9 *Chromosomes from a person with Down's syndrome (mongolism). This photograph shows a total of 47 chromosomes rather than the normal 46. Analysis of the smaller chromosomes shows that there is the normal number (two) of chromosome 22, but there are three representatives of chromosome 21. The unbalance of genes results in Down's syndrome.* COURTESY MURRAY BARR.

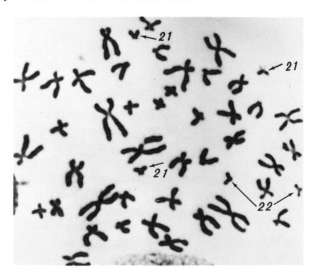

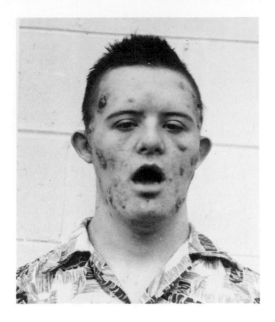

FIG. 15-10 *Down's syndrome (mongolism). This person would be normal except for the fact that he is trisomic for chromosome 21 which causes the mental retardation and other symptoms of this syndrome. He also has a skin infection which is not related to the syndrome.*

the larger chromosomes in the cell and the *Y*-chromosome is one of the smallest. The trisomic condition of the genes on chromosome 21 can upset the delicate balance of genes so that a child is produced with *Down's syndrome,* more commonly known as **mongolism** or **mongolian idiocy.** The latter names have been given because afflicted persons tend to have a round, full face and the upper eyelids tend to turn downward in a manner which is similar to the facial characteristics found in the Mongolian race. The condition appears in all races and has no relationship to the genes of the Mongolian race. In fact, Mongolian people often think that persons of their race who have this syndrome resemble Caucasians.

In addition to the facial characteristics, Down's syndrome produces typical palm prints, malformed ears, a creased tongue, possible congenital heart defects, high susceptibility to respiratory infections, no sexual maturity, and mental retardation. The exact level of mental development varies, probably because of inheritance of other genes which affect intelligence. The level ranges from the idiot to the high imbecile classification on the I.Q. scale. Rare cases of sexual maturity and fertility have been reported in women with Down's syndrome. Ten cases of females who have actually borne children have been found. Of a total of twelve children from these women, seven were normal and five had the syndrome, which is about as close to the 1:1 expected ratio as could be hoped for. Half of the eggs from such women would be expected to be diploid for chromosome 21 and the other half would be the normal haploid.

One important fact about Down's syndrome has been known for many years. The syndrome appears much more frequently in children born to women in the later part of their reproductive life. About 1 child out of every 500 born in America today has this syndrome. This means that the average risk of having such a child is about 0.2 per cent. For a woman between 40 and 45 years of age, however, the risk rises to about 1.25 per cent and over age 45 the risk rises to about 2.25 per cent. If the women of America could complete their families before they are 40, there would be about a 30 per cent reduction in births of children having Down's syndrome. The age of the father is not significant; the percentage of Down's syndrome in children from 20-year-old fathers is just as high as that found in children from 60-year-old fathers.

Why should age be a factor in non-disjunction of chromosome 21 in a woman and not in a man? A baby girl has developed her full complement of primary oocytes at birth, but these do not complete meiosis and form haploid ova until the time of ovulation, which may be from 15 to 45 years later. This aging of the oocyte appears to be the responsible factor. Such an assumption agrees with work done by the author on *Drosophila* in which virgin females were aged for varying periods of time before mating. They were then treated with X-rays which increase the amount of non-disjunction. The aged females showed a much higher percentage of non-disjunction in their progeny than did younger females, although the effect dropped off after about 8 days of aging because the females being held as virgins began laying many of their aged eggs and producing new ones. In the human female the aging of the oocytes plus the accumulating effects of natural background and other radiation could likewise account for the increase in non-disjunction. In the human male, on the other hand, sperm production is a continuing process, and fresh primary spermatocytes are generated continually during the reproductive life of the man.

Another question about Down's syndrome is frequently asked, "Is it hereditary?" This is a very important question to a married couple who have borne a child with this syndrome or to young couples who have close relatives who are afflicted. From studies so far, it might appear that heredity is not involved; how could heredity influence the chance of non-disjunction of a particular autosome? Sad experience, however, has shown that the expectation of a second mongoloid child, in a family already burdened with one, is greater than chance alone would indicate. The explanation of this fact has been found in chromosome studies which reveal that some persons with Down's syndrome have only the normal chromosome number of forty-six. However, a detailed study of these chromosomes shows that chromosome 21 has been translocated onto another autosome,

FIG. 15–11 *Effect of aging Drosophila females on the amount of non-disjunction of the X-chromosomes in response to high energy radiation. The flies were given 3,750 roentgens of X-rays to increase the frequency of non-disjunction. After 8 days, many infertile eggs were laid by the virgin females, so the eggs laid on the 9th and 10th days were actually not as old as those laid on the 4th to 8th days.* FROM WINCHESTER, JOURNAL OF HEREDITY.

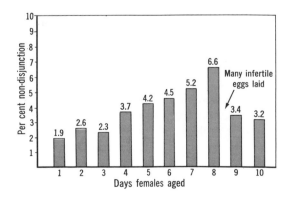

usually one of the longer autosomes, and also present is the normal pair of separate chromosomes 21. Thus, the person is triploid with respect to the genes found on chromosome 21, even though he has only a total of forty-six chromosomes. A study of the chromosomes of the parents of such persons usually shows a single chromosome 21 and a translocated 21 on another autosome in one of the two parents. Such a parent would be normal, but about one-fourth of his or her gametes will carry a single chromosome 21 and the translocated 21 as well. When such a gamete is fertilized by a normal gamete from the other parent, a trisomic set of genes (triplo-21) results. Other children of these parents will receive the translocated 21, but will have only one separate 21. These children will be normal, but they can transmit mongolism to future generations. By all means, then, a couple who have had one child with Down's syndrome should have chromosome studies made to determine whether the abnormality is due to a translocation and, therefore, is transmissible to future children or other more distant descendants.

Non-disjunction of chromosome 21 is just as likely to result in a monosomic zygote as a trisomic zygote. The fact that no cases of monosomy have been reported indicates that such a zygote is probably not viable.

Non-disjunction can occur with any chromosome. Human trisomy has also been observed in the *D* group of chromosomes. This group includes chromosomes 13 to 15, which are so much alike in size and shape that it is difficult to tell them apart; hence, we call the abnormalities resulting from trisomy of one of these, the *D* syndrome. Persons with this syndrome are severely defective in many ways. The exact combination of traits varies, but the following may appear: mental retardation, incomplete development of the eyes, cleft lip and cleft palate, extra fingers or toes, deafness, and congenital heart defects. Another autosomal trisomy occurs in the *E* group of chromosomes, chromosomes 16 to 18, and is generally considered to involve chromosome 17. Again the afflicted persons have a variety of serious defects which includes the following: mental retardation, low-set and malformed ears, improper development of the lower jaw, spastic muscles, flexed fingers, and heart defects.

That a complete set of 23 syndromes is not found in man, each caused by trisomy of each of the chromosome pairs, as in *Datura,* is probably because the trisomic condition of most chromosomes is so upsetting to the genic balance as to cause zygotic or early embryonic death. The monosomic condition, as in most organisms, is regularly lethal.

Polyploidy

This term refers to the condition in which organisms have extra haploid sets of chromosomes. When there are four such sets, rather than the normal two, in the cells of somatic tissue, the cells are said to be **tetraploid** rather than the normal diploid. Such a condition may arise as a result of **abnormal mitosis.** Mitosis will start — there will be duplication of the chromosomes in the prophase and arrangement on the metaphase plate — but then, for some reason, the process stops. This leaves a cell with double the normal chromosome num-

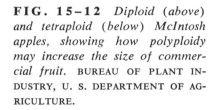

FIG. 15–12 *Diploid* (*above*) *and tetraploid* (*below*) *McIntosh apples, showing how polyploidy may increase the size of commercial fruit.* BUREAU OF PLANT IN-DUSTRY, U. S. DEPARTMENT OF AG-RICULTURE.

ber (tetraploid) at the time it goes into the interphase. Later the cell may undergo normal cell division and produce a group of cells which have double the normal chromosome number. If this occurs in the growing tip of a plant, a shoot may be produced which consists of cells that are exclusively tetraploid. This might be a branch of a tree which may be propagated by cuttings or through grafting to continue the tetraploid condition. Since the tetraploid is often larger and more vigorous than the diploid, it may be commercially valuable in the production of better fruit or other products. For instance, tetraploid grapes of the Portland and Fredonia varieties have been produced which are just about twice the size of the fruit borne on the diploid plants. Figure 15–12 shows the difference in the size of diploid and tetraploid McIntosh apples.

Many varieties of cultivated plants show various degrees of polyploidy when compared with wild forms. These include such plants as cranberries, blueberries, pears, cherries, and blackberries. No doubt some ancient ancestor of ours selected these varieties with chromosomal aberrations which occurred naturally but had no notion of the reason for their superiority. Today we do not have to wait for such changes to occur naturally — a method has been found to induce tetraploidy in plants. We will learn about it later in this chapter.

Polyploidy may also arise by **abnormal meiosis.** In the first division of meiosis the chromosomes may synapse and prepare for a normal reduction division, but fail to pull apart, so that one cell receives the entire diploid number and one receives no chromosomes at all. The latter type of cell dies in a short time, but the first type may go through a second meiotic division and produce diploid gametes. When such a gamete unites with a normal haploid gamete, a zygote is produced with three complete sets of chromosomes. This is called a **triploid.** The genes are in proper balance, but certain phenotypic effects result from this condition, just as from tetraploid organisms. When the time comes for reproduction, however, the triploids are sterile. The gametes usually receive two chromosomes of some types and only one of other types. When such gametes are used for fertilization, there is an unbalanced condition of the chromosomes which results in the early death of the zygote. Hence triploids among animals are more or less genetic freaks which cannot be maintained in a stable condition. In plants, however, triploids may be maintained through grafting, and by making cuttings and rooting them, or through propagation by bulbs and roots. Some triploid plants have commercial value and are maintained in these ways. The Baldwin apples and Keizer's kroon tulips are examples.

New Species Through Polyploidy

Whenever two diploid gametes unite, a **tetraploid** individual results. If this condition has arisen from a doubling of the chromosomes within one species **(autotetraploids),** the chromosomes will, to a certain extent, synapse in fours in meiosis. The irregular chiasma formation among the resulting eight chromatids causes an uneven distribution of the chromosomes to the gametes, and such tetraploids are, therefore, highly infertile. It is a different matter, however, when the tetraploids are produced in hybrids of different species **(allotetraploids).** In such cases the chromosomes synapse in normal pairs and tend to produce viable gametes and zygotes, whereas the diploid hybrid is typically sterile. Tetraploids of this nature have undoubtedly played an important part in the evolution of new species, especially in the plant kingdom.

An excellent example of this is furnished by the work carried out by a Russian geneticist, Karpechenko, before Lysenko came to power and ended scientific genetic research in Russia. He made intergeneric crosses between the radish, genus *Raphanus,* and the cabbage, genus *Brassica.* While these plants show considerable differences in their morphology, they each have a diploid chromosome number of 18 and can be crossed to yield a hybrid. The genes in the two are so different, however, that the hybrids are usually completely sterile because there can be no regular synapsis of chromosomes in meiosis. Karpechenko found that, out of a very large number of hybrids, there were a very few that were fertile. Cytological examinations of these fertile hybrids showed that they were tetraploid — they had thirty-six chromosomes. In these few plants there had been a doubling of chromosomes and this made it possible for normal synapsis to take place. This new form of plant bred true and when crossed back with either of its progenitors it gave highly infertile triploids. Thus, Karpechenko felt justified in calling this a new, artificially created genus, and he gave it the name *Raphanobrassica.*

There is much evidence to indicate that new species have arisen in this way naturally. As an example, there is a species of hemp nettle, *Galeopsis tetrahit,* which has thirty-two as the diploid chromosome number, but other species of this genus have only sixteen. It was postulated that this nettle arose as a tetraploid from a hybrid between two diploid species. To test this hypothesis, Müntzing selected two of the sixteen-chromosome species, which appeared most likely as progenitors of the tetraploid species, and crossed them together to obtain hybrids. The two species were *G. pubescens* and *G. speciosa.* From many hybrids he eventually found a tetraploid which was fertile and which proved to be indistinguishable from the original *G. tetrahit,* both in appearance and in crosses.

Thus, it appears rather certain that this is a case where man has duplicated the production of a new species which had previously taken place naturally.

Cytogenetic studies show that many of our cultivated plants are probably the result of such tetraploidy. The various species of cultivated cotton in America all seem to have arisen as tetraploids from hybrids of Old World cotton and New World cotton. Tobacco (*Nicotiana tabacum*) is also a fertile allotetraploid. Alfalfa appears to be a tetraploid from diploid species of clover. The dewberry is likewise an apparent tetraploid from raspberry hybrids.

Higher degrees of polyploids have been observed through continued doubling. In strawberries, for example, hexaploid (6n), octoploid (8n), decaploid (10n), and even 16-ploid (16n) types have been produced by methods outlined below.

Artificial Induction of Polyploidy

A number of chemicals have been found which can induce polyploidy in plants. The most efficient and most widely used of these is a quite poisonous chemical known as **colchicine,** which has long been used in minute quantities as a medicine for the treatment of gout, a disease characterized by a swelling of the legs and feet. This chemical, in proper concentration, will inhibit the formation of spindle fibers during mitosis, but will not inhibit the doubling of the chromosomes. Hence, when a normal diploid cell is so treated the chromosomes double, but when they are ready for the metaphase there is no functional spindle and they remain clustered as in the prophase — double chromosomes held together by single centromeres. The centromeres then duplicate and there is a tetraploid number of chromosomes in the cell. The cell does not divide, however, but instead returns to the interphase with its chromosomes doubled in number.

This effect can be demonstrated on an onion root tip. Place the tip in a solution of colchicine (about .03 per cent concentration) for several hours and then smear the tip in a nuclear stain such as aceto-carmine. If the tissue is then examined under the microscope, a very large number of prophases may be seen. The early stages of mitosis are not inhibited and cells continue to go through them, but they do not get beyond the metaphase. This also means that a great many more of the cells will be in mitosis than will be observed in untreated cells. There may be ten times as many cells in mitosis after treatment as before. If the treated cells remain on a living plant, they may recover from the effects of the colchicine and go into the interphase; but they will now have double the original chromosome number. These cells usually divide later by normal mitosis to yield more cells with the double number.

Colchicine is extremely valuable in horticulture, for plant tissues with double the normal chromosome number often yield products which are commercially superior. In practice, some colchicine is applied to the growing tip of a plant either in the form of a liquid or mixed with lanolin in a salve. The treated cells at the growing tip will die, but, when new tissue grows out from the cells just below the dead portion, it may contain tetraploid cells because of the

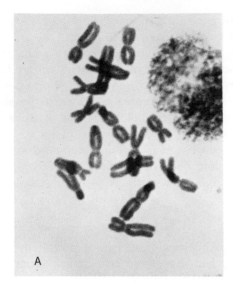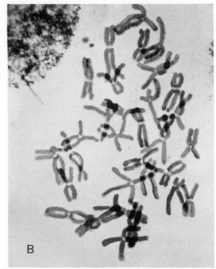

FIG. 15–13   *Effect of colchicine on the chromosomes of the onion root tip. A. The chromosomes of a diploid cell which has been treated. The chromosomes have doubled, but are held together by their centromeres, and mitosis has stopped at the metaphase. B. A tetraploid cell which has been treated. There are 28 rather than the normal 14 chromosomes and each has doubled, but they are still held together by the centromeres. This will result in a cell which is octoploid.* J. C. O'MARA IN JOURNAL OF HEREDITY.

effect of the colchicine. Cells of this new type can then be propagated through cuttings or grafts. It is also possible to treat the growing tips of plants having tissue of this new type and once again obtain a doubling of the chromosome number so as to get an octoploid; and a third treatment in some cases yields even a 16-ploid!

This technique is well adapted for use on plant tissues, since plants grow through the addition of new cells at the tips of their stems, but animals offer greater difficulties since they tend to grow in many different parts of the body at the same time. Chromosome doubling might be induced in various growing somatic cells, but it is generally not possible to propagate entire new animals from the altered portion of the body. It has been possible, however, to produce triploid rabbits by treating rabbit eggs with chemicals at the time of fertilization. Also, triploid and higher polyploid *Drosophila* have been produced by various treatments of the reproductive cells.

In conclusion, we can say that many chromosomal aberrations are distinctly harmful, but some have an important commercial value in cultivated plants and they appear to play an important part in the process of evolution through the origin of new species and through changes in the chromosome pattern within the species.

PROBLEMS

1. In *Drosophila* the genes for yellow body, miniature wings, and forked bristles are recessive to the wild type (gray body, long wings, normal bristles). These genes are sex-linked. A wild-type male is crossed to a female homozygous for these three genes. Among the offspring, the females are all of the wild type except one, which has miniature wings but a gray body and normal bristles. In the light of the facts studied in this chapter, suggest an explanation for this.

2. In maize the following genes are normally located on the second chromosome in this particular order: white sheath, glossy leaves, silkless, and chocolate pericarp. A certain strain was found, however, in which the order was: white sheath, silkless, glossy leaves, and chocolate pericarp. How would you explain this?

3. A second strain of maize was found in which the genes for silkless and chocolate pericarp were linked with a number of genes which are located on the third chromosome. Explain how this might have come about.

4. Cytological examination of meiosis of a certain maize plant shows that chromosomes 4 and 5 are involved in a pattern which includes a ring in the center. Explain this unusual synapsis.

5. In *Drosophila* yellow body and white eyes are recessive sex-linked genes. A gray-bodied, white-eyed male is crossed with a yellow-bodied, red-eyed female that has attached *X*-chromosomes. Show this cross by diagram and give the genotype and phenotype of the expected offspring.

6. If *Drosophila* super-females were fertile, what type of offspring would you expect when crossed with normal males?

7. When the cells of an onion root tip are treated with colchicine before being studied under the microscope, a great many more mitotic figures may be seen than in a root tip not so treated. Explain.

8. Just why is the hybrid between two species more likely to be fertile as a tetraploid than as a diploid.

9. Tetraploidy is more easily established in those plants which may be propagated by asexual means, such as budding or making cuttings. Why is this?

10. As an agricultural geneticist, suppose you are given the task of obtaining an improved species of squash, through induction of polyploidy, which can be propagated by seeds. Tell how you would proceed. (Note — you have several species of squash available to you for this undertaking).

11. A certain species of weed grows very abundantly in the midwestern plains. A careful examination of large numbers of these weeds shows that about one out of every 800 is a triploid. Explain how these might arise.

12. Will these triploids tend to upset the balance of the chromosome number in this species as they produce gametes with odd chromosome numbers? Explain.

13. In rare cases, tetraploid weeds are found among those described in the preceding problems. (Not weeds with only a part of the body tetraploid, but the entire plant tetraploid.) In the light of the frequency of the triploids, about how frequently would you expect these tetraploids?

14. Is there much chance that a new species might arise from these rare tetraploids? Explain your answer, in the light of facts brought out in this chapter.

15. In making preparations of human tissue for chromosome study colchicine is usually added to the culture several hours before the smear is made on a slide. Of what value do you think this treatment might be in the study of human chromosomes?

16. The cases of autosomal non-disjunction which have been found in man are limited to several of the smaller chromosomes. Does this mean that the larger chromosomes do not undergo non-disjunction? Explain.

17. Non-disjunction of the sex chromosomes in a woman can result in three types of sexual abnormalities and one lethal zygote. Non-disjunction of the sex chromosomes in a man, on the other hand, can result in only two types of sexual abnormalities. Explain.

18. Most persons with Down's syndrome have forty-seven chromosomes (including an extra number 21), but a few cases have been found with only the normal number of forty-six chromosomes. How can this be explained?

19. Some family pedigrees made of hundreds of persons include perhaps one or two persons with Down's syndrome. In other pedigrees the incidence is much higher; too much higher to be due to chance variation. How can this difference in the pedigree be explained?

20. It was once believed that Down's syndrome arose because of degenerative changes in the reproductive organs of older women who bore children. What observations led to this belief and how has modern knowledge dispelled this belief?

21. A young couple have just had their first child and it has Down's syndrome. They are anxious to know whether it would be advisable for them to have any other children. What could you tell them and what advice would you give them?

16

Giant Chromosomes

FOR OBVIOUS REASONS, the comparatively small size of chromosomes has been a great handicap to geneticists in their efforts to correlate genetic findings with cytological observations. Although *Drosophila* is almost ideal as an animal for use in making genetic crosses, its chromosomes were unusually small as they appeared in the gonad preparations which were used by geneticists for many years. Then, strange to say, it was discovered that some of the largest chromosomes known to exist may be found in *Drosophila,* though not in the reproductive cells. In 1931, B. P. Kaufmann made smear preparations of the salivary glands of *Drosophila* larvae and found that the cells of this tissue contained enormous chromosomes. They appeared as huge, worm-like bodies with alternating light and dark bands of various widths. Similar chromosomes had been reported by Balbiani in midge larvae as far back as 1881, and it is now known that all dipterans have them, but no one had yet realized the great significance of these structures. Dr. Kaufmann demonstrated his preparations at the annual meeting of the American Association for the Advancement of Science in New Orleans in late December, 1931. Present at that meeting was T. H. Painter, cytogeneticist at the University of Texas, who recognized the possibilities of these giant chromosomes as a tool for genetic study. Painter found that the bands on the chromosomes were constant in pattern and could be used as markers to help in the location of genes on the chromosomes by cytological methods. Since that time these chromosomes have been investigated extensively and much important genetic knowledge has resulted from the study of them.

STUDY OF SALIVARY GLAND CHROMOSOMES

Staining and "Smearing" Technique

The egg of *Drosophila melanogaster* is laid on moist food and, after a few days, hatches into a tiny, legless larva, or maggot, which crawls through the food, feeding almost constantly. This voracious appetite in the larva is accommodated by a relatively large pair of salivary glands. These glands are attached to its mouth and extend back into the body cavity for a distance equal to about one-third of the total body length. The glands can be studied most easily in the older larvae just before they transform into pupae, for these larvae are much larger than the tiny, newly hatched ones. The glands are dissected under the low power of a stereomicroscope. At this stage it can be seen that each gland is made up of about a hundred very large cells. The glands are then covered with a drop of acetocarmine or some similar stain, and a cover glass is applied. When examined at this stage of preparation, the chromosomes within the nuclei can be clearly seen, even though these cells do not undergo any division except in very young larvae. This is unusual, since chromosomes generally become so thin, as the matrix disappears and they uncoil in the telophase, that they are not discernible at all in the interphase.

We can learn more about these unusual chromosomes by use of the **smear technique.** When sufficient pressure is applied to the cover glass, the cell membranes burst and allow the chromosomes to float free in the crushed cell. We can now see the truly giant proportions of these chromosomes, for some of the arms are about 500 microns (0.5 mm.) in length, which is over a hundred times as long as the longest chromosomes of ordinary somatic or germ cells. These giant chromosomes have over a thousand times the volume of ordinary ones.

Appearance after Smearing

The stain affects certain parts of the chromosomes but not others. This gives the chromosomes a banded appearance, so that red bands of different

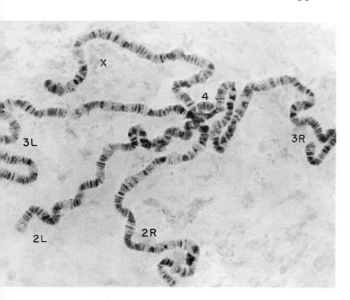

FIG. 16–1 *The chromosomes of a salivary gland cell of Drosophila after smearing and staining. The chromocenter, containing the centromeres and heterochromatin, can be seen in the center from which the euchromatin arms radiate. The numbers and letters identify the arms of the chromosomes.* BERWIND P. KAUFMANN, CARNEGIE INSTITUTION OF WASHINGTON.

FIG. 16–2 *A highly enlarged view of a portion of a cell from a salivary gland showing some details of the bands and other features of the chromosomes. This is a photograph of fresh, unstained chromosomes made with a phase contrast microscope.*

widths alternate with clear, nonstaining areas. With the chromosomes in this extended condition we can study them in more detail. All of them are attached to a mass of material called the **chromocenter** and extend outward from it. When we count these chromosomes, we note a puzzling thing. There are only six of them, five long ones and a very short one. How can we account for such an odd number as this when in this species of *Drosophila* eight is the characteristic diploid number? Only careful studies of the nature of these chromosomes will reveal the answer. First, let us reconsider the chromosomes of the more typical somatic cells of *Drosophila*.

Comparison with Typical Cells

There is one pair of very short rod-like chromosomes (designated number 4) which appear almost like dots. Then there are two pairs of very long chromosomes which have a V-shaped appearance in most preparations because the centromeres are attached near the center of each one. (These are designated numbers 2 and 3.) The other pair of chromosomes are the sex chromosomes (number 1). In the female these appear as two rather long rods, because of the terminal attachment of the centromere. In the male there is one such rod and another rod with a hook on one end. The latter is the *Y*-chromosome, which bears a subterminal centromere attachment that determines its shape.

Now let us correlate all this with the findings in the salivary glands. The chromocenter to which all the chromosomes are attached represents the centromeres of all chromosomes and certain inert portions of the chromosomes which are to be found in the neighborhood of the centromeres. These inert parts of the chromosomes contain few or no genes and are often called the **heterochromatin,** in contrast to the **euchromatin,** which is the typical gene-bearing portion of each of the chromosomes. Since the centromere lies near the center of each V-shaped chromosome, two arms should extend out from the chromocenter for each V-shaped chromosome. The short fourth chromosomes have a centromere which is almost terminal, and so they should show two very short arms. The *X*-chromosomes have terminal centromeres, and hence in the female are expected to appear as two fairly long single arms. This would give a total of twelve arms, whereas only six are visible in each smeared nucleus. This can be explained by a closer study of the chromosomes. A very thin longitudinal line can be seen running down each of the arms in a smear from the female salivary

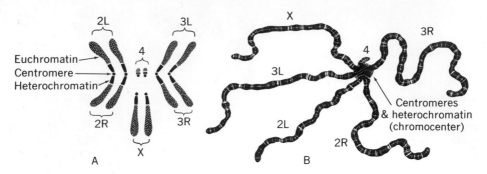

FIG. 16–3 *The relationship between the chromosomes of Drosophila melanogaster as seen in the metaphase of dividing cells (A), and the salivary glands of the larvae (B). The centromeres and heterochromatin of the chromosomes become incorporated into a chromocenter in the salivary gland cells and there is a pairing of the homologous portions of the euchromatin to give the six arms which are seen in the smears.*

gland cell. This suggests that each arm is double. Sometimes excessive pressure on the cover glass will cause the chromosomes to separate for a short distance along this line, affording further evidence in support of this supposition. This condition is brought about by somatic synapsis between homologous chromosomes. In Chapter 4 it was pointed out that like chromosomes bear a strong attraction for one another and that this causes them to pair (synapse) in the prophase of the first meiotic division. In most somatic tissue this attraction is quite weak, although homologous chromosomes sometimes lie side by side; but in *Drosophila* (and other dipteran larvae) the chromosomes do pair completely in the salivary glands and in the cells of some other somatic tissues. This explains the six elements seen in the cells of the salivary glands. Actually, there are only four pairs of chromosomes closely joined together, but two of the four pairs have two arms each. Sometimes the connections between the arms of the V-shaped autosomes can be seen through the chromocenter.

A male salivary gland smear will look very much like one from a female at first glance, and one may therefore wonder what has become of the *Y*-chromosome. Careful study will show that one of the long chromosome arms is only about one-half the thickness of the others. This is the *X*-chromosome; the male has only one of these — it is not paired like the other chromosomes in the cell. The *Y*-chromosome, being composed almost entirely of inert material (heterochromatin), is represented by just a few bands near the chromocenter. Thus we can understand the relationship between the six giant chromosome arms seen in salivary gland preparations and the eight actual chromosomes which compose them. We must examine further the formation of these chromosomes to learn why they are of such gigantic size.

FORMATION OF GIANT CHROMOSOMES

When ordinary somatic cells are not undergoing mitosis, the chromosomes consist of long and very thin gene strings or chromonemata. In the events leading

up to cell division the genes and the chromonemata which bear them are duplicated. The chromonemata then shorten by coiling, develop a matrix around
themselves, and become the typical sausage-shaped chromosomes of midmitosis.
This is followed by cell division, which separates the duplicated chromonemata
into two new cells. Thus there is one duplication of genes and chromonemata
for each division of a parent cell. Now let us compare this with the condition
in the salivary gland cells of *Drosophila*. When first formed, these cells contain
long, thin chromonemata just like other somatic cells, but the chromonemata become paired with their homologues as in cells preparing to undergo meiosis.
These then become duplicated to form four strands, again as if in preparation
for meiosis, but the process goes no further — there is no shortening of the
chromosomes through coiling and no matrix develops. There are now four
chromonemata representing each pair of chromosomes, but these paired, double
chromosomes remain extended, and there is no cell division. Some time later
there will be a second duplication of each of these chromonemata, making a
bundle of eight for each one of the paired chromosomes. A third duplication
increases this to sixteen, and so on. This continued duplication causes a thick-
ening of the chromosomes without any reduction in their length. There seem to
be several hundred chromonemata in the paired salivary gland chromosomes.
Figure 16–4 illustrates this process of formation.

FIG. 16–4 *Schematic representation of the possible method of formation of a salivary gland chromosome. First, there is somatic pairing of
the uncoiled chromonemata, followed by duplication of these chromonemata.*

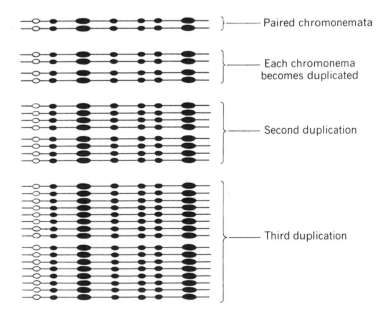

Cause of Banding

We may now seek an explanation for the banded nature of the giant chromosomes. We know from numerous studies of a variety of organisms that gene strings (chromonemata) are composed of the nucleic acid, DNA, closely associated with complex proteins. In certain parts of the chromonemata there are greater quantities of DNA, as shown by the fact that these parts take up stains which are specific for DNA more readily than other parts. In the opinion of some cytogeneticists this is due to the fact that the DNA ladder is more tightly coiled in these regions of the chromonemata. For whatever reason, these areas of nucleic acid concentration stain more heavily than other areas with the nuclear stains which are commonly used, for among the substances present the nucleic acid has the greatest affinity for the stain. As a result, a single chromonema often appears like a string of beads, the beads being the heavily stained regions rich in DNA. Such beads are called **chromomeres.** As such, a chromonema duplicates itself a number of times in the salivary gland cells of *Drosophila* and other dipterans. These chromomeres adhere side by side and in perfect alignment with homologous chromomeres so as to form dark bands, while the areas between the bands of adhering chromomeres will appear light. The chromomeres vary greatly in size; thus the bands vary in width. This causes a variation in the width of the entire chromosome.

Relation of Bands to Genes

The question now arises, "Just what relation do such bands bear to the genes?" "Do the individual chromomeres which form the bands represent single genes?" Bridges made a very careful study of the *X*-chromosome and identified 1024 crossbands. Slizynski, of the University of Edinburgh, found 137 bands in the short fourth chromosome. A survey of the maps of all four chromosomes yields a total band count of 5072. In some cases, however, it appears that two adjacent bands are formed by single chromomeres, and, if this is true, the total number of chromomeres is perhaps somewhat less than the total number of bands. These figures agree very closely with Muller's estimate of the number of genes in *Drosophila*, based on other techniques. The bands seem to be formed by 3795 paired chromomeres. Muller has estimated that the haploid gene number in *Drosophila* is about 5000, a very close correlation with the number of chromomeres. This would give some basis for a tentative conclusion that each band represents the location of a single gene. This is not to say that the chromomere itself is the gene; however, it represents a heavy concentration of DNA, and we now recognize DNA as the stuff of which genes are made.

The variations in the size and nature of the bands give a distinctive appearance to every part of the salivary gland chromosomes. A person who has studied these chromosomes extensively can recognize even a small portion of any of the chromosomes, even though the portion may be detached from its normal position. It is easy to see how such a means of recognition can be of inestimable value in making chromosome maps and in studying chromosomal aberrations.

CHROMOSOME MAPPING WITH GIANT CHROMOSOMES

Study of Small Deletions

With such large and clearly marked chromosomes, it has been possible to establish very accurate cytological chromosome maps which have thrown much light on the relationships between the chromosomes and the genes. Most of this work has been conducted by a study of small deletions. We have learned in Chapter 15 that small portions of chromosomes may often be deleted as a result of certain abnormalities in meiosis. This number may be greatly increased by treatment of the flies with X-rays. Such a study would be made on females bearing a rather large number of recessive genes on one chromosome; let us say genes *abcdef*. Females homozygous for these genes are mated to wild-type males which have been treated with X-rays in an effort to increase the possible number of deletions. All the normal offspring will be of the wild type, because of the presence of the dominant genes from the males. There may be one, however, which will show perhaps two of the recessive genes, e.g., *c* and *d*. This would probably be due to the fact that the portion of the chromosome from the male which carried genes *C* and *D* has been lost and the two recessive alleles now show up in the absence of these dominant genes. About one-half of the offspring of any such abnormal fly will receive the chromosome with the deletion. It is therefore an easy matter to study this deletion in the salivary glands of those numerous larvae which descend from this fly and which show the deletion. Figure 16–5 shows how the deletion would look. The chromosomes will be perfectly paired except in the region concerned. Here the normal chromosome shows a loop, or buckle, corresponding to the deleted portion of the chromosome with which it is paired. We can conclude that genes *c* and *d* lie somewhere in this loop. This gives us a rather rough location of them.

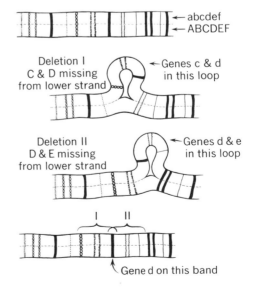

FIG. 16–5 *Locating genes on the salivary chromosomes through a combined study of breeding results and cytology. Breeding tests show that genes c and d are missing in deletion I which includes four bands of the chromosome. Tests also show that genes d and e are missing from deletion II. These two deletions overlap on only one band which must be the band which contains gene d.*

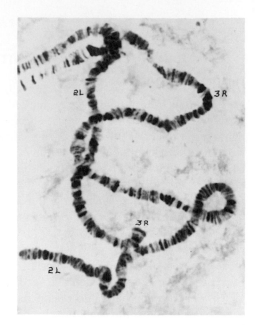

FIG. 16–6 *Pairing brought about by a reciprocal translocation between the left limb of the second chromosome and the right limb of the third chromosome. These may be identified by the numbers.* BERWIND P. KAUFMANN.

After extending our study, we may find a deletion which would allow the recessive genes *d* and *e* to show in a heterozygous individual. This would indicate a deletion which had removed the two dominant genes *D* and *E*. A salivary gland study will show that this deletion includes a different series of bands from the previous study. Figure 16–5 shows that it includes only one band which was present in the previous deletion. Since both deletions include the gene *d,* we can assume that this gene is associated with this particular band. In most cases, of course, we would have to study more than two deletions in order to narrow the gene down to a location on one band, but this illustrates the way in which it can be done. Extensive chromosome maps have been made by this technique, many of them by Bridges.

Other Aberrations

The salivary gland chromosomes are also of great value in studying chromosomal aberrations other than deletions. For many years geneticists speculated on the configurations which would be produced when a chromosome bearing an inversion or a translocation underwent synapsis with a normal chromosome. With the discovery of the giant chromosomes of *Drosophila,* it became possible to see such figures clearly, and they were found to show remarkable similarity to the theoretical figures which had been drawn before this discovery (see Figure 16–7).

A study of the bar-eye locus of *Drosophila* reveals some very interesting things about chromosomal aberrations. This is a sex-linked character which, when heterozygous, causes the eye to be elongated and bar-shaped. In homozygous females and in males, which carry only one *X*-chromosome, the eyes are still narrower. Hence it was formerly thought that bar-eye resulted from a gene intermediate in dominance. But cytological examination of the salivary gland

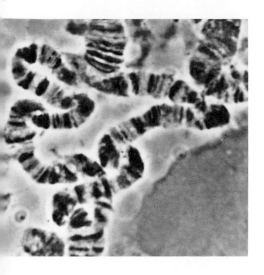

FIG. 16-7 *Appearance of an inversion in the paired salivary gland chromosomes of Drosophila. This photograph shows how the inverted portion of one chromosome is repelled by the normal portion of the other, thus giving an area of separation, but at the top and bottom of the picture there is a matching of the genes and the synapsed chromosomes appear as one.*

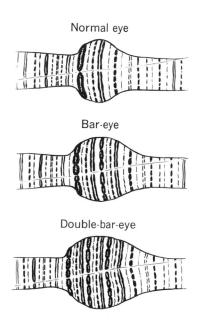

Normal eye

Bar-eye

Double-bar-eye

FIG. 16-8 *A small portion of the X-chromosome of Drosophila as seen in a salivary smear. It can be seen from this that bar-eye is actually due to a duplication of a small section of the chromosome carrying four or five bands. This is thought to be due to unequal crossing over. In double-bar, there is an additional duplication to give three sets of bands in this small region of the chromosome.*

chromosomes of heterozygous females has shown that bar-eye is produced by a duplication of a very small segment of the chromosome including only four or five bands (see Figure 16–8). In some cases crossing over will occur in homozygous bar-eyed females in such a way as to give a chromosome with three of these segments. In the male or in the homozygous female this chromosome produces an even more extreme bar-eye, for the eye is reduced to a tiny slit of eye tissue which is known as **ultra-bar.** This unequal crossing over will also result in a chromosome which has reverted to the normal condition (without this duplicated portion), and this type of chromosome will produce normal eyes. Hence, we can conclude that this is a case of position effect, where the surroundings of the gene influence its phenotypic expression, though the gene itself is not changed, for when it is returned to its normal surroundings it functions normally again. Various combinations of chromosomes further support this conclusion. A female which is homozygous for bar-eye will have two chromosomes with duplicated segments (four segments). A female which is heterozygous for ultra-bar-eye will have three segments in one chromosome and one in the other (again four segments). Still, the second female will have a narrower eye than the first. This again illustrates the possible effect of the position of genes on the chromosomes. The two females have exactly the same number of genes located in this region, but they are distributed differently.

INTERSPECIFIC HYBRIDS

There are a number of different species of *Drosophila,* and a study of the salivary gland chromosomes is often of great value in establishing possible phylogenetic relationships. For instance, *Drosophila pseudo-obscura* and *Drosophila miranda* are two species which are so similar in appearance that it is quite difficult to distinguish them. A comparative study of the oogonial chromosomes is of little help, for we find that both species have five pairs. Before the discovery of the giant chromosomes it was necessary to cross the two in order to demonstrate the species distinctions. When crossed, these two species produce hybrid offspring which are sterile. Why should they be sterile when the two parents seem to possess the same kind of genes — as far as can be told by their appearance — and the same number and kind of chromosomes, as evidenced by cytological preparations? Salivary gland preparations made from the tissue of the hybrids reveal the answer. Their chromosomes show all sorts of inversions and translocated segments which make it impossible for normal synapsis to take place; thus no viable gametes can be produced. It is reasonable to conclude that these two species have descended from a common ancestor, but, owing to a long period of separation, various chromosomal aberrations have brought about differences in the internal nature of the chromosomes which cause the hybrid to be sterile even though there are no superficial marks of these changes.

GIANT CHROMOSOMES IN OTHER ORGANISMS

The great genetic discoveries which have resulted from the thorough study of salivary gland chromosomes in the larvae of *Drosophila* and other dipteran insects have stimulated investigation of large chromosomes in other forms of life. Belling has found very long chromosomes with distinct chromomeres in the early prophase of meiosis of a lily, *Lilium paradalinum.* He has concluded that in this species there are about 2000 to 2500 chromomeres in a haploid set of chromosomes and is inclined to believe that there is an equal number of genes. Hence it is possible that each chromomere represents a single gene locus. In maize also the chromosomes are very long in the early prophase of meiosis and, like the salivary chromosomes of *Drosophila,* each chromosome may be identified cytologically by its length, the position of the centromere, and the arrangement of chromomeres of various sizes on the chromosome. The study of giant chromosomes has opened new vistas to the geneticist and holds promise of still further discoveries which as yet perhaps have not even been surmised.

PROBLEMS

1. The salivary gland chromosomes of dipterans are much longer than the chromosomes as seen in a metaphase plate of an oogonial cell, yet the number of gene loci in the two is the same. Explain.

2. The salivary gland chromosomes are also wider than the metaphase chromosomes of an oogonial cell. Explain.

3. There are eight chromosomes in the diploid cells of *Drosophila melanogaster,* yet salivary gland smears reveal only six chromosome-like bodies extending out from a chromocenter. Explain.

4. What parts of the chromosomes are found within the chromocenter?

5. What is the nature of the bar-eye "gene" as evidenced by the salivary gland studies?

6. Microscopic examination of a *Drosophila* salivary gland smear reveals a distinct buckle out to one side of a chromosome. How would you interpret this in terms of chromosome synapsis?

17

The Nature of Genes and Their Actions

GENES ARE OF VITAL CONCERN TO GENETICISTS since they are the ultimate units of heredity. Thus far in our study, we have learned that genes are extremely small in size, that they are made of a chemical material known as DNA, that they form an integral part of chromosomes, and that they can be located with great accuracy on the chromosomes of many species of plants and animals. Also, we have studied the effects which they bring about in the development of organisms. In recent years genetic research has been directed toward an understanding of the nature of genes and just how they bring about their effects. In this chapter we shall attempt to evaluate some of the discoveries which have been made.

THE NATURE OF A GENE

As our knowledge of DNA and its relationship to the chromosome has increased, one theory has been proposed that genes do not even exist as discrete particles, but that what we call a gene is only a part of a continuous chain of genetic material which forms a chromosome, and that what we have been calling gene mutations are, in reality, only small substitutions, deletions, and rearrangements within the chromosome. The fact that some genes change their phenotypic expression when translocated from one part of a chromosome to another (position effect) is one of a number of factors which lend support to this theory. Recent detailed studies of crossing over, mutation, and chromosome rearrangements, however, support the concept of the gene as a discrete unit of the chromosome, although a unit that is closely related chemically with its surroundings.

From the standpoint of evolution this would appear to be necessary. Only a limited amount of evolutionary change would be possible from the various translocations, deletions, and other rearrangements of parts of the chromosome, but mutations of particulate genes would provide the unlimited possibility of change which would be necessary if evolution on earth has been as extensive as the evidence leads us to believe.

The Physical Basis of the Gene

A chemical analysis of chromosomes shows that they are made of at least two kinds of protein closely associated with two kinds of nucleic acid, deoxyribonucleic acid (DNA) and ribonucleic acid (RNA). It was formerly postulated that the genes were made of this same combination of protein and nucleic acid, and that genetic diversity was due to the variations of the protein component. Large protein molecules may show an almost infinite variety in structure, and it was not supposed that all the many different kinds of genes could be explained by variation in the DNA. Each of the many different forms of life on the earth has its own peculiar complement of genes which number in the thousands in all but perhaps the very simplest living organisms. And, within each species, great variation among the genes confers distinctive characteristics upon different individuals within the species. We have seen how varied these can be in a study of human blood, which is just one part of the human body. Thus, it seemed that DNA could show comparatively little diversity in comparison with the protein molecules and, therefore, would not be a likely substance for the genetic material.

All of the evidence of recent genetic research, however, points to DNA as the primary substance of which genes are composed. Genetic diversity is possible because the DNA molecule is very long and capable of many different sequences of its four elementary components along its length. Before going further into the nature of the gene, however, let us explore some of the evidences that DNA is the actual material of which genes are made.

Evidence from Studies of Cell Division

When chromosomes are clearly seen in the early prophase of mitosis, they are already double in nature. Thus the actual gene duplication must take place some time before this state. Daniel Mazia of the University of California studied the DNA content of human cells in tissue culture. He found that the DNA quantity remained rather constant for about 8 hours after a cell had divided and then the quantity of DNA was doubled. About 5 hours after this duplication of DNA, the first evidence of a prophase of mitosis was visible under the microscope. The protein portion of the chromosome, on the other hand, showed considerable variation in quantity and no such regular doubling.

Evidence from Bacterial Transformation

Bacteria have the unique ability to take up genetic material from their surroundings and thereby become transformed in their genetic properties. One strain of pneumococcus, the bacterium which causes bacterial pneumonia, can be ground up finely and the remains added to the food upon which another strain is growing. Soon some of the living bacteria will show genetic traits which were a part of the bacteria that were destroyed in the grinding process and will pass on the altered traits to their offspring. Evidently the living bacteria have taken up some of the genes which were a part of the bacteria that were destroyed. It is not necessary to add the entire remains of the ground-up bacteria. An extract of the latter can be so treated that practically pure DNA is obtained. When this DNA is added to the medium in which the living bacteria are growing, the genetic transformation of some of the organisms occurs. But when, on the contrary, the DNA is destroyed (by the enzyme deoxyribonuclease), and proteins and various smaller molecules are not harmed, no transformation occurs. Thus, convincing evidence that DNA makes up the genes was obtained.

Typical pneumococci are encased in capsules and, as they are grown on a plate of solidified nutrient medium, they form colonies with a smooth, shiny surface. A frequent mutation in this organism, however, provides strains that do not form capsules and consequently makes the surface of the colony irregular and rough in appearance. The capsulated strain is highly virulent, for apparently the capsules help to protect the organisms from the defenses of the body, but the unencapsulated strain lacks the protective envelope and does not cause infection. The body can bring the latter under control before they multiply to the point of doing severe damage and causing pneumonia. Transformation can take place while the bacteria are growing in a living organism. Mice injected with the unencapsulated strain remain healthy, but, if they are also injected with the DNA from a capsulated strain, they will become sick and die. Capsulated bacteria can then be recovered from the dead mice. Therefore, it follows that some of the living bacteria within the body of the mouse must have taken up genes in the form of DNA from the capsulated strain and in turn have become capsulated and virulent.

The transformation of pneumococcus can also be demonstrated in culture dishes, as shown in Figure 17–1. DNA extracted from the smooth colonies of capsulated bacteria is incorporated into the food medium upon which bacteria from rough colonies have been placed. Some of the bacteria grow into smooth colonies, thus showing that transformation has been accomplished by the uptake of DNA. When this experiment was first done, some persons suggested that the change from rough to smooth might be due to mutations which would have taken place anyway in the "rough" cells without any DNA from the smooth bacteria being used. Studies on the frequency of such mutations, however, showed that the number of transformations obtained was many thousands of times greater than the number which could be expected from mutation. Thus, there is little room for doubt that the bacteria can actually take up DNA from their environment and incorporate it into their own genetic systems. This change, of course, would make the bacteria temporarily diploid for the genes taken up, since they already have a full set of genes, including alleles of all those genes

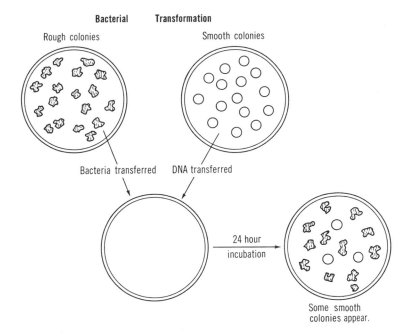

Bacterial   Transformation

Rough colonies   Smooth colonies

Bacteria transferred   DNA transferred

24 hour
incubation

Some smooth
colonies appear.

FIG. 17–1   *Bacterial transformation provides convincing evidence that DNA is the material making up the genes. When pneumococcus bacteria from a rough colony are grown on a nutrient medium containing DNA from a smooth colony, some smooth colonies will develop. The bacteria were transformed by absorbing DNA from the medium.*

differing in the external DNA and the recipient cell. The haploid state is ultimately restored, however, when the bacteria undergo divisions. Excess genes are gradually thrown off during the division process, and these genes can be either those newly acquired or those which were an original part of the recipient bacteria.

Evidence from Virus Research

Viruses are the smallest units of matter which may be said to be living. They can be seen only with the great magnification and resolution possible with the electron microscope. Since viruses exhibit detectable variations which are transmitted in a genetic manner, independent of each other, they too must have "genes" in their make-up. An individual virus particle is composed of an outer protein coat and a central core of nucleic acid. In most viruses this nucleic acid is DNA, but in some, especially those which infect higher plants, the nucleic acid is the closely related nucleic acid, ribonucleic acid. A typical virus of the latter type is the tobacco mosaic virus which infects tobacco plants. The protein coat of this virus is in the form of a cylinder with a hollow core that is filled with the nucleic acid. H. Fraenkel-Conrat of the University of California

FIG. 17-2 *Removal of protein overcoat from tobacco mosaic virus.
On the left is shown an intact virus particle; in the center the central core
of RNA can be seen partially extruded from the overcoat; at the right
is an end view of the overcoat showing the hole in the center after the
RNA has been removed.* ELECTRON PHOTOMICROGRAPH COURTESY R. G.
HART.

has been able to separate these two components by osmotic shock, as shown in
Figure 17–2. The parts can then be brought together again, and they will re-
combine to make complete virus particles which are fully as efficient in causing
infection as they were before their separation and subsequent recombination. It
was even possible to bring about a combination of the protein coat from one
strain and the nucleic acid core of another. The strains show some variation in
the degree of virulence, some of them causing quick infection which spreads
rapidly while other strains cause a low degree of infection which does not greatly
damage the tobacco plant. When the central core of nucleic acid from a highly
virulent type was combined with a protein coat from a virus of low virulence, a
highly virulent virus resulted. The reverse combination gave a virus of low
virulence. This again is evidence that it is the nucleic acid rather than the protein which determines the genetic properties of an organism.

Further evidence is revealed by the method of infection and reproduction of
a virus which infects the common bacterium found in the human colon, *Escherichia coli* (see Figure 17–3). A virus which infects bacteria is known as a **bac-**

FIG. 17-3 *The bacteriophage which infects the colon bacillus. The
picture at left shows the T_2 phage as normally seen with the electron-micro-
scope. After osmotic shock (at right) the DNA has escaped, leaving only
the protein overcoats or "ghosts."* FROM R. M. HERRIOTT IN THE CHEMICAL
BASIS OF HEREDITY, JOHNS HOPKINS PRESS.

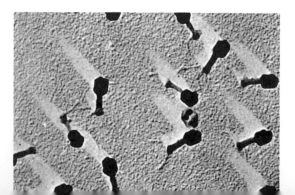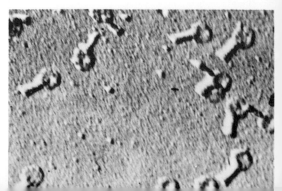

teriophage or, in shortened form, simply phage. Electron microscope studies show that this phage is shaped like a hexagon with a protruding tailpiece. Without cytoplasm or other parts found in cells, besides the chromosomes, this particle has no power to take in food, to grow, or to reproduce. However, whenever it comes in contact with the proper host, a colon bacillus, its tail becomes attached to the cell and it empties its DNA into the bacterium. The protein overcoat remains outside. We have clear evidence that DNA and not protein enters the bacterial cell from experiments with radioactive isotopes done by Hershey and Chase. Sulfur is found in protein but not in DNA. When radioactive sulfur was incorporated into the protein coat of the bacteriophage, it was found that the interior of an infected bacterium did not become radioactive. Hence, there could be no transfer of protein from the phage to the cell. On the other hand, phosphorus is a constituent of DNA and, when radioactive phosphorus was incorporated into the phage, the interior of the bacterium was found to be radioactive after infection. These experiments thus confirm the visual evidence that the protein coats of the phage remain on the outside of the bacteria after the latter have been infected.

When a bacterium has been infected, there is no evidence of the infection within the cell until about 15 to 30 minutes later when 50 to 100 or more complete new virus particles can be seen within the cell. The cell wall is then **lysed** (dissolved), and these particles escape. Each phage is now free to infect another bacterium and to repeat the entire process. The infecting DNA thus appears to be able to utilize the components of nucleic acid produced by the bacterium to replicate its own genes, and also produces the protein overcoat and the tail from the amino acids within the cell. The entire complex phage particle is then assembled. The evidence is clear that different genes of the injected DNA control separate steps in DNA synthesis, coat protein synthesis, tail synthesis, and assembly of the phage particle.

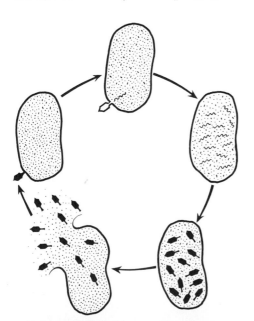

FIG. 17-4 *Cycle of bacteriophage infection. This shows a phage particle attached to the cell membrane of the bacterium on the left. Then the DNA is emptied into the cell, but the protein coat remains outside. There follows a vegetative stage in which the DNA multiplies and then becomes surrounded with new protein coats. Finally, the cell ruptures, releasing the phage particles which are free to infect other cells.*

Furthermore, this virus DNA shows properties very much like the genes of larger forms of life. Both virus DNA and the DNA of typical cells multiply only within living cells, and both produce specific effects on certain cells but not on others. The genes for melanin production in man are present in all cells of the body; yet these genes are active and function only in the pigment cells of the skin, iris of the eye, and roots of the hair. This condition can be compared with that produced by the genes of the virus which causes rabies. These genes function and cause degenerative changes only in certain cells of the nervous system. The virus which causes warts affects only the epithelial cells of the skin and imparts a characteristic effect.

Both the DNA of the viruses and the genes of cells retain their individuality through repeated multiplications; yet both do mutate in rare cases. It is the mutations in viruses which cause so much concern to the medical men of today. We have developed vaccines for the three common variants of the virus of polio and thus have made great progress toward the eradication of this disease; but, should a mutation for a new type appear, we might find that the present vaccines do not confer immunity toward this mutant form of the disease. The great influenza epidemic of 1918 appears to have started as a mutant form of a comparatively mild influenza virus. Fifteen million people died as a result of infection with this mutant form in one of the most deadly plagues in the history of mankind from the standpoint of the total number of people who died. In more recent times a new strain of influenza virus, one which originated in Asia, has swept over the world, but fortunately this mutation did not produce a very deadly form of the disease.

Another bit of evidence related to viruses is also pertinent. Whenever two phages with different characteristics infect the same bacterium, the phage progeny which are produced may show genetic recombinations of the characteristics of the two original strains. These results indicate that we are dealing with the same sort of genetic material in the viruses as are found in higher forms of life. The few viruses which have RNA as their nucleic acid rather than DNA, of course, are exceptions. RNA is not self-replicating as it exists in cellular forms of life; it is produced by the DNA of the cell. The method by which the RNA viruses achieve replication is still one of the unsolved problems of genetics at this time. We do know that such RNA replicates only when it is in cells containing DNA. Perhaps these viruses somehow have the power to call upon the DNA of the host cell to achieve replication of the virus RNA.

Evidence from the Amount of DNA in Interphase Cells

The amount of DNA is remarkably constant in the different cells of an animal or plant body, even though some of the cells may show rather great differences in size and degree of specialization. Sperms and eggs, on the other hand, have only about one-half as much DNA as the other body cells; this is in agreement with the fact that these gametes have only half as many genes as are found in somatic cells.

There is a major exception to this rule in the case of the very large salivary gland cells found in fly larvae. In mature *Drosophila* glands, for instance, the amount of DNA per cell is approximately 1,024 times greater than the amount found in haploid sperm or egg cells. Furthermore, if we study the amount of DNA in these cells from their earliest beginnings in the embryo, we find that they start with the normal diploid quantity, but, as the cells grow and there is repeated duplication of the gene strings, the quantity of DNA increases in step-wise fashion from 2 to 4 to 8 to 16 to 32, etc. There is no such regular and stepwise increase in the proteins. This stepwise increase in the amount of DNA not only furnishes substantial evidence that DNA is the genetic material, but also is a good indication of the number of single gene strands which combine to constitute the salivary gland chromosomes.

The Chemical Nature of the Genes

Since there is overwhelming evidence that DNA is the material of which genes are made, let us learn something more about the chemical nature of this nucleic acid that plays such an important role in the life of all organisms. We know that the DNA molecule has a very high molecular weight and is made up of combinations of simpler organic compounds. One is a **pentose sugar** (a sugar with five carbon atoms) which is always linked to an inorganic phosphate (that is, the sugar is phosphorylated). The sugar in DNA is **deoxyribose** while that of RNA is the very closely related **ribose.** The chemical structure of these two sugars is given in Figure 17–5 in their phosphorylated form.

The second sort of organic substance which goes into the formation of nucleic acids is a variety of ring-compound bases known as **purines** and **pyrimidines.** The purines in DNA are **adenine** and **guanine;** the pyrimidines are **thymine** and **cytosine.** The exact way in which these component parts are assembled into the large DNA molecule has been the object of intensive research.

FIG. 17–5 *Structures of deoxyribose phosphate and ribose phosphate.*

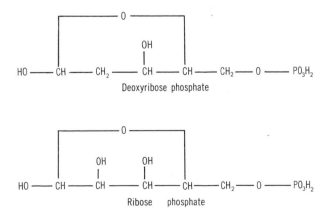

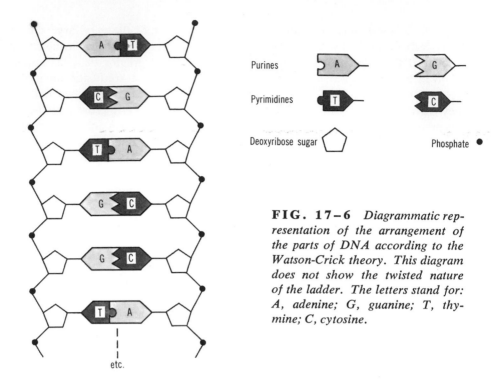

Purines

Pyrimidines

Deoxyribose sugar

Phosphate

FIG. 17–6 *Diagrammatic representation of the arrangement of the parts of DNA according to the Watson-Crick theory. This diagram does not show the twisted nature of the ladder. The letters stand for: A, adenine; G, guanine; T, thymine; C, cytosine.*

This is very important, for it is here that genetic diversity must exist, since the sugar-phosphate-sugar-phosphate backbone of the DNA molecule is everywhere the same. Consequently, the differences between all forms of life on earth must lie in the different sequences in which these bases are arranged. We might compare it to the great variety of words which are possible through different arrangements of letters, or to the Morse code which, by using only two symbols — the dot and the dash — can spell out any word in any language.

A theory to explain this structure in a way which accounts for all the known properties of the DNA molecule has been proposed by Watson and Crick. According to their theory, the molecule consists of a very long double helix (two strands twisted about one another much as two pieces of string would be when put together and twisted). The two helices consist of deoxyribose sugar molecules connected by their phosphate bonds to form the long strings. From these strings the purine and pyrimidine bases project inward, and the two strings are held together by bonds between these projections. These are only weak hydrogen bonds, but, though very weak, they are sufficiently numerous to hold the helix in shape. Also, according to this theory, a gene would consist of a segment of this double helix. An analysis of DNA obtained from different organisms shows that the percentages of the four base substances vary considerably, yet the number of adenine groups is always equal to the number of thymine groups and the number of guanine groups is always equal to the number of cytosine groups. This is explained by the fact that, whenever there is a projection of adenine

280

FIG. 17-7 *Structural chemical formulas of two of the bases
which make up a single nucleotide
pair of the DNA molecule. This
shows the two-ring purine, adenine,
joined with the single-ring pyrimidine, thymine, by means of weak
hydrogen bonds. The pentagons
labeled D are deoxyribose sugar
molecules.*

from one strand, it will fit sterically only with a thymine projecting from the
partner strand. The same is true of guanine and cytosine.

This theory may be carried a step further in an attempt to explain the
method of gene duplication. When time for replication comes, the two strands
of the double helix break apart at their weak hydrogen bonds and unwind into
single strands, not greatly different from a zipper becoming unzipped. Then
each single strand of DNA attracts to itself complementary parts from the store-
house of chemicals within the cell. Adenine will attract to itself a **nucleotide** consisting of thymine and a deoxyribose sugar molecule together with an attached
phosphate group. (A nucleotide is a sugar molecule with its phosphate group
together with its attached purine or pyrimidine base.) Guanine attracts cytosine,
etc. Thus, two complete strands are formed, and the gene duplicates itself.

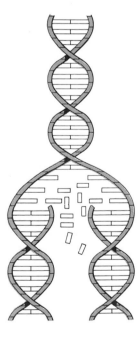

FIG. 17-8 *Diagram of the
method of gene duplication according to the Watson-Crick theory.
The two strands become separated
at the weak hydrogen bonds which
join the purines and pyrimidines.
Each strand then picks up the missing parts from the surroundings and
two complete helices are formed.*

The Fine Structure of a Gene

Now that we have discovered that DNA is the genetic material, the question arises as to the relationship of the genes to the long double coils of DNA. How much of the helix makes up a gene? Our classical concept of the gene has held it to be the smallest unit which can be recognized for recombination, mutation, and function. Modern research, however, has shown that these are not necessarily the same units. Pseudoalleles (see Chapter 12) show us that recombinations can take place within a region which we have called a gene because it is a single functioning unit. In microbial genetics it has been possible to show that the units of recombination, mutation, and function are different. Three terms have been coined to designate these three units where such distinctions are necessary.

A **recon** is the smallest element which is interchangeable through genetic recombination. Extremely delicate studies of recombination in microbes indicates that a recon consists of not more than two pairs of nucleotides, maybe only one.

A **muton** is the smallest part which, when altered, can give rise to a mutation. We have already learned that an alteration of a single nucleotide pair can result in a mutation.

A **cistron** is the term applied to a functional unit and conforms more closely to what we commonly think of as a gene. Careful study indicates that there can be over a hundred points within a functional unit wherein a mutation can take place and cause a detectable phenotypic effect. This means that a cistron is over a hundred nucleotide pairs in length and there is good evidence that some cistrons may be as long as 30,000 nucleotide pairs.

With this information, it is easy to understand how pseudoalleles are merely recombinations of recons within cistrons and how multiple alleles are easily possible whenever changes occur in different mutons within the same cistron. These new discoveries and new terms do not mean that we will henceforth abandon the use of the term "gene," but rather that we now understand the gene structure much better than previously. In general, we refer to the functional unit (cistron) when we use the word "gene" without qualification.

Artificial Synthesis of DNA

The formation of nucleotides from simpler compounds and the assemblage of the nucleotides into DNA are carried out by certain enzymes in the presence of energy sources within the cells. A number of workers have recently reported laboratory synthesis of nucleotides, and Arthur Kornberg has reported the actual synthesis of DNA from nucleotides. He was able to extract and purify two enzymes from the bacterium, *E. coli,* which are a necessary part of the synthesis within the cell. When these two enzymes were mixed with the nucleotides, together with a small amount of DNA as a "primer" and ATP (adenosinetriphosphate) as an energy source, he obtained DNA. The next step in this research might well be the synthesis of DNA which can be incorporated into a living system through transformation. Such an accomplishment would actually mean that man had achieved the creation of living matter from non-living chemical matter.

GENES IN ACTION

Now that we have some knowledge of the chemical nature of genes, let us turn our attention to a consideration of the ways in which they operate to bring about the phenotypic effects. For instance, why should a person with two genes for albinism fail to develop pigment in his skin or hair, or the irises of his eyes, whereas another person with one or two dominant alleles of this gene will have normal pigmentation? What is there about the dominant allele which causes the formation of the pigment? How can one gene produce a person with normal mentality, while a variant of this same gene produces a person who is mentally defective? Until recent years geneticists were content to learn what effects the genes produce without great concern about the way in which these effects are brought about. Learning such facts is somewhat like learning which trains from the Grand Central Station in New York City arrive at specific cities in the United States. We may learn, for instance, that if we board a certain train and stay on it long enough, we will eventually arrive in Miami, Florida. This is very nice to know, especially if we should ever want to go to Miami; but it is far from the whole story, and if we travel much, we will soon want to know more. What states does the train pass through? What kind of power draws the train? What switches must be thrown to insure that we arrive in Miami rather than Chicago, Indianapolis, or Atlanta? The answers to these and other questions are vital to one who would like to know the whole picture. Likewise the geneticist soon wants to know more than what genes cause what effects. Our genetic knowledge of the ultimate result of the presence of genes is now very extensive. Furthermore, recent genetic research has begun to give us some insight into the mode of action of the genes in the process of producing these results.

Eye Color in Drosophila

Many, if not most, genes exert their influence by means of enzymes whose specificity they control. In *Drosophila,* for instance, the work of Beadle and others indicates the presence of substances which are concerned with the development of eye pigments. The eye color of the normal wild-type *Drosophila* is a particular shade of red which is produced by a blending of an orange-red and a brown pigment in the facets of the eye. A different substance is necessary for the production of each of these pigments. Whenever flies are homozygous for a certain recessive gene, **vermilion,** one of these substances, the one which forms the brown pigment, is not produced. As a result, the eye is orange-red, since there is no blending with the brown.

This concept can be tested by **transplantation studies.** It is possible to remove pieces of tissues from the larvae of *Drosophila* and transplant these into the body cavity of other larvae where they will grow. A bit of tissue destined to produce an eye may be removed from a larva carrying the dominant normal

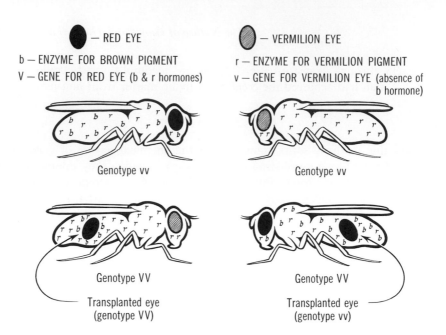

— RED EYE   — VERMILION EYE

b — ENZYME FOR BROWN PIGMENT   r — ENZYME FOR VERMILION PIGMENT
V — GENE FOR RED EYE (b & r hormones)   v — GENE FOR VERMILION EYE (absence of
                                                              b hormone)

Genotype vv   Genotype vv

Genotype VV   Genotype VV

Transplanted eye   Transplanted eye
(genotype VV)   (genotype vv)

FIG. 17–9 *Eye-color hormones in Drosophila. The wild-type red eye is produced when enzymes for both brown and orange-red pigments are present. Vermilion eyes result from a gene which fails to produce the enzyme for the brown pigment. When an eye is transplanted from a fly with red genotype into one with vermilion genotype, the eye is still red because the genes in this region produce both enzymes. When an eye with vermilion genotype is transplanted into a fly with red genotype, the eye becomes red because of the presence of the enzyme for brown pigment in the body into which it is transplanted.*

genes and transplanted into the body cavity of a larva which is homozygous for vermilion eyes. After metamorphosis this insect will have vermilion eyes on its head as expected, but when the body cavity is opened, a wild-type red eye will be found inside it. This result indicates that the cells which came from the first larva produce all that is necessary for the formation of the brown pigment in spite of the fact that it was growing within a body which contained only cells without such a capacity. Now if the procedure is reversed and the eye bud from a larva carrying two vermilion genes is transplanted into the body cavity of a larva carrying the wild-type genes, an interesting result is obtained. The adult which comes from this larva has the normal wild-type red eyes, of course, but when we open the body cavity we find the same wild-type red eye inside. This striking event bears out our concept. The cells of the wild-type larvae must secrete a diffusible substance which penetrates the transplant and permits the production of the brown pigment, even though the cells of the transplant itself do not have the power to produce such a substance.

The many variations in the eye color of *Drosophila* result from variations in the quantity of one or both of the two pigments described, variations in acidity which alter the color of the red pigment, and various degrees of oxidation or reduction of the brown pigment which cause changes in its color. Enzymes, under the control of genes, govern these various reactions.

Genes and Enzymes in Neurospora

Much of our knowledge of the biochemical action of genes has come from studies of the mold, *Neurospora*. This mold has the ability to live on a culture medium with only the barest of nutrients. A **minimal medium** containing a few simple salts, sugar, ammonia, and one of the B vitamins, biotin, is all that is necessary for the wild type of this mold to live. Does this mean that *Neurospora* does not need any amino acids or any of the other vitamins? Not at all; an analysis of this mold shows that it contains about twenty amino acids, all of the vitamins that are needed in human nutrition, and the purines and pyrimidines needed for construction of DNA and RNA. Since these substances are not provided in the minimal medium, it is obvious that the mold must synthesize each of these substances from the materials that are present in the minimal medium. This synthesis is brought about by enzymes.

We can learn about these enzymes and how they function through studies of what happens in their absence. If a mutation occurs in a gene which produces an enzyme necessary for the synthesis of the amino acid arginine, we find that arginine cannot be produced and the mold cannot grow on a minimal medium. The medium must be supplemented with arginine before this mutant form of the mold can grow. The detection of such mutations is relatively simple. The mold can be treated with X-rays to increase the mutation rate and then, when spores are formed, each spore is placed in a separate tube on a medium containing arginine. After the growth has become established, transfers can be made to tubes with media not containing arginine. All transfers which do not grow represent a mutation of some gene which produces an enzyme needed for arginine production.

Further study showed that not one but at least seven enzymes are involved, each produced by a separate gene. These are shown in Figure 17–10. Note that one gene produces each enzyme of the series and a mutation of any one of these genes will prevent arginine synthesis. We can tell just where the mutation has occurred by placing the mold on media with different intermediate products. If the mold grows on a medium with citrulline, we know that gene 7 is not the one which has mutated, for the enzyme which converts citrulline into arginine is still

FIG. 17–10 *How Neurospora converts sugar and ammonia into the amino acid arginine through a series of enzyme-stimulated reactions. Each enzyme is produced through the action of a single gene and a mutation of any one of the genes in the series can block the formation of arginine.*

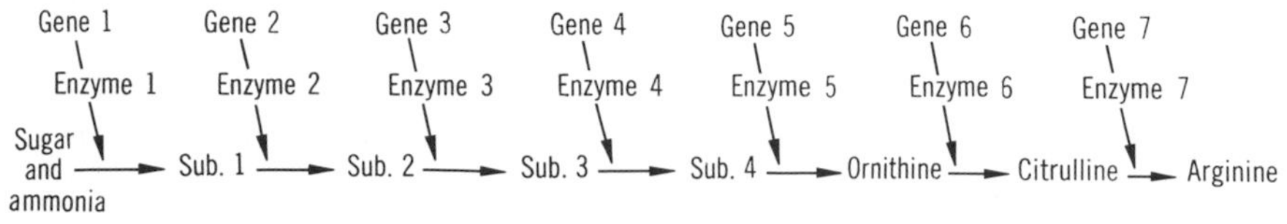

present. If the mold grows on ornithine, but not substance 4, we know that the mutation is in gene number 5. A similar chain of reactions is necessary for each of the amino acids and the vitamins which are synthesized. Thus, it is apparent that the number of genes and enzymes involved in the synthesis of the many products in the cell must be very great.

Does the mutation of a gene actually prevent the formation of an enzyme, or does it only change the nature of the enzyme so that it is still present, but so altered that it no longer accomplishes its former function? This question has been answered by ingenious techniques involving antigen-antibody reactions. One of the enzymes produced by the wild type of *Neurospora* is known as **tryptophan synthetase.** This enzyme functions in the production of the amino acid, **tryptophan,** from certain intermediate products synthesized by other enzymes. A mutant strain of the mold lacks the ability to make this synthesis and must live on a medium containing tryptophan. Purified tryptophan synthetase can be extracted from the wild-type *Neurospora* and injected into a rabbit. The rabbit responds by producing antibodies which are specific for this enzyme. When blood serum from such a rabbit is mixed with an enzyme extracted from the tryptophan-requiring strain of *Neurospora,* there will be an antigen-antibody reaction in some cases. This shows that a protein is present which is very similar to tryptophan synthetase, because it reacts to the antibody in the same way, yet it must be slightly altered because it cannot synthesize tryptophan. In other cases, however, there is no antigen-antibody reaction, for the mutation has so altered the gene that it no longer produces any protein at all, or at least none with the same antigenic properties as tryptophan synthetase. These results indicate that there can be more than one kind of mutation of a gene to result in a tryptophan-requiring strain. One is a very slight change, but there is at least one other kind of change that causes a greatly altered product or perhaps no product at all.

Gene Action in Man

Many people believe that the origin of the concept of enzymes carrying out the directions of the genes began with the work on *Neurospora.* Actually, the concept began with studies of human heredity. In 1909, an English physician, Archibald E. Garrod, published a book called *Inborn Errors of Metabolism.* In this he described various physiological abnormalities of man that appeared to be inherited. He proposed the hypothesis that some of these abnormalities developed because of the absence of specific enzymes which were present in normal persons. His hypothesis went further to propose that genes produce enzymes and that mutant forms of the genes do not produce the enzymes. The work of Garrod was somewhat like that of Mendel; the scientific world was not ready for his discoveries and they went largely unnoticed. Geneticists of Garrod's time were more interested in studying genetic ratios and in determining gene locations. Few of them were interested in conducting studies of the ways in which genes work. However, with the research of George Beadle and E. L. Tatum on the enzymes of *Neurospora* beginning in 1941 (for which a Nobel prize was awarded in 1958), interest in genes and enzymes developed and this area became a popular field of investigation. It was soon found that there were chains of enzyme-mediated reactions in man as there are in the mold. One of these which has been studied most thoroughly involves the breakdown of the amino acid, **phenylalanine.** This amino acid is a part of almost all protein foods we eat. Digestion breaks down the proteins into their component amino acids and frees the phenylalanine in the digestive tract. It is absorbed, along with the other soluble food products, and distributed over the body. It can pass into cells of the body by means of diffusion and other means of transport across the cell membranes.

Phenylalanine Pathways

Once it is in the cell the phenylalanine may follow one of three courses, depending upon which enzyme acts upon it. It may be converted into cell protein; it may be converted into another amino acid, **tyrosine;** or it may be converted into phenylpyruvic acid. Figure 17–11 shows the three pathways open to phenylalanine. This diagram also shows where various steps in the chain of reactions

FIG. 17–11 *The chains of reactions involving the amino acids phenylalanine and tyrosine in man. Each arrow indicates a change brought about by an enzyme. Letters within the arrows indicate places where genes are known which can break the chain of reactions at these points. See text for further details.*

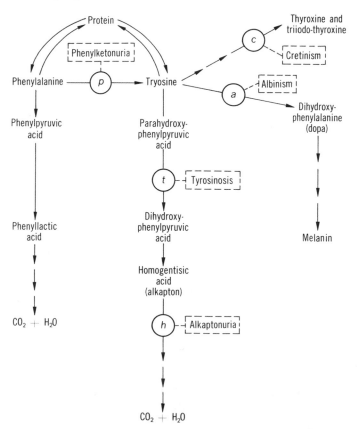

may be interrupted by mutations which interfere with the production of certain enzymes. The recessive gene *p,* for instance, when homozygous, causes a failure in the production or functioning of the enzyme which converts phenylalanine into tyrosine. Denied this outlet, the phenylalanine tends to accumulate in abnormal amounts and there is also an abnormal amount of phenylpyruvic acid, since some of the excess phenylalanine will be converted into this compound. The excess of these two substances diffuses into the blood, and some of it will be excreted in the urine where it can be detected by simple chemical tests. Persons with this characteristic are said to have **phenylketonuria,** frequently abbreviated to PKU. Whereas normal persons will have a plasma phenylalanine level of 1 to 2 mg. per 100 ml., a person with PKU will have 15 to 63 mg. per 100 ml. The urine of a normal person may have as much as 30 mg. per 100 ml., whereas a person with PKU will have 300 to 1000 mg. per 100 ml. More important, there is great mental retardation among those who have PKU. There is some variation in mental level, possibly because of the presence of other genes influencing mentality, just as there is variation in the mental level of the general population. About 64 per cent of those with PKU will have an I.Q. below 20, the idiot level; about 96 per cent are below 50, the imbecile level; and a bare 4 per cent range between 50 and 70 on the I.Q. scale. It seems that unusually high quantities of phenylalanine in the body tissues interfere with the development of the brain and cause the mental retardation.

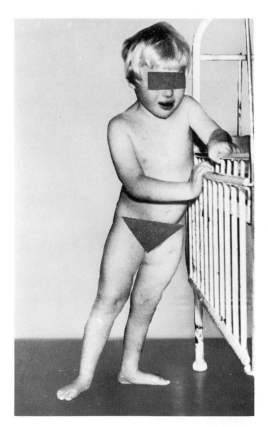

FIG. 17–12 *A child with phenylketonuria (PKU). Not only is the child mentally retarded, but shows structural weakness in the feet and ankles which makes standing difficult. Note the very fair hair which often accompanies this affliction.* COURTESY CARL A. LARSON, UNIVERSITY OF LUND.

The discovery of the nature of phenylalanine metabolism has led to the finding of a way to prevent much of the effect of PKU on the brain. It is possible to remove most of the phenylalanine from protein food by first treating the proteins with enzymes which break the proteins down into their constituent amino acids. The phenylalanine can then be extracted and destroyed and the child can be fed the residue of amino acids from the protein with only a small amount of phenylalanine included. They must have this small amount since phenylalanine is an amino acid essential to the building of cellular proteins. If babies are put on a diet very low in this amino acid when they first begin showing symptoms of PKU, the brain will develop more normally and the person will grow up with a mentality lying within the range of normalcy. It appears that the damage is done during the first 5 years of life, so the diet can be discontinued at about 5 years of age, since the parts of the brain are fully formed by that age. PKU is comparatively rare — only about one baby out of each 25,000 born in the United States is homozygous *pp*, yet the results of untreated cases are so tragic that many physicians advise that all babies be given the test for PKU. Formerly this was done by testing the urine for phenylketones which come from the improperly metabolized phenylalanine. Unfortunately, however, the test does not show a positive until several weeks after birth, and by this time the brain may already have suffered partial damage. It is estimated that 5 I.Q. points will be lost for each 10 weeks that treatment is delayed. A more recent test, the Guthrie test, is effective on the newborn baby. This test is based upon the principle that inhibition of growth of the bacterium, *Bacillus subtilis,* by thienylalanine in a minimal culture medium is prevented by the addition of phenylalanine to the medium. The bacteria are placed on the medium containing the inhibitor. Then small filter paper discs impregnated with blood from a skin puncture are placed on the medium along with control discs. If the blood is rich in phenylalanine there will be an abundant growth around the discs containing the blood as well as around the control discs which contain a high concentration of phenylalanine. The phenylalanine has prevented the inhibitor from functioning and the bacteria grow. Normal blood will not have sufficient phenylalanine in it to allow growth to occur. If the Guthrie test shows a presumptive positive, a sample of the baby's blood is removed and subjected to a quantitative blood assay to be sure that the baby has PKU before the diet is started.

Further possibilities for the control of mental retardation because of PKU now exist. It is now possible to detect the carriers of the gene because of a small heterozygous effect. To make this test an excess of phenylalanine is given to a person and, after several hours, the plasma is checked for the level of both phenylalanine and tyrosine. The phenylalanine level will be high and the tyrosine level will be low in the persons who are carriers of the gene for PKU. Thus, we see that it is not a completely recessive gene. Heterozygous persons produce less of the enzyme needed to convert phenylalanine into tyrosine than persons who are homozygous for the normal allele. It appears as if the lowered enzyme production of the heterozygotes is sufficient to take care of the normal intake of phenylalanine in the diet and they experience no difficulty with an excess of the amino acid in the blood. The body produces the enzyme with a factor of excess of at least two when the person is homozygous normal. If a young couple find that they both have PKU in their relatives they may wish to take the test, and if they find that both of them are carriers of the gene for PKU they may wish to adopt children rather than run the risk of having the abnormality in some of their children. This could reduce the incidence of the genes in the population, but this might be counterbalanced by those with PKU who have developed normally because of the special diet early in life and who have children. Every person who is normal owing to eating the special diet should be made to understand that his genes have not been changed and that he is a homozygous carrier.

Tyrosine is the second substance in the enzyme series that we shall consider. This may be produced from phenylalanine or it may be obtained directly from the digested proteins which are eaten. There are four pathways which tyrosine may take, depending upon which enzymes act upon it. Tyrosine is an amino acid and it may become a part of the cellular proteins. Second, tyrosine may be combined with iodine in the thyroid gland to produce thyroid hormones, **thyroxine** and **triodothyronine.** These hormones are important as regulators of body metabolism and are also necessary for normal mental and physical development. A pair of recessive genes, *cc,* results in a failure to produce the necessary enzymes for normal production of the thyroid hormones and the condition known as **genetic goitrous cretinism** develops. This is characterized by severe physical and mental retardation and a hypertrophy (overgrowth) of the thyroid gland. This is another inherited human affliction which can be prevented because of our knowledge of its cause. If thyroid hormones are administered to a baby as soon as it shows signs of cretinism, and the administration continued throughout life, a person with normal physical and mental development will result.

The third pathway open to tyrosine leads to the formation of **dihydroxyphenylalanine,** which in turn leads to the formation of **melanin** in a series of steps. Melanin is the pigment which is found in varying quantities in the skin, hair, and iris of the eyes of the great majority of people. The presence of a pair of recessive genes, *aa,* however, blocks the production of the enzyme needed for the conversion of tyrosine into dihydroxyphenylalanine and, in the absence of this precursor, no melanin is produced. The condition known as **albinism** then results. So far, no way has been found to prevent or correct this trait. You might wonder about the feasibility of injecting the missing enzyme, but you must remember that we are dealing with enzymes within the cells in all of this series. Enzymes are proteins and proteins cannot pass into cells; the plasma membrane will not admit protein particles. If the enzyme is broken down into smaller molecules, which can pass into the cell, it no longer has its enzymatic properties. Hence, the possibility of treating afflictions resulting from cellular enzyme deficiencies by injection of the enzymes seems to be impossible. It is interesting to note that persons with phenylketonuria also frequently have a low melanin deposit. Less tyrosine is available for the formation of melanin.

Much of the tyrosine in the cells follows the fourth pathway open to it and is finally reduced to carbon dioxide and water along with nitrogenous wastes with

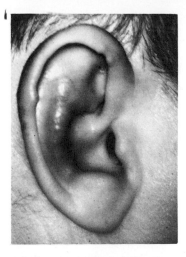

FIG. 17–13 *Cartilage discoloration in the ear of a person with alkaptonuria. The accumulation of homogentisic acid (alkapton) in the cartilages of persons with this inherited condition sometimes results in a reaction which produces pigment in the cartilages. This is most noticeable in the ears as it shows clearly through the thin skin in this region.* W. K. HALL, MEDICAL COLLEGE OF GEORGIA.

a release of energy as a normal part of cellular metabolism. This degradation proceeds through a gradual series of enzyme-mediated steps. The first compound formed in this series is **parahydroxyphenylpyruvic acid.** The next step results in the formation of **dihydroxyphenylpyruvic acid.** A rare recessive allele of the gene which produces the enzyme which mediates the latter reaction causes a failure of the production of this enzyme. When a person is homozygous for this allele, *t,* there is no conversion of dihydroxyphenylpyruvic acid and this acid builds up in the system. Also the amount of tyrosine, denied this important outlet, will increase. These two substances will be present in the urine and the person is said to have **tyrosinosis.** There seem to be no other serious symptoms.

The next step in the chain results in the formation of **homogentisic acid. Maleylacetoacetic acid** is next in the series, but this conversion may be blocked by the presence of a gene, *h,* in a homozygous state. The extra homogentisic acid, also called **alkapton** (Gr., black), which accumulates will be excreted in the urine and an afflicted person is said to have **alkaptonuria.** There is no difficulty in recognizing this condition; a mother knows if her child has it soon after birth, because the homogentisic acid is oxidized upon exposure to air and turns first an amber, then brown, and finally a black color. This is clearly evident upon the baby's soiled diapers. Many persons with alkaptonuria suffer no apparent ill effects, but in later life some suffer from degenerative arthritis. This appears to be caused by crystallization of some of the acid in the cartilages of the body with advancing age. There is also likely to be a progressive darkening of the cartilages of the body, most noticeable in the ears and the nose since at these points the cartilage is covered only by the outer skin. In a few cases it spreads to the fibrous tissue of the skin and the sclera (white) of the eye, where it produces disfiguring discolorations. Still other enzymes reduce the maleylacetoacetic acid to carbon dioxide through a series of steps, but as yet we have not discovered any breaks in this enzyme series.

Through a study of the possible pathways of the breakdown of a single amino acid, we have seen how complicated an enzyme series can be and how abnormalities can develop when one link in the chain of reactions is broken. Equally complex pathways and abnormalities, no doubt, exist in the metabolism of other amino acids.

PROBLEMS

1. If we were to accept the theory that the chromosome is one continuous unit of genetic material and that mutations merely represent rearrangements of small parts of this material, what problems would arise as we attempted to explain the many mutations which would be necessary for evolutionary development?

2. In Pneumococcus there is a strain that is resistant to the antibiotic streptomycin — it can live in media containing this antibiotic. Another strain is sensitive — it is killed by streptomycin. Describe the procedure you would use to induce transformation of a sensitive strain into a resistant strain.

3. How does transformation in bacteria support the concept that it is the DNA rather than the protein component of the chromosomes which is the material of which genes are made?

4. Different strains of polio virus vary greatly in their virulence. Certain strains are so non-virulent that they appear never to cause any severe symptoms of the disease. Sometimes epidemics of paralytic polio appear in an isolated region without any evidence of outside contact with a virulent virus. What could be a possible explanation of this?

5. What evidence from the bacteriophages supports the concept of DNA as the genetic material?

6. If practically all genes are formed of the same material, DNA, then how can we achieve the great variation in the genes which now exist in the many different forms of life on the earth?

7. Let us assume that a single strand in a certain small section of the DNA molecule has its bases in the following order: adenine, thymine, adenine, guanine, cytosine, guanine. In what order would the bases occur in the partner helix?

8. Tell how the gene for albinism in man causes its effects according to the enzyme hypothesis.

9. Describe some human characteristic, other than those discussed in this chapter, which you think might be the result of the presence or absence of intracellular enzymes. Tell why you select this characteristic.

10. If the eye bud from a fruit fly larva, homozygous for the wild-type red eyes, is transplanted into the body of a larva that is homozygous for white eyes, what type of eye would you expect to develop from the transplanted bud? Explain.

11. What great advantage does the mold, *Neurospora,* have over *Drosophila* as an experimental form for genetic studies?

12. Trytophan synthetase is an enzyme found in the wild-type *Neurospora.* It can combine serine and indole glycerol phosphate to produce tryptophan, an essential amino acid. A mutant strain of the mold will not grow on a minimal medium, but will grow if either tryptophan or indole glycerol phosphate is present. How would you analyze the enzyme deficiency of this mutant strain?

13. One strain of *Neurospora* cannot convert citrulline into arginine. How could you determine if the enzyme involved is lacking or present, but in a slightly altered form?

14. Describe the evidence which indicates that the genes for enzymes in man are actually somewhat intermediate in their expression even though they appear to be strictly recessive in terms of whatever abnormalities they cause?

15. Certain experiments have shown that the compound, serotonin, is necessary for the normal development of the brain tissue. Tests on patients with phenylketonuria indicate a deficiency in serotonin, and administration of extra serotonin

to phenylketonuric mice prevents impaired mental development. What relationship do you think might exist between these results and what you have learned about the cause of phenylketonuria in this chapter?

16. Why is it not possible to cure enzyme-deficiency diseases by the simple administration of the deficient enzymes by injection into the body?

17. Persons with phenylketonuria usually have very fair skin and hair regardless of their genes for density of pigment deposits. What explanation can you offer for this? Would you expect the pigmentation to be more dense in those which show a normal mental development because they have been on a low phenylalanine diet? Explain.

18

Gene Control of Cell Activity

THE CELL IS THE UNIT OF ACTIVITY IN ALL FORMS OF LIFE, with the exception of the viruses which must enter cells to carry on their growth and reproduction. It is in the cell that new protoplasm is produced from the raw materials which enter as food. Likewise, it is in the cell that enzymes, which mediate the many reactions which are a necessary part of living, are produced. The seat of the synthesis of the structural proteins of the protoplasm and the protein enzymes lies in the small granules in the cytoplasm known as **ribosomes.** These ribosomes are a part of a network, the **endoplasmic reticulum,** which forms an important part of the cytoplasm. It is only with the aid of an electron microscope, with its extremely high magnification and resolution of the cell parts, that we can see the endoplasmic reticulum together with the ribosomes attached to it. The genes which determine the nature of the proteins to be synthesized by the ribosomes lie within the nucleus of the cell. They have the information, the genetic code, for the construction of the proteins. How do they transfer this information to the cytoplasmic ribosomes which carry out the actual construction? There must be some kind of messages which are sent from the genes to the ribosomes.

GENE MESSENGERS

The messages sent by the genes to the ribosomes are in the form of molecules of RNA, the same substance which is the sole nucleic acid component of some viruses, namely those which infect higher plants, as well as some animal viruses, such as those which cause mumps, Newcastle disease, and influenza in man. RNA has certain basic differences from DNA. First, as we have learned in Chapter 17, RNA has a ribose instead of a deoxyribose sugar in its molecule.

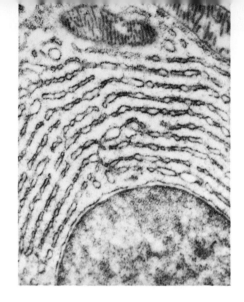

FIG. 18-1 *Endoplasmic reticulum and ribosomes in a cell from the pancreas of a guinea pig. This electron photomicrograph shows the nucleus at the lower right and the endoplasmic reticulum appears in the cytoplasm. The ribosomes are the small spherical bodies attached to the endoplasmic reticulum.* COURTESY GEORGE E. PALADE.

Second, the pyrimidine, uracil, is substituted for thymine. Thus, thymine is the only one of the four nucleotides which is found in DNA but not in RNA. Third, the messenger RNA molecules are much shorter than the very long DNA molecules. Finally, there is some evidence that the RNA molecule is a single strand rather than being a double-stranded helix like DNA.

The production of messenger RNA from a portion of the DNA molecule appears to be as follows: The DNA molecule separates temporarily at one part of its length. RNA nucleotides are then arranged in a complementary order on one of the two strands of the DNA. If the DNA strand reads adenine, thymine, guanine, the RNA formed will read uracil, adenine, cytosine. Ribose sugar and phosphate bonds complete the RNA molecule. The two strands of DNA then become rejoined. The following question might have arisen in your mind as you read this account: what determines which of the two strands of DNA are to be used to mold the RNA? The answer is that we do not know at the time of this writing. This is an important question, for the two strands could yield different types of RNA and one would be complementary to the other, but we must await further research for an answer. Once the RNA is formed, it is ready to be passed into the cytoplasm. There is some evidence that the nucleolus can serve as a storage chamber for the RNA. The nucleolus is very rich in RNA and is closely associated with the chromosomes. Also, it disappears during mitosis when the DNA is not actively synthesizing RNA, but reappears as soon as mitosis is completed and the RNA synthesis is resumed.

Once in the cytoplasm, the messenger RNA goes to the ribosomes. The ribosomes are made of protein and RNA, ribosomal RNA, in about equal proportions. They serve to receive the messages incorporated in the messenger RNA.

Evidence That Messenger RNA Carries the Gene Code

Substantial evidence to support this pattern of code transfer from gene to ribosome is provided by studies that have been made on the one-celled protozoan, *Amoeba*. Amoebae are first put into a medium containing radioactive phosphate, some of which is taken up by the cells and used in the construction

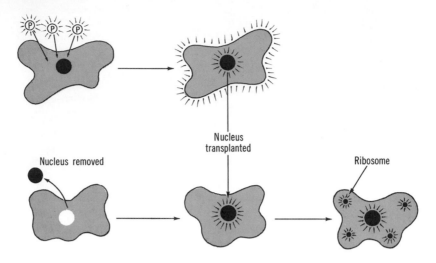

FIG. 18-2 *Experiment to demonstrate the passage of material from the nucleus to the ribosomes. An amoeba is placed in water containing radioactive phosphorus and both nucleus and cytoplasm become radioactive. The radioactive nucleus is then transplanted to the enucleated amoeba and soon the ribosomes show radioactivity.*

of phosphate-containing compounds within the cells (see Figure 18–2). Since DNA is one of these compounds, the DNA soon becomes radioactive as it undergoes duplication during the period preceding cell division. Also the RNA produced by the DNA becomes radioactive since it also uses phosphorus. By means of delicate micropipettes, it is possible to remove the radioactive nucleus from one of these amoebae and transfer it to a non-radioactive amoeba from which the nucleus has been removed. In a short time this second amoeba will reveal radioactive cytoplasm, and, on separating the cytoplasmic parts, the radioactivity is found to be concentrated in the region of the ribosomes.

Another experiment can show that DNA itself does not leave the nucleus (see Figure 18–3). Radioactive hydrogen, tritium, can be incorporated into thymidine (the nucleotide containing thymine as its base). When amoebae are placed in water containing such thymidine, the latter is taken up and incorporated into duplicating DNA within the organisms. When a nucleus containing the radioactive thymidine is transplanted into a non-radioactive amoeba, the radiation remains within the nucleus. Since RNA does not contain thymine, it does not transfer the radioactivity to the cytoplasm.

Having seen how the genetic code can be transferred from the genes in the nucleus to the ribosomes where the actual protein construction takes place, let us next turn our attention to the methods of construction of proteins.

PROTEIN SYNTHESIS

Amino acids are the building blocks of proteins. Animals may eat whole proteins, but these are broken down into their component amino acids which are taken up by the cells and reassembled to make proteins characteristic of the

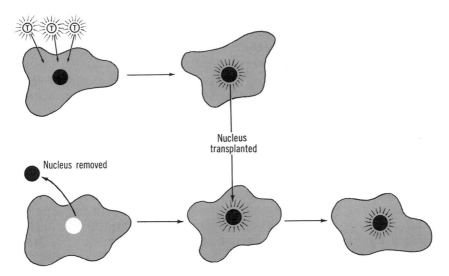

FIG. 18-3 *Experiment showing that DNA does not leave the nucleus. An amoeba is placed in water containing radioactive thymine, which is soon taken up by the DNA. No other part of the cell uses thymine, so the radioactivity is concentrated in the nucleus. When this radioactive nucleus is transplanted to an enucleated amoeba, the radioactivity remains within the nucleus and is not transferred to the cytoplasm.*

cell in which they are constructed. This is also true of all plants which do not manufacture their own food, but the green plants, on the other hand, manufacture their own amino acids from the sugars which they produce, plus some nitrogen salts they absorb from the soil. Enzymes within the cells in both groups of organisms cause amino acids to be linked together by making the carboxyl group of one amino acid combine with the amino group of the next amino acid. This forms what is known as a **peptide bond.** Thus chains of amino acids, known as **polypeptides,** are formed; and it is these polypeptide chains which form protein. Some protein molecules consist of a single polypeptide chain, folded and cross-linked to form a more stable structure. Other protein molecules may be **polymers,** made up of two or four polypeptide chains which may or may not be identical. Protein synthesis, therefore, consists of (1) an end-to-end union of amino acids to form the chains which make up the protein molecules; (2) folding and crosslinking of the chains; and (3) often a combination of several such folded chains into a compound structure, a polymer. The great variety and complexity of protein molecules are due to variations in the sequences of amino acids in the chains, as well as to the length and relationships of the chains.

We can trace the source materials of protein synthesis through studies of the uptake of amino acids and their utilization within the cell. Radioactive amino acids placed around a cell are soon absorbed. If the cell is killed a short time later, radioactive proteins will be found around the ribosomes, but nowhere else in the cell. If a longer time is allowed to elapse before the cell is killed, the radioctivity can be found in the protein of other parts of the cell. These results

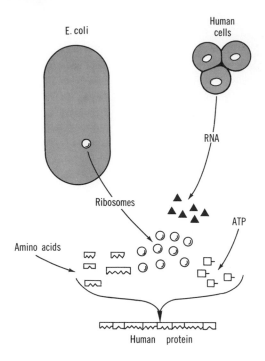

E. coli

Human cells

RNA

Ribosomes

ATP

Amino acids

Human protein

FIG. 18-4 *Diagram to illustrate experiments which show that RNA is the messenger material which codes the production of proteins by the ribosomes. Ribosomes from bacteria (E. coli) will produce human protein when RNA from human cells and assorted amino acids as well as ATP are present.*

supply rather convincing evidence that the major site of synthesis of proteins is at the ribosomes.

Further experiments confirm the concept that the ribosomes are the centers of protein formation in accordance with messenger RNA from the genes. By delicate centrifugation it is possible to separate the various parts of cells. Thus, ribosomes and the fraction of the cytoplasm containing the RNA can be extracted from bacteria, for instance, and proteins will be formed outside the cells. Amino acids can be added to such a mixture, and, since energy is required for protein synthesis, the energy-yielding compound known as *adenosine triphosphate (ATP)* must also be added. ATP is produced by the mitochondria in the cytoplasm and supplies the energy for most cellular activities. In a short time whole proteins will be found in the mixture, and these proteins will be of a specific nature characteristic of the cells from which the ribosomes and RNA were taken. To show that it is the RNA and not the ribosomes which determine the specificity of the proteins, it is possible to add RNA extracted from human cells to the ribosomes taken from bacterial cells. In such a mixture, the proteins which are formed will be of the same nature as those found in the human cells from which the RNA was extracted.

RNA Storage in Non-Nucleated Cells

If the nucleus is removed from a cell, the local source of messenger RNA to the ribosomes is cut off, but the ribosomes can continue to synthesize proteins for a short time because of the existing supply of messenger RNA which was already in the cytoplasm. In time, however, the synthesis stops since messenger RNA is destroyed after its use in protein synthesis and eventually the cell dies. Since human red blood cells are not nucleated as they normally exist in the blood stream, we may wonder how they can continue to live without genes to direct their cell activities. These cells are produced in the red bone marrow and do contain nuclei in their immature stages. Sufficient messages are sent to the ribosomes and sufficient protein is produced to keep the cells alive and active after the nucleus has been expelled. The red cells do not continue to grow and synthesize new protein, however, after they are released into the blood stream, and their life span is limited to about 120 days.

White blood cells retain their nuclei after they are released into the blood stream and can continue to grow and synthesize protein. This is demonstrated by the photograph shown at the left in Figure 18–5. Radioactive amino acids were placed in a small sample of human blood, and, after several hours, smears of the blood were made on microscope slides. The smears were then covered with a thin photographic emulsion which was allowed to stay in contact with the smear for several days. The slides were then placed in a photographic developer which caused a darkening of the silver grains of the emulsion wherever it was exposed to radiation. You will note that there are dark grains within the cytoplasm of the white blood cells, a fact indicating that they have taken up the radioactive amino acids and incorporated them into proteins within the cells. No such grains show in the normal red cells. In the abnormal blood there are some red cells, which were released prematurely while still nucleated, and in one of these, shown at the right of Figure 18–5, there has been some uptake

FIG. 18–5 *Autoradiograph showing that protein is manufactured in the nucleated human white blood cells, but not in the non-nucleated red blood cells. The photograph on the left is of normal blood and shows one white blood cell and several red blood cells. The photograph on the right contains some nucleated red blood cells, which have been released into the blood prematurely. The nucleated cells, both white and red, show protein synthesis, but the non-nucleated cells do not. See text for details.*
COURTESY C. P. LE BLONDE.

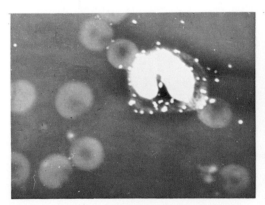
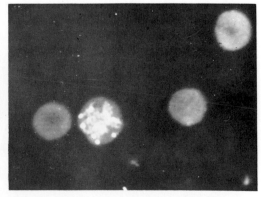

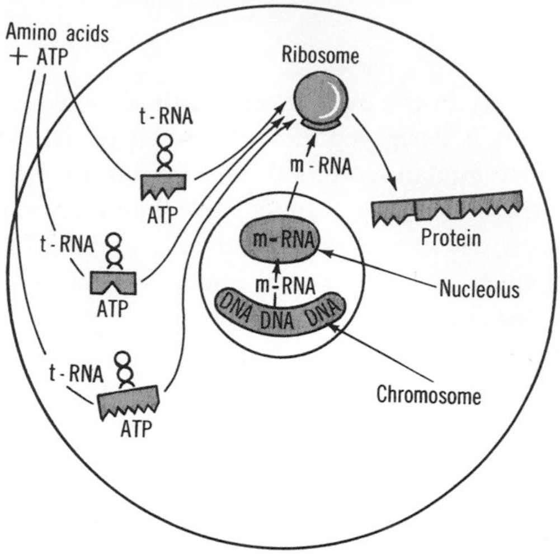

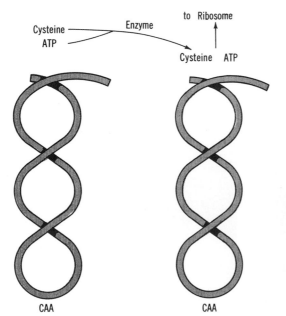

FIG. 18–6 *Diagram to show how the genes control protein synthesis through messenger RNA which moves out to the cytoplasmic ribosomes.*

of amino acids and some protein synthesis. A person with such cells is anemic because red cells with nuclei have less volume available for hemoglobin and cannot carry oxygen as efficiently as the normal, non-nucleated cells.

Transfer RNA

There is another aspect of protein synthesis remaining to be explained. This is the method of transfer of amino acids to the ribosomes. The transfer is accomplished by a cytoplasmic RNA, known as transfer RNA (t-RNA). It is an RNA of low molecular weight and is made up of a relatively short, single strand, which is probably bent back upon itself somewhat like a twisted hairpin (see Figure 18–7). There appear to be only about 90 nucleotides in the t-RNA

FIG. 18–7 *Details of the transfer RNA activity in transporting a particular amino acid. A transfer RNA molecule with the code CAA at one end picks up the amino acid cysteine after the amino acid has united with the energy-yielding compound ATP.*

molecule, but actually only three of these carry the message for a particular amino acid, a point to be discussed later in this chapter.

Each transfer RNA molecule is coded to move only one kind of an amino acid to a ribosome. Since human proteins are made up of about 20 different kinds of amino acids, we can expect at least 20 different kinds of transfer RNA molecules to be in the cytoplasm of human cells. Actually, there are considerably more than this because it is known that some amino acids can be moved by several different combinations of RNA nucleotides. According to the information available at present, the method of transfer of amino acids is somewhat as follows: To be transported an amino acid must first be activated and this occurs when the amino acid joins with a molecule of ATP, the energy-yielding compound. The activated amino acid then combines with a particular transfer RNA molecule which carries it to the ribosome. The messenger RNA appears to move across the ribosome somewhat like a coded magnetic tape of a computer machine. The ribosome seems to "read" the first **codon** (set of three nucleotides) and the proper amino acid is deposited from the transfer RNA. The second codon of the code on the messenger RNA is then "read" and the second amino acid is deposited from a second transfer RNA molecule and linked up with the first. Thus, the polypeptide chain grows according to the code sequence contained in the messenger RNA. Each transfer RNA molecule, after depositing its amino acid, is free to pick up another amino acid of the same type and deposit it when the code calls for this amino acid again.

Advantages of Use of RNA in Protein Synthesis

The foregoing discoveries have provided a rather complete picture of the way in which genes control protein synthesis in the cell. The genes themselves serve as a mold upon which rather short sections are transcribed into the messenger RNA which is then transferred to the cytoplasm, possibly by way of the nucleoli. The messenger RNA then goes to the ribosomes. Amino acids, in activated form, are then brought to the ribosomes by the transfer RNA and put together in the form of a polypeptide chain in accordance with the code in the messenger RNA. The polypeptide chains make up the proteins which become either important structural or functional proteins such as hemoglobin, or enzymes which mediate important cell reactions that in turn make the cells grow and act according to their heredity.

The advantage of such an indirect functioning of the genes is obvious. The genes remain protected within the nucleus and are less subject to the destructive forces of respiratory metabolism in the cytoplasm. Destruction of the RNA messages is not of serious consequence because more messages can be produced by the genes. As an analogy, let us study the method of building a house. An architect produces a master copy of the plans giving detailed instructions for constructing the house. These plans (genes) can then be used to produce blueprints (messenger RNA) which are sent out to the workmen (ribosomes). The workmen, however, must have materials with which to build the house. The lumber, bricks, cement, nails, and other materials (amino acids) are brought to the workmen by delivery trucks (transfer RNA), each truck bringing only one kind of material as it is needed. Thus, the house is constructed according to the plans which are kept safely in a vault at the architect's office. The workmen can construct other houses according to the same plans, but the blueprints become destroyed through use on the job. New blueprints can be printed from the original plans, however, so there is no permanent loss of the information needed for the construction of this particular type of house.

HUMAN HEMOGLOBINS

We have learned much about the synthesis of proteins and the gene code through studies of the hemoglobin found in the red blood cells of man and other animals. Hemoglobin, a protein with a great affinity for oxygen, serves as the major oxygen-transporting medium in the blood. The existence of different kinds of human hemoglobins became evident in 1910, when blood examinations of people with a certain type of anemia revealed the presence of red blood cells that were elongated and shaped somewhat in the form of sickles. The anemia was called sickle-cell anemia because of this unusual shape of the red blood cells. These cells are poor carriers of oxygen, and the anemia is so severe that it often brings about the early death of afflicted persons; at best, they are severely handicapped throughout life. Studies of the blood showed that the sickling occurred when the cells were exposed to conditions of low oxygen concentration, such as would be found in the veins after the oxygen has been given up in the capillaries.

Inheritance of Sickle-Cell Anemia

In studies of pedigrees which included persons with sickle-cell anemia, it was found that the condition was transmitted by a gene. The anemia develops

FIG. 18–8 *How a deficiency of oxygen in the blood causes a sickling of red blood cells in a person homozygous for the gene for sickling. A normal blood smear is shown on the left and a smear of sickled cells on the right.* SICKLED CELLS COURTESY JAMES V. NEEL.

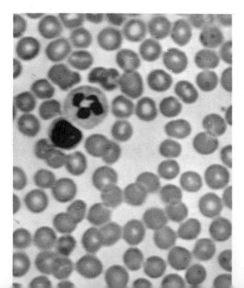

in those persons who are homozygous for the gene involved, but the heterozygous carriers of the gene also show some degree of expression of the characteristic. The red blood cells of these heterozygous persons will undergo sickling when exposed to conditions of extremely low oxygen concentration. Within the veins the oxygen concentration is still high enough ordinarily to prevent sickling, but, if the blood is removed and placed in a container where there is a complete absence of oxygen, sickling will take place. These heterozygous people usually do not suffer from anemia, but if they were to travel to a high altitude, where the oxygen concentration of the air is low, some sickling will occur and an anemia might develop. Also, the impairment of circulation to some body part will bring about sickling in such persons. These heterozygous persons are said to have the sickle-cell trait.

The gene for sickle-cell anemia is most prevalent in large parts of Africa, Sicily, and southern Italy, Greece, the Near East, Arabia, and certain areas of India, Pakistan, and Burma. The prevalence of the gene seems to be correlated with the incidence of a particular form of malaria in these regions of the world. We shall learn the reason for this in Chapter 25.

Chemical Analysis of Hemoglobins

It was not until 1949 that the chemical basis for the sickling of the red blood cells was demonstrated. Linus Pauling, a Nobel prize winner at the California Institute of Technology, first removed the hemoglobin by placing the cells in water. This caused hemolysis of the cells; i.e., they burst from internal pressure as they absorbed water by osmosis. Some of the extracted hemoglobin was then placed in the center of a paper wet with a weak salt solution, and positive and negative electrodes were attached at opposite ends of the paper. The hemoglobin migrated toward the positive pole, which indicated that the hemoglobin must have a negative charge. This was called **hemoglobin A,** normal adult hemoglobin, possessed by the great majority of people. Hemoglobin from a person with sickle-cell anemia, **hemoglobin S,** showed a different reaction during electrophoresis. It migrated toward the positive pole also, but at a rate which was slower than that of hemoglobin A (see Figure 18–9). When the hemoglobin from a heterozygous person, one with the sickle-cell trait, was tested, it was found to separate into two components, one component migrating more rapidly than the other. These reactions indicated that both hemoglobins A and S were present.

V. M. Ingram of Cambridge University, England, took up the study and succeeded in finding the chemical difference between the two hemoglobins. The hemoglobin molecule is made of two symmetrical halves, each half containing two different polypeptide chains, known as the alpha chain with 141 amino acids and the beta chain with 146 amino acids. The molecule can be broken down into smaller parts by exposing it to trypsin, a digestive enzyme which breaks the chains wherever the amino acid lysine or arginine is present. By using this technique, Ingram obtained 28 amino acid groups, known as peptides, from each half molecule. He was able to separate these peptides from one another

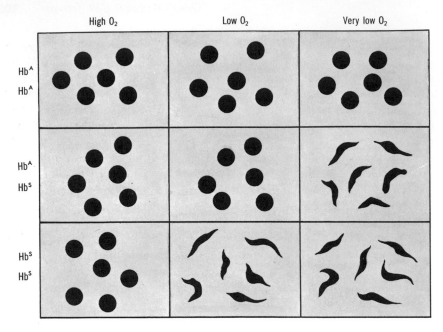

High O₂   Low O₂   Very low O₂

FIG. 18–9 *Reaction of human hemoglobins to variations in oxygen concentration. A person homozygous for hemoglobin A has cells which are not affected. Cells from a person heterozygous for A and S will sickle under very low oxygen concentration, but not under the concentrations which are normally found in the veins. Cells from a person homozygous for S will show sickling even in the veins.*

because each peptide has a different rate of migration on paper that is subjected to electrical fields and various solvents. Hemoglobins A and S showed identical patterns on the paper, with one exception — one peptide out of the 28 was in a different position.

FIG. 18–10 *Migration patterns of hemoglobins when placed on buffered paper and exposed to an electrical field. Hemoglobins A and S move at different speeds and can be separated out from the blood of heterozygous persons by this method.*

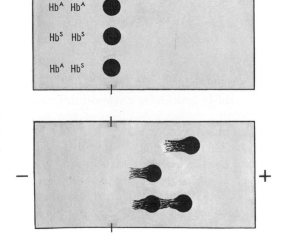

Ingram cut out the areas of the paper which showed this difference and analyzed their amino acid content. He found that the peptides differed with respect to only one amino acid. In hemoglobin S, valine had been substituted for glutamic acid. This one substitution, in one out of 287 amino acids, accounts for the entire difference in the two kinds of hemoglobins. Glutamic acid has a negative charge and valine has no charge; hence the charge of the entire molecule is altered. That difference in charge accounts for the difference in the reaction to an electrical field. The difference in electrical charge also accounts for the sickling of the entire cell under conditions of low oxygen concentration. Under such conditions the hemoglobin crystallizes, and the lengthwise orientation of the molecules results in the distortion of the red blood cells. Such crystallized hemoglobin cannot pick up and transport oxygen. In persons with the sickle-cell trait the cells contain only from 25 to 45 per cent of hemoglobin S, an amount not sufficient to permit crystallization and sickling except under extraordinarily low concentrations of oxygen.

Homozygous persons, i.e. those with sickle-cell anemia, were at first supposed to have only hemoglobin S, but analysis of blood samples from such persons shows that some of them have as little as 70 per cent of hemoglobin S. The remaining 30 per cent is not hemoglobin A; it could not be, if our ideas are correct, for these people do not have one of the genes needed to make hemoglobin A. Instead, this 30 per cent is hemoglobin F, fetal hemoglobin, which is produced in all persons during the fetal period of prenatal life. Hemoglobin F normally persists in diminishing quantity until about 6 months of age and by then is replaced by the adult type; but small amounts may be present in some adults. The larger quantities of hemoglobin F found in some cases of sickle-cell anemia can be explained by a partial restoration of the activity of genes that normally function only in the fetus and infant. This restoration is probably an adjustment of the body to the chronic low oxygen level of the blood that occurs when many of the cells are sickled.

Other Hemoglobins

The discoveries of hemoglobins A and S led to further research, and other human hemoglobins have been found. **Hemoglobin C** was first detected in an American Negro, and it was then found that almost 2 per cent of American Negroes have the gene for hemoglobin C. In Africa the distribution of the gene for this type of hemoglobin is limited to a circumscribed region of West Africa. Hemoglobin C is formed by an allele of the gene for hemoglobin S and is characterized by a substitution of lysine for the identical glutamic acid that in hemoglobin S is replaced by valine. Persons homozygous for hemoglobin C have anemia, but it is not as severe as the usual sickle-cell anemia. Heterozygous C and S individuals have sickle-cell anemia.

At least 40 types of hemoglobins have now been found, all produced by individual amino acid substitutions. At least five loci on the chromosomes are involved. These variations in hemoglobin can be related to the DNA chain of

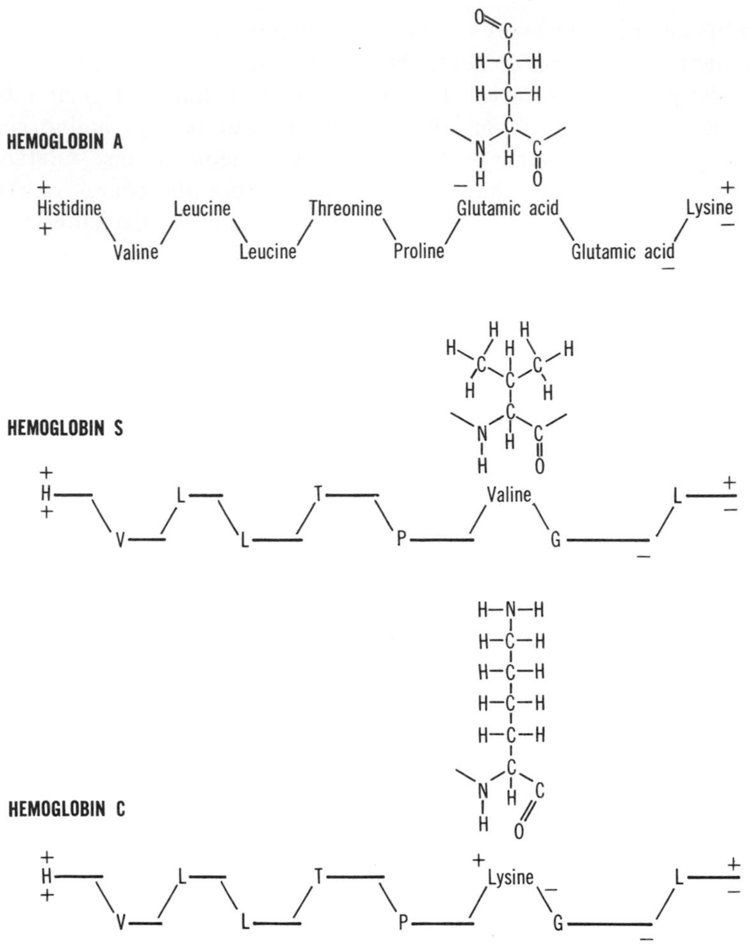

HEMOGLOBIN A

+
Histidine   Leucine   Threonine   Glutamic acid   Lysine
+   −
     Valine   Leucine   Proline   Glutamic acid
                                                                   −

HEMOGLOBIN S

+
H——   L——   T——   Valine   L—— +
+   −
   V——   L——   P——   G——

HEMOGLOBIN C

+
H——   L——   T——   Lysine   L—— +
+   −   −
   V——   L——   P——   G——

FIG. 18–11 *The structural formulas of the amino acids which vary in hemoglobins A, S, and C are shown. Note that in hemoglobins S and C valine and lysine replace the glutamic acid present in hemoglobin A.*

the genes involved in hemoglobin production. According to the theory that the base sequences of the DNA are lined up in an order corresponding to the amino acids of the hemoglobin, it is easy to see how a change in a single nucleotide pair of the DNA might lead to the substitution of a different amino acid at a particular point in the chain of amino acids of the hemoglobin molecule. We can understand this better after a look at the nature of the genetic code.

THE GENETIC CODE

Several theories on the method of coding have been proposed to explain how a difference in sequence of bases in DNA, when transcribed as complementary bases in RNA, can determine just which amino acids will be placed in specific positions on the protein molecule. The theory which has gained the greatest acceptance was the one proposed by F. H. C. Crick, coauthor of the DNA theory of gene structure. This theory holds that there is a triplet code — that a sequence of three bases codes for a single amino acid. Such a triplet of bases is called a **codon.** Starting at a set point on the gene, the triplets of bases in order determine which specific amino acids are to be assembled in this same order in the ribosomes. Likewise, the amino acid which is picked up by each kind of transfer RNA is determined by three bases at the loop of the "hairpin." For instance, research indicates that the combination AAA picks up the amino acid phenylalanine and places it on the RNA at the ribosome wherever the UUU sequence is found. The corresponding unit of DNA which coded the messenger RNA would be AAA. If through mutation the DNA sequence were changed to read AAT, then the messenger RNA would carry the message UUA to the ribosome and this particular amino acid in the protein would be changed to tyrosine. The change in the hemoglobin molecule from type A to S, according to this theory, would require the substitution of only one base in the messenger RNA. When AGU, which is an RNA codon for glutamic acid, is changed to read UGU, which is a codon for valine, the substitution would be made.

Discovery of the Genetic Code

How has it been possible for geneticists to work out the detailed information needed to establish a genetic code? Several techniques have been used. One of these was first reported in 1961 by M. W. Nirenberg and J. H. Matthaei of the United States National Institutes of Health. These men succeeded in synthesizing a synthetic RNA which contained only one base, uracil. To test the effectiveness of this synthetic RNA, it was necessary to place it in a system which could synthesize protein. Such a system was obtained from cells of the bacterium, *Escherichia coli,* by first grinding them to destroy the cell membranes and then treating the cell extract with the enzyme deoxyribonuclease, which destroys DNA. This treatment removed the source of fresh messenger RNA

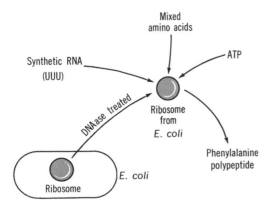

FIG. 18–12 *Technique of determining the genetic code. When synthetic RNA containing only uracil is added to extracted ribosomes in a mixture of amino acids and ATP, a polypeptide chain is formed which contains only phenylalanine. Hence, it appears that a code sequence for phenylalanine is UUU.*

to the ribosomes, but left the ribosomes themselves intact. In time the messenger RNA in the cell extract was depleted. To this mixture were then added the various amino acids, the synthetic RNA containing only uracil, and the necessary ATP. Polypeptide molecules were formed, and they were found to contain only phenylalanine. Hence, it was concluded the UUU is the codon for phenylalanine on the basis of a triplet code.

It was not long until other RNA combinations were obtained, some of them by chemical synthesis and some by extraction from natural sources, such as the RNA of certain viruses. RNA containing only cytosine, for instance, was found to produce a polypeptide containing only the amino acid proline; thus CCC was accepted as the codon for proline. Then combinations of different bases were used, and eventually all of the possible combinations of bases were tested and codons for all of the 20 common amino acids determined, although many of these remain tentative because of difficulties in determining the sequence of the bases within each codon. Table 18–1 shows the codons for each of the 20 common amino acids as they have been worked out on this tentative basis. Note that all of the amino acids except two can be designated by several base combinations. This multiple system of coding, known as a **degenerate system,** provides a protection to organisms against many harmful mutations. Should a mutation of one base of the DNA molecule change the codon for an amino acid, it is possible that the mutant codon would still correspond to the same amino

TABLE 18–1

Genetic Codons for the Amino Acids*

Amino acid	RNA codes
1. Alanine	CUG CAG CCG
2. Arginine	GUC GAA GCC CGC
3. Asparagine	UAA CUA CAA ACA
4. Aspartic acid	GUA CGA ACA
5. Cysteine	GUU UUG UGG
6. Glutamic acid	AAG AUG AGU AGA
7. Glutamine	AGG AAC
8. Glycine	GUG GAG GCG UGG
9. Histidine	AUC ACC
10. Isoleucine	UUA AAU
11. Leucine	UAU UUC UGU GUU
12. Lysine	AUA AAA AAC AGA
13. Methionine	UGA
14. Phenylalanine	UUU UUC
15. Proline	CUC CCC CAC CCU
16. Serine	CUU CCU ACG UCG
17. Threonine	UCA ACA CGC CAC CAA
18. Tryptophan	UGG
19. Tyrosine	UUA UAU
20. Valine	UUG UGU

*Some of these codons are tentative with respect to the order in which the bases occur, except those in which the three bases are the same.

acid, and no change in the polypeptide chain would then result. There are 64 possible combinations of four bases taken in triplets, so it is easily possible for several different triplets to code for the same amino acid. All 64, however, do not code for amino acids. Some of the possible triplets do not correspond to any kind of transfer-RNA in the cell. These would be **"nonsense" codons** and would not place any amino acid on the polypeptide chain.

An understanding of the nature of the genetic code makes it possible to postulate methods by means of which mutations of the genes may occur. A mutation of a gene might take place whenever there was a change in the code. Such a change could come about through the addition, subtraction, or substitution of a single nucleotide in the DNA, or by rearrangements of two or more nucleotides. We might compare mutations of the genes to typographical errors made in printing words. For instance, suppose a typist is copying some material and strikes the wrong key. Such an error could change a word and alter the meaning of an entire sentence. A single letter change could cause the word "hot" to read, "pot," "hop," "lot," "not," "hit," or "hut." Should the typist accidentally insert an extra letter the word might read "hoot" or "holt." Frequently errors of either kind result in nonsense words such as "hxt," hotp," "mhot," or "hht." These letter combinations spell nothing and might simply destroy the meaning of an entire sentence. Likewise, substitutions in the genetic code might make altered sense, lead to the substitution of one amino acid for another, and thus form a polypeptide chain, but one which is different from that formed before the change. Or, changes in the DNA bases could result in a nonsense codon and block the synthesis at a particular point. This could mean that the ribosomes might not be able to make an entire protein if this amino acid were essential. If this missing protein were an enzyme, there would consequently be a loss of some important cell function. For example, if it were an enzyme necessary for the synthesis of melanin in man, its absence would cause albinism.

Sense and Nonsense Base Combinations

Let us illustrate the possibility of a mutation through substitution in a single one of the RNA combinations that signifies alanine. The code sequence we will use is CAG, which we will assume is found at some place on a messenger RNA molecule. There are nine possible variations of this code sequence which can result from the substitution of a single letter in this code. Some of the substitutions will produce base combinations which make sense, and a different amino acid will then be substituted for alanine at one point on the polypeptide chain which forms the protein. As a result, the protein may show important alterations of its properties because of the substitution of this one amino acid, even though this may be only one amino acid out of hundreds in the entire molecule. Other base substitutions will make nonsense, so that no amino acid at all will be deposited at this particular place in the protein molecule. This deficiency could result in a changed protein or it might block the synthesis at this point and lead to an incomplete protein which cannot function. No doubt, many lethal

T A B L E 1 8 – 2

Variations in an Alanine Codon Through a Single Base Substitution*

Codon for alanine	Variations possible from single base substitutions
CAG alanine	CAU nonsense — no amino acid CAA sense — asparagine CAC sense — proline CCG sense — alanine (no change) CGG nonsense — no amino acid CUG sense — alanine (no change) UAG nonsense — no amino acid GAG sense — glycine AAG sense — glutamic acid

* This chart is given as an illustration of how variations in one letter can alter the amino acids in proteins. It should be kept in mind, however, that the sequence of many of these is tentative.

gene mutations come about because of such changes. Table 18–2 shows the various combinations which result in single letter substitutions in CAG and indicates which ones make sense and which ones make nonsense, according to present knowledge of the code.

REGULATION OF GENE ACTIVITY

It is obvious that all genes do not function at all times. You have genes for the development of eyes in all of the nucleated cells of your body, yet you do not have eyes on the bottom of your feet or at other equally ridiculous places. Why do genes function at the time and place where they are supposed to function and suspend their operations at places where they are not supposed to function? We certainly do not have all of the answers to these questions today, but we do have some small inkling of the mechanism involved in such gene control.

The Puffing Phenomenon of Giant Chromosomes

Studies of the giant chromosomes from the salivary glands of *Drosophila* and other fly larvae have been important in studies of gene activity. At certain regions of these chromosomes there are swellings or puffs which look distinctly different from the other parts of the chromosomes. Photographs of such puffs, as seen in fresh unstained preparations, are shown in Figure 18–13. Analysis of these puffed regions shows amounts of RNA much larger than those found at the non-puffed regions, thus indicating that the puffs must be the places where RNA is being synthesized in high quantities. We can, therefore, postulate that the genes in the puffs are those connected with the activities of the salivary glands, whereas the genes in the more condensed regions of the chromosomes are related to activities in other parts of the body. This hypothesis is supported by the fact that the puffs are not constant in position, but instead first one region shows the puffing phenomenon and then other regions as the glands develop and alter their activities. These facts give us a clue to the activity or inactivity of the genes on the chromosomes but leave a big question unanswered: what factor controls the puffing or non-puffing of certain regions of the chromosomes in particular cells and at particular times?

Stimulation and Repression of Gene Activity

There is some evidence that the protein histones, which are a part of the chromosomes along with the DNA, act as inhibitors of gene activity. James Bonner of the California Institute of Technology has used chemical treatment to remove much of the protein histones from the small embryos of germinating garden peas. The genes then produced five times as much messenger RNA as they did previously — evidence indicating that the histones must act to suppress the output of RNA from the genes. When extracted histones from one set of embryos were added to the growing embryos of other plants, there was a reduction in the amount of RNA released.

Antibiotics have also been found to inhibit RNA output from the genes, and this observation may well explain the inhibiting effects of antibiotics on specific types of cells. Antibiotics help bring diseases under control by inhibiting the metabolic activities of bacteria, thus slowing their growth and reproduction. The antibiotic, actinomycin, when added to the salivary gland cells of flies, was found to inhibit the puffing phenomenon and to repress RNA production. This possibly explains the growth-inhibiting action of the antibiotic — when there is a reduced flow of messenger RNA from the genes, there is likewise a reduction in protein synthesis. Since metabolic enzymes are proteins, there will be a corresponding reduction in metabolism, including growth through synthesis of new protoplasm.

Just the opposite effect is found when certain hormones are applied to growing cells. When the female hormones, the estrogens, of higher animals are

FIG. 18–13 *Puffs on the chromosomes from the salivary glands of Drosophila. Analysis shows that this puffing phenomenon is accompanied by a high output of RNA. These photographs are of fresh, unstained preparations made with phase contrast.*

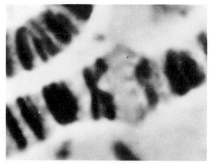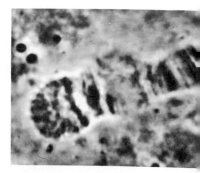

added to cells of the female uterus growing in a culture, they cause a marked reduction in the ratio of histones to DNA and a rise in the output of messenger RNA. Likewise, the male hormones, the androgens, have a similar effect on tissue from a prostate gland. In plants, the flowering hormone, florigen, when applied to plant buds will greatly increase protein synthesis and soon flowers bloom out of the buds. In insects, the presence of the molting hormone, ecdysone, causes a stimulation of chromosome puffing and of RNA output.

Embryonic Organizers

These discoveries fit into our knowledge of the pattern of embryonic development. A certain region of frog and salamander embryos is known to produce a chemical organizer that diffuses into the surrounding cells and causes the genes to function at the proper time and place. This region, known as the dorsal lip of the blastopore, can be removed from one embryo and transplanted to the side of another embryo, and another set of organs then starts to develop. A two-headed frog or salamander may result. The organizer is a chemical substance which can retain its organizing capacity even when extracted and applied to developing embryos.

Once embryonic cells have been exposed to the organizer, and we may presume that the proper genes have been stimulated to produce their messenger RNA, those cells will continue their line of development even though transplanted. For instance, a region of a frog embryo that is destined to form an eye may be transplanted to the side of another embryo and it will still make an eye, even though there was no sign of differentiation before the part was transplanted (see Figure 18–14).

Histones in Salivary Gland Puffs

From these various studies of the methods of regulation of gene activity, the pattern of gene control would appear to be due to the movement of histones from one region of the chromosome to another, suppressing activity in regions

FIG. 18–14 *A three-eyed frog. While still a young embryo, this frog received a transplant of tissue from the eye region of another embryo. Even though the transplant showed no differentiation at the time, the genes within the cells had already been stimulated to produce the messenger RNA necessary to form eye tissue and a fully developed eye appeared at the site of the transplant on the side of the body.* COURTESY C. L. MARKERT.

where they are most abundant and allowing greater RNA production in regions where they are least abundant. Various outside chemical substances, such as hormones and embryonic organizers, coming into the cell seem to influence this movement of the histones and thus play an important part in the activity of the genes. An actual analysis of the chromosome puffs of salivary gland chromosomes of *Drosophila* by E. C. Horn and C. L. Ward, however, does not support this hypothesis. These men used selective stains and found that the density of the histone stain in areas of the puffs indicated that the total amount of histone was just as great after a puff developed as before a puff developed. Both the DNA and the histone appeared just to spread out in the puffing phenomenon. The proportion of one to the other did not seem to change. It is quite possible, however, that there could be changes in the quantity of histones in relation to DNA which are below the level of detection by judgment of the degree of staining. Further research must be done to give the answer.

Structural, Regulator, and Operator Genes

Other work which explains the control of gene activity has been done by F. Jacob and J. Monod. As a result of experiments on the bacterium, *E. coli,* they formulated a theoretical model for the control of RNA synthesis by the genes. Their model assumes that there are three kinds of genes: **Structural genes** are those genes of classical genetics which produce the messenger RNA which, in turn, governs protein production by the ribosomes in the cytoplasm. In addition, there are **operator genes (operons)** which act as switches to stimulate the synthesis of RNA by the structural genes. In most cases a single operon governs the activities of a cluster of nearby structural genes, perhaps through histone inhibition if histone is truly related to RNA synthesis. Finally, there are **regulator genes** which produce substances which act on the operons to turn them off so that they do not stimulate the structural genes. Regulator genes are probably sensitive to factors from the outside, such as hormones or organizers, which have previously been mentioned.

It should be clear after studying this chapter that the discoveries relating to gene control of cell activities have really just given us a partial insight into this important topic. Without doubt, many important new discoveries will be made in this field by the time this book is available to the student.

PROBLEMS

1. Why is the synthesis of proteins within cells of such great importance to geneticists?

2. Would the transfer RNA found in the brain cells of a horse be expected to be the same as the transfer RNA found in the brain cells of a monkey? Explain.

3. Would the messenger RNA found in the brain cells of a horse be expected to be the same as the messenger RNA found in the brain cells of a monkey? Explain.

4. If chemical analysis of a certain enzyme shows it to be a polypeptide protein chain of 1200 amino acids, how many nucleotide pairs would you expect to find in the gene which control the synthesis of this enzyme? Explain.

5. Describe the nature of a hemoglobin molecule and tell how an alteration of one amino acid can alter its electrophoretic reactions.

6. How can substitution of a single base in the DNA of a gene bring about an extensive change in the organism?

7. How have the studies of human hemoglobins contributed to our understanding of the genetic code?

8. Explain what is meant by sense and nonsense genetic codons and how these are related to mutations.

9. Give evidence for and against the role of histones as controlling factors in gene activity.

10. Suppose a portion of a chromosome has nucleotides that produce messenger RNA with the code sequence CAGUGG. As two triplets, these signify that alanine and tryptophan should be deposited side by side when the protein molecule is formed in the ribosome. In the center of this code sequence, however, there is a sense sequence of GUG which signifies glycine. Formulate a theory to explain why glycine is not sometimes inserted in the protein instead of alanine and tryptophan.

11. It has been found that certain chemical mutagenic agents alter one base of DNA, but not others; for instance, nitrous acid changes adenine, but does not affect the other three nucleotides. Some have said that this knowledge opens the way for man to direct the course of mutation. Evaluate such a statement in the light of what you have learned in this chapter.

19

Gene Mutations

IN THE YEAR 1791 A THRIFTY NEW ENGLAND farmer, Seth Wright, noticed a peculiar male lamb in his small flock. This lamb had unusually short legs which were somewhat deformed, so that it had a rather awkward gait. With typical Yankee shrewdness Mr. Wright thought that it would be nice to have an entire flock of sheep like this one, for the New England pastures were usually inclosed by rather low stone fences, and farmers had considerable trouble with the sheep jumping the fences and damaging crops under cultivation in adjacent fields. Sheep like this new one, however, would give no such trouble, since they could do little leaping. Mr. Wright's flock consisted of fifteen ewes and one ram. Upon the advice of his neighbors, he killed his ram and allowed this new ram to mature for breeding. Of fifteen lambs produced during the first season two were of the new type. In the following years others were produced. By breeding these to one another Mr. Wright obtained a pure breed. He first called this the Otter breed, because of its fancied resemblance to an otter — and because many persons thought that this condition had arisen because the mother of the first lamb had been frightened by an otter during her period of gestation. Later the name **ancon** was given to the breed, a name from the Greek "elbow," because the crooked condition of the forelegs makes them resemble a human elbow.

This new breed of sheep arose as the result of a recessive mutation which had occurred somewhere in the recent ancestry of the first lamb which showed the condition. The crosses indicate that the ram was homozygous for the mutant gene and several of the ewes carried it in a heterozygous state. It then became possible, by breeding the mutant phenotypes to one another, to establish the ancon breed of sheep. This same mutation appeared in a flock of sheep in Norway in 1925, and a separate breed of ancon sheep was established from it.

315

FIG. 19-1 *The ancon mutation in sheep. The ram on the right and the ewe in the center are homozygous for a mutant gene which causes certain skeletal changes including a shortening of the legs. The normal ewe on the left is heterozygous.* COURTESY LIFE MAGAZINE, © 1947, TIME INC.

A similar development occurred during the latter part of the nineteenth century, when a worker in South America discovered a branch on an orange tree which bore a peculiar type of orange. At one end of the orange there was a shriveled, indented portion which resembled a human navel. The pulp of such an orange contained no seeds. A study of the method of growth showed that the flowers bloomed and were fertilized normally and that the fruit began its development in the usual way but soon became abortive. Around this abortive fruit a second fruit developed. All the potential seeds remained in the aborted fruit, leaving the fleshy part of the orange free of seeds. This condition resulted from a mutation in a cell of the meristematic region of a bud which grew to form this branch of the tree. Through subsequent budding and grafting, the mutation was propagated. An American tourist brought a twig back to California, and thus the great navel orange industry of America was founded.

These illustrations show how mutations may arise and how some of them may be of sufficient advantage to man to warrant their selection and continued propagation. In a similar way, it is easy to see how mutations may arise in a natural population and how some may be of sufficient advantage to become established through the process of natural selection. We must be careful, however, not to confuse the appearance of hidden recessive traits with the occurrence of mutations. We have already learned that recessive genes may be carried for generations without any phenotypic expression. The appearance of an aberrant form in a population may be only the expression of a long established gene which has become homozygous through the mating of heterozygous parents. Careful breeding tests or studies of the ancestry are necessary to determine whether a particular case is really a recently mutated gene.

The possibility that new types of inherited characteristics may appear suddenly, without any previous indication of their presence in the race, was first suggested by Hugo De Vries in 1901 as a result of his experiments and observations on the evening primrose, *Oenothera lamarckiana.* He showed that from plantings of very large numbers of these primroses a certain small proportion of the offspring would show variant characters which would be transmitted to future generations. Most of the variations he described were due to rare cross-overs between translocated chromosomes rather than to changes in the genes, but De Vries deserves credit for the formulation of the concept of mutation and its importance from an evolutionary point of view.

Discovery of Mutations

The scientific study of true mutations did not begin until 1910. During that year Thomas H. Morgan, working at Columbia University, reported the discovery of a few *Drosophila* males with white eyes among hundreds of flies with the normal red eyes. Recognizing the potential significance of this discovery, he bred these males with their red-eyed sisters and obtained some white-eyed females, a result which indicated that some of the females in the original culture were heterozygous. By mating white-eyed males to these white-eyed females, he established a pure-breeding race of white-eyed flies. This change in eye color proved to be produced by a single gene mutation which had occurred in an *X*-chromosome of a germ cell of a normal fly. Literally millions of fruit flies were carefully scrutinized during the next few years in an effort to find other mutations. This tedious search bore fruit, and about five hundred mutations were found during the next 17 years as a result of the combined efforts of geneticists throughout the world. Flies were found with pink eyes, purple eyes, rough eyes, small eyes, and no eyes; they were found with curved wings, bent wings, blistered wings, outstretched wings, miniature wings, and no wings; they were found with yellow bodies, black bodies, ebony bodies, and misshapen bodies; one was even found with legs growing out of its head in place of antennae. The discovery of these and many other mutations made possible the extensive genetic studies which have been carried out on the fruit fly.

FIG. 19–2 *A bud mutation in grapes. The Emperor seedless grape on the left arose as a mutation on one branch of a vine bearing Emperor seeded grapes, which are shown on the right. This is a mutation which is beneficial to man, but would be harmful to a plant not grown under cultivation.* H. P. OLMO IN JOURNAL OF HEREDITY.

THE NATURE OF MUTATIONS

The term mutation is sometimes used to refer to all forms of changes which result in altered patterns of heredity. Since this usage often causes confusion, we shall restrict our use of the word to changes of individual genes, which are sometimes called **gene mutations** or **point mutations,** since only one point or locus on the chromosome is involved in each case, in contrast to chromosomal aberrations which affect larger portions of the chromosomes.

How Mutations Occur

Gene mutations occur suddenly. At one instant a certain human gene may be of such a nature as to contribute to the formation of a normal iris of the eye, but in the next second it may be so altered in its pattern that it no longer has the power to carry out its function in a normal manner. It has now acquired a new structure and will reproduce itself in this new pattern — in other words, it has mutated. Should this mutation occur in the germ cells of a man, it may be transferred to a zygote which will grow into another human being. Through countless thousands of mitotic divisions the gene will faithfully reproduce itself in its new form. In all other body tissues it may exert no noticeable effect, but,

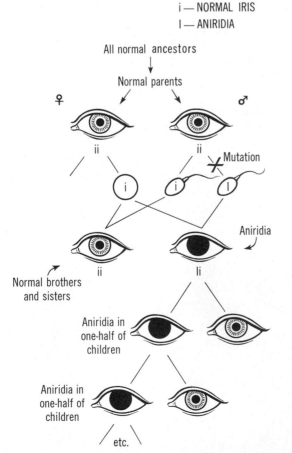

FIG. 19–3 *The origin of a mutation for aniridia (absence of the iris of the eyes) in man.*

TABLE 19–1

Direct and Reverse Mutations Obtained by Irradiation of Adult Males
(Dosage of 3975 r)

Gene	Direct mutations		Reverse mutations		Per cent mutations	
	No. flies examined	Mutations	No. flies examined	Mutations	Direct	Reverse
Y	11,620	3	69,923	1	0.0258	0.0014
sc	11,620	6	101,042	3	0.0516	0.0030
w^a	11,620	0	69,302	0	0.0000	0.0000
ec	11,620	18	57,323	0	0.1548	0.0000
ct	11,620	0	57,323	1	0.0000	0.0017
v	11,620	6	61,119	1	0.0516	0.0016
m	11,620	3	39,923	2	0.0258	0.0050
g	11,620	6	57,323	4	0.0516	0.0070
f	11,620	6	130,421	11	0.0516	0.0084
car	11,620	0	69,302	1	0.0000	0.0014
Total	116,200	48	713,001	24	0.0413	0.0034

This table shows a comparison of the rate of direct mutation (from the dominant wild-type gene to a recessive allele) and reverse mutation (from the recessive allele back to the dominant wild-type) in *Drosophila melanogaster*. The rate of reverse mutation is less in all cases except two, but the fact that such mutations occur at all indicates that mutations are not irreversible changes of the genes. X-rays were used to increase the rate of mutations for these studies. Compiled from the data of Winchester, Moore, and Johnson in *The American Naturalist*.

when the time comes for the development of the iris of the eye, something goes wrong. This is a dominant gene and, even though there is a normal allele, the iris cannot be formed properly and the child is born without any irises in the eyes. We say that it has *aniridia*. The mutant gene will also be included in the germinal epithelium from which the germ cells are to be derived when the new individual becomes mature, and one-half of the germ cells of such a person will carry the mutant gene. This means that on the average one-half of his children will show the same condition.

Reverse Mutations

A mutant gene will be propagated through future generations, although there may be rare cases in which it reverts to normal. Let us assume, for instance, that the mutant gene in man has passed through many generations and has appeared in perhaps thousands of afflicted persons. Perhaps in one germ cell of one afflicted man the gene may suddenly mutate to the normal recessive allele and continue to propagate itself in its new reverted state. This is known as a **reverse mutation.** The occurrence of reverse mutations shows us that mutations are not permanent mutilations, losses, or inactivations — whatever the nature of the change, it is reversible. The author has made an extensive study of a number of recessive genes in *Drosophila* which had mutated from the normal or wild type. It was found that reverse mutations of these genes occurred much less frequently than direct mutations, but, when a gene mutated back to the normal condition, it regained all the characteristics of a normal gene at this locus which had never undergone the double change.

Reverse mutations may be demonstrated in *Neurospora* with considerably less effort. A mutant form of this mold has appeared which requires the presence of adenine in the food — it has lost the power to produce one of the enzymes necessary for the synthesis of this substance. To test for the presence of reverse mutations in this strain it is only necessary to put large numbers of them on a plate of agar which does not contain adenine. If any colonies grow, it would appear that they must come from reverse mutations which have regained the power to produce this enzyme. About 600 million spores of this mold were placed on each of these agar plates not containing adenine. The spores on one plate were treated in various ways which might increase the rate of mutation. There were 47 colonies on the treated plate, but only one on the untreated plate.

Thus we may conclude that genes, in some cases at least, do not undergo any irreversible changes in the process of mutation. Logical reasoning will show that this must be true. Mutations play an important part in evolution in that they provide new characteristics upon which natural selection can operate. Any gene as it exists today must have undergone mutation many times in the course of the development of the species. It would indeed be difficult to explain how the complex forms of life on the earth today could have arisen from a series of destructive changes in their individual genes. We learned in Chapter 17 that an infinite variety of genes can exist because of the infinite variety of combinations of bases in the DNA molecule, and mutations can appear when there are small rearrangements or substitutions of the bases. Such slight deviations may be very far-reaching in their effects, as we have already learned. Within the framework of such a concept, it is easy to see how reverse mutations may occur through some change of the altered parts back to the condition which existed before mutation took place.

THE FREQUENCY OF MUTATIONS

So far in our study of genetics we have emphasized the stability of the gene, although we have recognized the fact that mutations do occur. H. J. Muller has estimated that there is an average chance of less than one in a million of any given gene undergoing a mutation in *Drosophila* from the time it is formed by a replication until it divides again. This estimate indicates the very great stability of genes. For at this rate there will be a mutation of any particular gene in *Drosophila* only about once every 40,000 years. This may seem too infrequent to have much evolutionary significance, yet when we compare such a mutation rate with the total number of genes per individual, we find that there will be some sort of mutation in about every twenty germ cells on the basis of

Muller's estimate. This figure is higher than the rate which is commonly assumed, because it includes the mutations which do not produce any visible effects but only certain small effects on the viability or fertility of the flies which receive such mutant genes. The great majority of mutations are of this type, as we shall learn later in this chapter.

Variations in Mutation Rates

This discussion is not intended to give the impression that all genes mutate with equal frequency or that any one gene mutates at the same frequency under all conditions. There is a considerable range of variation among the genes of different species and among different genes of the same species, as well as a difference in the mutation rate of any one gene under different conditions. The work of Haldane has indicated that, in man, the average gene is considerably more stable than that in *Drosophila*. The mutation rate in *Drosophila* per generation is much greater than the rate found in man. On account of the great disparity in the lengths of generations in the two species, however, the average rate of mutation in *Drosophila* for any given period of time during any generation is greater than the rate in man for the same period of time. This lower rate in man is to be expected because most mutations are harmful, and if man had the same mutation rate per unit of time as is found in *Drosophila,* he would accumulate such a large number of mutations in his lifetime as to lead to serious genetic incapacity in the race.

Haldane estimates that a single human gene has a life expectancy of 2,500,-000 years without change. This estimate was obtained partially from observations of mutations which occurred in large hospitals in proportion to the total number of births. It is quite likely, of course, that this study includes genes which are among the most mutable in man, and it is possible that the average mutation rate for genes in mankind as a whole is even lower than Haldane figured. As an example, from records of 128,763 children who were born in two large hospitals in Copenhagen, it was found that fourteen of the children were **chondrodystrophic dwarfs.** In this type of dwarfism the head and trunk are of a normal size but the arms and legs are extremely short. This condition is known to result from the presence of a dominant gene. Three of these children had parents who were also afflicted with this condition, but the others were from normal parents. We can assume that this last group resulted from mutations in the germ cells of the parents. This gives a rate of one mutant dwarf in about every 11,500 births, or one mutation for every 23,000 genes at this locus in the children. This also means one mutation for each 23,000 parents. If we assume that the average age of the parents was 30 years, then we can estimate that this particular gene mutates once every 690,000 years, which is somewhat more frequently than the estimate for mutations as a whole — which was an average of one mutation in each 2,500,000 years for any one gene locus. Of course, such a method is subject to a number of possible sources of error, but a statistical study of the number of chondrodystrophic dwarfs in Norway in proportion to the total population yields similar results. Such a study was made by

FIG. 19-4 *Chondrodystrophic dwarfism. This condition is brought about by a gene mutation which has been studied extensively in an effort to determine the mutation rate in man.* COURTESY C. NASH HERNDON.

comparing the number of such dwarfs to the number which would be expected as descendants from afflicted parents. The difference between the two figures represents the number which arose as a result of a mutation in the germ cells of normal parents. The figures obtained in this study agree with the results obtained from the hospital records.

Similar studies have been made for other genes. The results of some of these are given in Table 19–2. These figures show variation in the mutation rates of different genes, as would be expected, but from them we can gain some idea of the average mutation rate in man.

Effect of Temperature, Age, and Sex

Temperature and age are factors which influence the frequency of mutations. Animals raised under a high temperature will yield more mutations than animals of the same breed raised under a low temperature. This applies mainly to animals that do not maintain a constant body temperature, for the germ cells of warm-blooded animals remain at a rather constant temperature regardless of the surroundings. Also, in *Drosophila* it has been shown that sex is a factor in the mutation rate. Studies of the mutation rate in the *X*-chromosome indicate that sex-linked mutations occur considerably more often in the males than in the females. A study of the mutation rate of hemophilia in man indicates a similarly higher rate in males.

Thus we see that there are many factors involved in the frequency of mutations, and no definite figure can be set which will apply to all genes under all conditions. Nevertheless, this is what would be expected on the basis of our knowledge of the nature of the gene. Just as various conditions affect the rate of chemical reactions, we would expect a variation in the mutation rate of a gene under different conditions.

TABLE 19-2

The Mutation Rate of Ten Genes in Man
(From Figures by Haldane, Neel, Schull, and Falls)

Condition caused by the gene	Effect on the body	Appears once in every	Mutation rate in per cent
DOMINANT			
Pelger anomaly	Abnormal white blood cells, tends to reduce resistance to disease	12,500 gametes	0.0080
Chondrodystrophic dwarfism	Shortened and deformed legs and arms	14,300 gametes	0.0070
Retinoblastoma	Tumors grow on retinas of the eyes	43,500 gametes	0.0023
Aniridia	Absence of iris of the eyes	200,000 gametes	0.0005
Epiloia	Red lesions appear on face; later tumors develop in brain, kidneys, heart, etc.	83,333 gametes	0.0012
RECESSIVE (autosomal)			
Albinism	Melanin not developed in hair, skin, or iris of the eyes	35,700 gametes	0.0028
Amaurotic idiocy (infantile)	Born normal, deterioration of mental facilities during first months of life	90,909 gametes	0.0011
Ichthyosis congenita	Rough, scaly skin at birth, may cause infant death	90,909 gametes	0.0011
Total color blindness	Difficulty distinguishing colors	35,700 gametes	0.0028
RECESSIVE (sex-linked)			
Hemophilia	Blood clots very slowly	31,250 gametes	0.0032

TYPES OF MUTATIONS

Most of the mutations which have been used as illustrations in earlier chapters in this book have been those which involve clearly distinguished phenotypic effects. These, of course, are the ones which are most easily found and most easily studied. Not all mutations, however, produce changes in the organism which will be visible upon cursory examination. Many will result in changes in physiological states which will have no recognizable effect on the structure of the body. A mutation may influence the viability of the organism or its fertility, and while either of these kinds of change might be of great evolutionary significance, neither may give any external indication of its presence.

A very careful analysis of mutations in *Drosophila* has been made to determine the relative proportions of the different kinds of mutations. From this study it was found that most of them (80 per cent or more) are **detrimental** mutations. Such mutations produce no visible effect, but have an influence on the viability of the fly. **Lethals** were the second most common type of mutation. These are genes, you will recall, which have such an extreme effect on the body that all or most organisms which are homozygous for them will die. Somewhat less than 20 per cent of all mutations fall in this category. This leaves less than 1 per cent of the total mutations in the group which produce **visible** effects. Thus the mutation rate is actually far greater than the frequency of visible mutations would lead us to suppose. We have mentioned earlier in this chapter that there would be about one mutation in every twenty *Drosophila* germ cells, but this would mean only one mutation in every 2000 germ cells which could be detected phenotypically. And even these mutations could not be detected in the first generation unless they happened to be dominant. Thus it becomes evident that any studies which attempt to determine the total number of mutations that occur naturally or that may be induced by any artificial agent must include techniques for the detection of all types of mutations.

THE DETECTION OF MUTATIONS

Dominant Visible Mutations

These are the easiest of all to detect, for they express themselves in individuals which carry them either in the heterozygous or in the homozygous state. Unfortunately, from the standpoint of detection, these occur much less frequently than recessive visible mutations. This statement may not seem to hold true for man, since the majority of human mutations which have been detected are dominant. In other forms of life which have been studied genetically, we find that the great majority of mutations are recessive. This state of affairs is due not to a difference in the types of mutations, but to the difference in the method of detection. Since we cannot use experimental breeding to detect recessives in man, and since close inbreeding normally does not occur, we are limited in our ability to detect recessive mutations. Hence, the majority of mutations which are detected in man are dominants.

Intermediate Visible Mutations

The mutations which fall into this category are also easily detected because they have a phenotypic effect when heterozygous. Evidence is accumulating that many genes — possibly most — which were formerly thought to be entirely recessive actually produce some detectable phenotypic effects when heterozygous, and that would place them in this category. Many of these effects are small and can hardly be said to be intermediate between the full expression of the characteristic and expression of the homozygous allele, but they would lie in this category. The fact that there is any effect at all is of great value in trying to detect mutations, especially in man where the breeding techniques for the detection of mutation are not possible. For instance, there is a gene in man which, when homozygous, causes a condition known as **xanthomatosis.** This is characterized by a defect in fat metabolism and results in localized accumulations of excess lipids (fatty substances). Fatty tumors develop in various parts of the body as a result of this defect. Formerly thought to be entirely recessive, it has recently been found that the heterozygous carriers of the gene have hypercholesterolemia (an excess of cholesterol in the blood). Thus, through blood analysis it is possible to detect this gene in the first generation after it arises through mutation. Some other examples of incompletely recessive genes in man are given in Chapter 5.

Sex-Linked Visible Mutations

The detection of this type of mutation is comparatively simple, since all sex-linked genes act as dominants in the heterogametic sex, which is the male in most animals. An examination of such males will reveal any visible mutations that have occurred on the X-chromosome of their female parents. The recessive mutations which arise in the germ cells of the male parents will not show in the females which descend from them, but can be detected in the offspring of these females. This would also be true of recessive mutations which occur in the X-chromosome of a female but are transmitted to female offspring — the mutation will show in the second generation males. In *Drosophila* it is even possible to short-circuit this two-generation cycle and detect sex-linked mutations in the male offspring of the males in which they occur. This has been made possible by the discovery of females with attached X-chromosomes. In such females the two X-chromosomes which are attached do not separate in meiosis. Hence some eggs are 2 X and some have no X-chromosomes. Males, therefore, come from the no-X eggs fertilized by X sperm. Hence, a mutation on the X-chromosome of a male can be expressed in one of his male offspring. It is interesting that most of the sex-linked mutations of man which have been detected are recessive. This is understandable in the light of the ease with which recessive, sex-linked mutations are detected.

Autosomal Recessive Visible Mutations

Special breeding tests are necessary, as a rule, to detect this type of mutation. It is simplest in monoecious organisms (those that produce both types of gametes in one individual). The garden peas which yielded such good results for Mendel are excellent in this respect, for they are not only monoecious but also self-fertilizing. Corn produces both gametes on one plant, but it is usually cross-pollinated by the wind, and to make a genetic analysis it is necessary that self-pollination be carefully done by the experimenter. This is a simple procedure, however, and corn is therefore used extensively in genetic studies. If a recessive mutation occurs in a reproductive cell, the offspring coming from that cell will as a rule be heterozygous. When this organism is self-fertilized, the recessive mutation will show in one fourth of the offspring.

Practically all animals which are being studied genetically are dioecious, and it is not possible to apply to them the comparatively simple method which can be used for most plants as described above. At least three generations of crosses are usually necessary after the origin of the mutation before an autosomal recessive mutation can be discovered. Should such a mutation occur in a sperm cell, the offspring from this cell, after union with a normal egg, will be heterozygous and indistinguishable from its homozygous siblings. This heterozygous individual must be mated to a homozygous normal individual, with the result that the second generation will also appear to be normal. But one-half the offspring of such a cross will be heterozygous, and when these are crossed freely among themselves about one-fourth of the matings will occur between two heterozygous individuals. Thus about one-sixteenth of the total offspring in this third generation will show the mutation. To accomplish this in three generations, however, it is necessary to isolate the offspring of each first generation pair. If there is indiscriminate mating of the second generation individuals, the recessive gene may be carried for six, seven, or many more generations before two heterozygous individuals happen to breed and produce offspring which show the mutation. There will be many cases in a natural population where an autosomal recessive mutation will arise and will never be expressed. It may become eliminated without ever becoming homozygous in any individual.

There is another method of detecting recessive visible mutations which is sometimes used in *Drosophila* and other animals for which quite a number of genes are now known. This consists of crossing the normal or wild type with individuals carrying a number of recessive genes. The first generation following such a cross will show any mutations which have occurred in the wild-type individual at any of the loci which are being studied. For instance, suppose we cross wild-type *Drosophila* males with females which are homozygous for sepia eyes, curled wings, ebony body, and rough eyes. These are all recessive genes which are located on the third chromosome. Now, should a mutation at any one of these four loci occur in the germ cells of the wild-type males, a fly will be produced in the offspring which will show the mutant character. Such a method would show only those mutations which appeared at one of these four loci. These would represent a very minute proportion of the total number of possible mutations, but the method is practical when one wishes to determine the mutation rates of specific genes. Figure 19–5 illustrates diagrammatically this method of detecting mutations.

Lethal Mutations

The lethals which give a visible effect when heterozygous, such as the bulldog gene in cattle which was discussed in Chapter 11, are easy to detect. When two heterozygous individuals are crossed, one obtains the unusual ratio of 2 : 1 among the surviving offspring. Most lethal mutations are recessive, however, and the heterozygous individuals cannot be recognized. These are more difficult

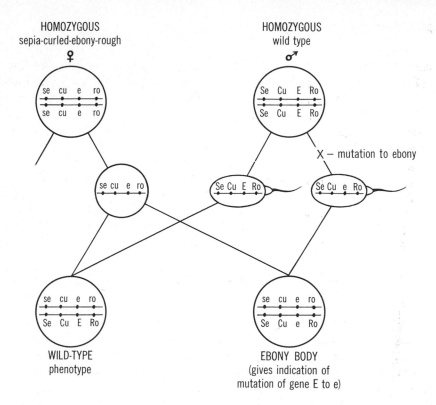

FIG. 19-5 *How mutations of specific loci in Drosophila may be detected by the use of recessive genes in one of the parents.*

to detect, but they can be located by a method somewhat like that used for the detection of visible recessive mutants. If a lethal mutation appears in a germ cell of an animal, then any offspring developing from this germ cell will be heterozygous. When bred with a normal individual, this one will produce more heterozygous offspring. A careful study of the progeny of these in turn will reveal the presence of this lethal. If the animal lays eggs, approximately one-fourth of the eggs from heterozygous parents will fail to hatch. If the animal is a mammal, about one-fourth of the offspring will die in various stages of embryonic development, at birth, or shortly thereafter. Among seed plants about one-fourth of the seeds will fail to germinate or will produce plants which die shortly after sprouting. This type of study is very tedious but will show the presence of lethal mutations.

In *Drosophila* a more rapid technique has been worked out using chromosomes marked with dominant genes. When lethals appear, there will be an absence of certain expected classes of offspring. One such technique is described in the next chapter in connection with the artificial induction of mutations.

Detrimental Mutations

Detrimental mutations are the most common yet most difficult of all types of mutations to detect. Let us assume that a recessive mutation of this nature appears on the *X*-chromosome of *Drosophila* and that it causes a reduction in viability of 20 per cent. One-half of the male offspring of heterozygous females will receive this gene, and 20 per cent of these will die. This means that there will be an over-all reduction of 10 per cent in the male offspring. To detect the presence of this gene, it is necessary to examine a much larger number of flies than must be examined in order to detect a recessive, sex-linked lethal, which causes a 50 per cent reduction in the number of male offspring. To detect a decrease in fertility, it is necessary to breed each male independently and to note the percentage of infertile males. Likewise, autosomal detrimental mutations may be detected in somewhat the same manner as autosomal lethals, but again it is necessary to study larger numbers of offspring in order to detect a smaller decrease in numbers than is true for lethal genes.

MUTATIONS IN HAPLOID ORGANISMS

Many of the difficulties in the way of detecting mutations are avoided in those forms of life which are haploid in the major part of their life cycle. Such an organism is the mold *Neurospora,* which we have already discussed in some detail. Among animals there are certain insects, such as male honey bees, which have only one of each type of chromosome in their bodies. Unfortunately for genetics, the breeding habits of the bees and the problems of studying them do not make them favorable subjects for investigation. However, other members of the order *Hymenoptera* make very satisfactory forms for genetic breeding and study. One is the small parasitic wasp, *Habrobracon juglandis,* which has been investigated extensively by P. W. Whiting who has found that females carrying recessive genes yield male offspring which show these genes regardless of what type of male the female is mated to. This is due to the fact that the males develop from unfertilized eggs. These tiny wasps are parasitic on the meal worm, *Tenebrio,* and may be raised very conveniently in the laboratory with a little corn meal and the *Tenebrio* worms.

In all haploid organisms every mutation behaves as if it were a dominant, and hence is just as easily detected as dominance in a diploid organism. Because of this valuable characteristic, many important genetic discoveries have been made on such organisms during recent years. An example of how *Neurospora* can be used will illustrate how mutations with a biochemical effect can be detected. We start with a wild-type mold growing on a minimal medium. Asexual spores from this culture are transferred to a complete medium containing vitamins and amino acids. These spores may be treated with X-rays or in other ways to increase the mutation rate. From this tube single spores of sexual fruiting bodies are transferred to tubes containing complete media. Then from these tubes transfers are made to tubes which are deficient in specific vitamins and amino acids. Whenever a tube is found with no growth, that is a sign that a mutation has occurred in some gene which is necessary to synthesize the specific vitamin or amino acid which is lacking in this particular tube. Figure 19–6 shows how this would come about if there was a mutation which affected the synthesis of pantothenic acid (one of the B vitamins).

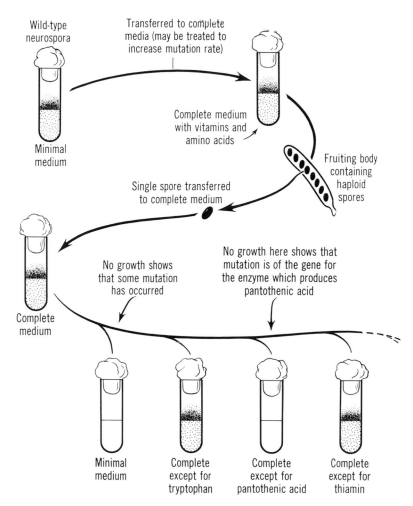

FIG. 19-6 *How mutations with a biochemical effect may be detected in Neurospora. We start with wild-type mold growing on a medium which contains the minimum needs for growth. Asexual spores from this are transferred to a complete medium containing vitamins and amino acids. These spores may be treated with X-rays or in other ways to increase the mutation rate. From this tube single spores of sexual fruiting bodies are transferred to tubes containing complete media. (Only the tube is shown here.) From this tube, transfers are made to tubes deficient in certain vitamins and amino acids. In this illustration, there is no growth in the tube without pantothenic acid, thus indicating a mutation of the gene which produces the enzyme necessary for the synthesis of pantothenic acid.*

Gametic and Zygotic Lethals

Lethal genes are not only easier to detect in haploid organisms, but it is easier to learn something about their effects if they are early-acting lethals. Dominant lethals sometimes arise in diploid organisms, but these can be detected only by noting a decrease in the number of offspring produced, and one fails to distinguish between those lethals which cause inactivation of the sperm and those which cause early death of the zygote. In *Habrobracon*, however, inactivated sperm would mean fewer fertilized eggs and hence more male offspring, because males hatch from unfertilized eggs. Zygotic lethals carried by the sperm will result in the death of the zygotes carried by fertilized eggs. There will hence be a reduction in the number of both males and females.

HARMFUL MUTATIONS

Mutations are, without doubt, an essential part of the process of evolution, by means of which all forms of life are constantly changed and adapted to changing environmental conditions. In spite of the fact that it is mutation which makes evolution possible, the great majority of the mutations which occur are harmful to the organism that expresses them. Consider a few of those in man which we have used as illustrations — albinism, aniridia, alkaptonuria, amaurotic idiocy — these genes are definitely harmful to the persons affected. Many other mutations in man are neutral in survival value; very few can be classified as of positive value to the persons showing them. Why are most mutations harmful, and how can mutations be of benefit to the race or species in which they appear if the majority of them are harmful and if very few actually possess any survival value?

To obtain answers to these questions, we must first understand that the living organisms on the face of the earth today represent a highly selected group. Only a small fraction of the total number of species of living things that have inhabited the face of our globe since life began upon it are now in existence; the great majority have failed in the struggle for existence and have become extinct. Those that continue to inhabit the earth, we may be sure, are highly selected and efficient organisms, for they have survived where most have failed. None is perfect. There is still room for improvement, and there must be improvement if the species are to continue to survive; but even so, they are quite efficient as they are.

We may compare this constitution to that of a highly efficient, but not perfect, mechanical instrument — say a watch. Suppose you have a watch which keeps fairly good but not perfect time. Knowing nothing about watches, suppose you open it and push a lever without any knowledge of what that lever is for or what it does. It is, to be sure, possible that pushing the lever in ignorance of what it may do will make the watch keep better time than before. There is a much greater chance, however, that the change will make the watch keep poorer time, and if the change is too extreme, it may stop running altogether. Like a watch, a living organism is highly efficient but not perfect, and mutations represent random changes in the controlling mechanism of the living organism. As with the watch, a random change (a mutation) may just happen to result in an effect which is beneficial, though the chances are against it. It is much more likely that such a change will make the organism less efficient, and, if the change is too extreme, it may cause life to stop altogether. Even so, the few

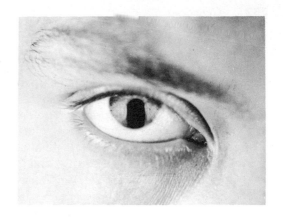

FIG. 19-7 *Coloboma iridis, a sex-linked, recessive, visible mutation which can appear in the son of a woman in which it occurs. It results in a fissure in the iris which causes the pupil to extend out to the edge. Since this reduces the power of accommodation to light intensity, it is one of the many possible mutations which produce small but harmful effects on the body.*

beneficial changes which do occur provide a sufficient basis for evolutionary advance. Through selection, the many harmful mutations are held in check and eliminated, while the advantage held by the organisms with beneficial changes will enable them to become established and to multiply their kind. To summarize, most mutations are harmful because they represent changes in an already highly efficient organism; yet mutations are necessary to the continued survival of a species, for without them there would be no beneficial changes to afford a basis for natural selection.

A highly adapted species might very well go on existing indefinitely if it were fortunate enough to live in an environment which never changed from one generation to the next. But the conditions of life do change, and the pages of geologic history amply reveal the fact that species must in most cases readapt to altered circumstances or perish. Thus mutations may not be essential to the survival of the species in certain limited conditions, but for most forms of life, most of the time, mutations are essential. To survive, a species must strike an equilibrium between the harmful effects which result from too many mutations and the insurance of continued existence which is provided by the occurrence of some mutations.

SOMATIC MUTATIONS

Usually the mutations emphasized in the genetic studies are those which occur in the germ cells, for mutations which occur in somatic tissues generally perish with the individual in which they occur and have no lasting significance from the standpoint of heredity. Conceivably, a gene for brown eyes may mutate to a gene for blue eyes in a somatic cell of the skin on the bottom of a person's foot, but it lives and dies without ever making its presence known. Many mutations, no doubt, arise in this manner with never a chance to express themselves. Occasionally, however, somatic mutations do have a phenotypic effect. If the mutation occurs early enough in the embryonic development of the organism, it may express itself in the individual in which the mutation occurs. Suppose a mutation from brown to blue eyes occurs in one of the cells of a very young embryo that is heterozygous for the brown and blue. This will give one cell that is homozygous for blue. Now, let us further assume that one side of the head develops from the cell which mutated and the other side of the head develops from one of the unmutated cells. The eye which contains the gene for brown will be brown since brown is dominant, while the eye containing the genes for blue will be blue. Thus a person can have one brown and one blue eye. A popular magazine recently printed a colored picture of a beautiful young lady with this unusual combination of eye colors. When asked how she thought such a thing came about she provided an eloquent although somewhat ungenetic explanation — "I guess my genes just got all loused up."

Mutations which occur early in embryonic life may produce mosaic effects on the body if the gene influences the entire body. Thus a fruit fly may be produced which is yellow on one side of the body and gray on the other. Should the mutation occur in the four-cell stage, only one-fourth of the body may be yellow. A mutation from blue to brown in later stages of embryogenesis may explain a spot of brown in the otherwise blue iris of a human eye.

Cancer, that dread affliction of mankind which seems to yield so slowly to the progress of scientific research, may be related to somatic mutations. It may be that the mutation of a gene which is related to the correlation of cell growth and activity suddenly changes the cell containing this gene into an outlaw cell, a cell growing wild, living parasitically on the tissues of the body of which it once was a normal part. We know that inheritance plays an important part in a person's predisposition to cancer. Perhaps such inheritance consists of the relative instability of some gene related to cell correlation; it may be that such a gene mutates more easily in persons predisposed to cancer. There are a good many "if's" involved in this question, but such a concept is in line with the known facts relating to other genes. There are a number of cases in both plants and animals where genes are transmitted which have a predisposition to frequent mutations. The variegated pattern on the leaves of some plants is produced by such a gene.

PROBLEMS

1. Why is it easier to establish mutations in plants than in animals?
2. Show the genotype of the ancon ram and the normal ewes which produced the ancon lambs in Seth Wright's flock.
3. Most of the mutations which were discovered in *Drosophila* in the early days of mutation study were on the *X*-chromosome. What is the reason for this?
4. What evidence do we have that mutations are not always destructive changes in the genes?
5. In most forms of life which are studied genetically the known recessive mutations far outnumber the known dominant mutations, but in man the condition is reversed. What is the probable explanation for this?
6. Among the sex-linked mutations known in man, the recessive genes are more numerous than the dominant. Explain this situation in the light of Question 5.
7. Why are sex-linked lethals easier to detect than autosomal lethals?
8. What is the great advantage of *Habrobracon* as an experimental animal in genetics?

9. What is the effect of a somatic mutation on the appearance of an individual In which it occurs?

10. We say that somatic mutations are not of great genetic significance in animals, yet they may be in plants. Explain.

11. Most mutations seem to be harmful rather than helpful to the organism in which they are expressed. Why?

12. What difference will there be in the way in which recessive mutations and dominant mutations, which occur in the germ cells, come to be expressed phenotypically?

13. Suppose, as of this moment, there were no more mutations of genes. Describe the effects which would be expected in the future as a result of this change.

20

The Artificial Induction of Genetic Changes

MUTATIONS ARE EXTREMELY VALUABLE to the plant or animal breeder who is constantly seeking new variations of existing types of cultivated plants or domestic animals in an effort to improve them. Great things can be accomplished through selection, but selection can only go so far without mutation. Also, mutations are extremely valuable to the experimental geneticist, for genes can be recognized and studied only when a variant allele is available. Hence, the more mutations that can be found, the more can be learned about genes and their action. Unfortunately, from this point of view, the natural rate of mutation is exasperatingly slow. Many thousands of fruit flies must be carefully examined in order to find one visible mutation. Until 1927, geneticists could only work, hope, and (if they were of a religious nature) perhaps pray that mutations would occur in the forms of life with which they were working. Nevertheless, many had dreamed of the possibility of speeding up the mutation rate. The poor *Drosophila* had been subjected to almost every conceivable maltreatment in an effort to force the stubborn genes into a more rapid rate of mutation. Poisons in sublethal doses, supersonic vibrations, centrifugal force, and electricity were among the many treatments which were tried with uniform failure.

THE USE OF X-RAYS

On Drosophila

It was in the twenties that a geneticist at The University of Texas, H. J. Muller, who had come from the stimulating atmosphere of the Columbia University group trained by T. H. Morgan, began some experiments on the effects

334

of **X-rays** on *Drosophila*. He reasoned that the high energy radiation from the X-ray tube could penetrate the cells and perhaps cause mutations. His experiments were successful, and with very heavy doses of X-rays he found that the mutation rate could be increased about a hundredfold over the rate that occurred in his controls, which were not treated with X-rays.

On Barley

Simultaneously with Muller's work, although reported a year later because of the longer time required to raise and analyze the material, L. J. Stadler demonstrated that X-rays produce gene mutations in barley. Barley is a good plant for the study of the effects of irradiation on the production of mutations, because each seed produces several tillers and each tiller bears flowers and seeds. Each tiller comes from an independent primordium in the seed, so that a mutation which is induced in one primordium will affect only one tiller and its flowers. Moreover, the flowers are self-fertilized, and recessive mutations will tend to show in the second generation plants. It is possible, therefore, to irradiate barley seeds which contain the young embryos and to detect within two generations mutations which may thus have been induced. Stadler's work demonstrated the important fact that X-rays are just as effective as mutagenic agents in plants as in animals.

Significance of Early Discoveries

Here at last geneticists had a tool with which to speed up the process of genetic change. Muller reported his findings before the International Congress of Genetics in Berlin in the fall of 1927. Since that time X-rays and other sources of radiation have become standard equipment in genetics laboratories throughout the world, and countless important discoveries in this field have resulted from the use of radiation. Muller was awarded the Nobel prize in 1946 for his brilliant achievements in this field. He succeeded in his experiments because, like Mendel, his work was carefully planned and was executed on a

FIG. 20-1 *Dr. H. J. Muller in his laboratory at the University of Indiana. Dr. Muller was awarded the Nobel prize in 1946 in recognition of his discovery of the effect of radiation on the production of mutations. This discovery opened great avenues of genetic research and has greatly expanded our knowledge of heredity.*

sufficiently large scale to produce statistically significant results. Other workers had performed hybridization experiments with plants before Mendel's time, but they failed to detect the principles of heredity because they did not plan their experiments with the meticulous care that characterized Mendel's work or did not carry them far enough and analyze them mathematically. Likewise, other workers had tried X-rays before Muller's discovery, but they failed because their experiments were not planned with the scientific exactness of Muller's work so as to make the results clear-cut. Muller deserves credit, therefore, not only for his selection of X-rays as a very likely mutagenic agent, but even more for devising a scheme to detect, in accurate numbers, any mutations which were induced.

The ClB Method of Detecting Lethal Mutations

Since lethal mutations occur with a much greater frequency than visible mutations, any study of the rate of mutation can proceed much faster through a study of lethals, provided a method can be devised for easily detecting them. Muller worked out what is known as the *ClB* method for detecting lethals, a method which is illustrated in Figure 20–2. For this work a special type of female is used, carrying one normal *X*-chromosome and one *X*-chromosome in which the central portion has been **inverted** (*C*). This inversion prevents it from crossing over with the normal *X*-chromosome. The inverted chromosome also carries a dominant gene for **bar-eye** (*B*), in which, as we have seen, the eye is bar-shaped rather than rounded in outline. This gene acts as a marker and makes it possible to recognize flies having this particular type of *X*-chromosome. Finally, this unusual chromosome carries a recessive **lethal gene** (*l*). Females of this type are called *ClB* females — *C* for crossover suppressor; *l* for lethal; and *B* for bar-eye.

Females of this type are mated with normal males which have been treated with X-rays. The viable offspring of such crosses will be normal males, normal females, and bar-eyed females in a ratio of about 1 : 1 : 1. The other class of males (bar-eyed males) will die because of the lethal gene in their one *X*-chromosome and will be missing among the living offspring. From these flies the bar-eyed F_1 females are chosen to continue the test. These females, which carry the inverted *X*-chromosome and the *X*-chromosome from a male that has been treated with X-rays, are mated to the normal males which hatch in the same vial. Only one female is placed in each vial for the final test. One-half the male offspring of these F_1 females will die because of the presence of the lethal gene on the *X*-chromosome which carries the gene for bar-eye. The other half will live if there is no lethal in the other *X*-chromosome. If a lethal mutation has appeared in the *X*-chromosome derived from an X-rayed male, however, this other half of the F_2 males will also die, and there will be no males at all among the offspring. Thus it is possible to detect lethal mutations by a quick examination of the vials in this second generation. Each vial which contains only females represents one lethal mutation in a germ cell of one of the original males which were treated with X-rays. A large number of such vials

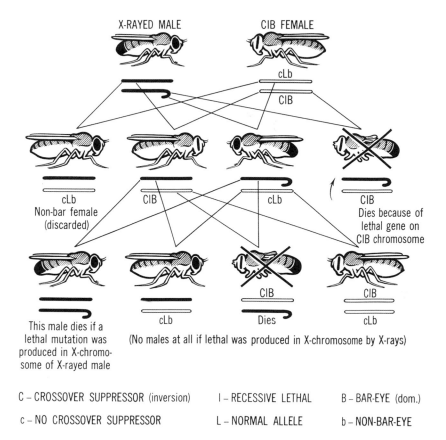

X-RAYED MALE   CIB FEMALE

cLb

CIB

cLb
Non-bar female
(discarded)

CIB

cLb

CIB
Dies because of
lethal gene on
CIB chromosome

This male dies if a
lethal mutation was
produced in X-chromo-
some of X-rayed male

cLb   Dies

CIB

CIB

cLb

(No males at all if lethal was produced in X-chromosome by X-rays)

C – CROSSOVER SUPPRESSOR (inversion)	I – RECESSIVE LETHAL	B – BAR-EYE (dom.)
c – NO CROSSOVER SUPPRESSOR	L – NORMAL ALLELE	b – NON-BAR-EYE

FIG. 20-2 *The ClB method of detecting lethals produced on the X-chromosome of Drosophila males. By this method there will be no males at all in the second generation if a lethal is produced in the X-chromosome of the treated male.*

can be prepared conveniently, and a statistical study of the number of lethal mutations produced on the *X*-chromosome can be made. Of course, a similar number of vials must be prepared containing untreated males to act as controls, for a certain proportion of the mutations would have occurred anyway, and we cannot distinguish the effects of X-rays unless we know how many mutations would occur under similar conditions without the X-ray treatment.

After the report of Muller's discovery, other sources of high energy radiation were tested for their mutagenic property. It was found that the radiation from radioactive isotopes of chemical elements would likewise increase the mutation rate. Since some understanding of the nature of high energy radiation is necessary for an evaluation of its mutagenic effects, let us digress from this discussion long enough to survey the different forms of high energy radiation and how they are measured.

THE NATURE OF HIGH ENERGY RADIATION

Discovery of X-rays

X-rays were discovered in 1895 by the German physicist, Konrad Roentgen, a professor at the University in Würzburg. He constructed a vacuum tube through which he passed a high voltage direct electric current. The electric current jumped a gap in the tube and struck a platinum target. The impact of the electric spark against the target caused much of the energy of the electricity to be given off as short wave, high energy rays. These rays cannot be seen with the naked eye, but Roentgen found that they caused fluorescence of certain minerals in rocks on a nearby table when the tube was turned on in a darkened room. He called these emanations from the tube "X-rays," "X" standing for the unknown.

Roentgen was greatly interested in photography and had the brilliant idea that these rays, although invisible to the human eye, might affect the sensitive emulsion on photographic plates in the same way as visible light rays. He tested this hypothesis by placing metal keys over the plates, still in their light-tight holders, and exposing this set-up to X-rays. When the plates were developed, he saw a clear outline of his keys on an otherwise blackened plate. The rays had penetrated his film holder and exposed the plate within, but the metal keys were of a sufficient density to absorb some of the rays, so the emulsion was exposed less underneath the keys.

Roentgen became so engrossed with his work that he spent most of his time in the laboratory. His wife became concerned about him, and, in order to convince her that he had not gone insane, he invited her to visit the laboratory and see his remarkable new discovery. While she was there, Roentgen asked her to hold her hand over a photographic plate while he turned on the electric current. Development of the plate revealed an outline of the bones of her hand and her rings but very little flesh showing. Roentgen reported and demonstrated his findings in a lecture at the University in January, 1896. His discovery was

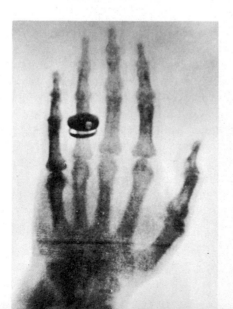

FIG. 20-3 *One of the first X-ray photographs to be made. This was made by Roentgen during the early days of experimentation with X-rays and their penetrating qualities.* BY KONRAD ROENTGEN.

hailed by physicians all over the world; now they had a way to "see" inside the human body. Broken bones could be set with greater precision, and intestinal obstructions and lung infections could be located. The discovery was as sensational in its time as the discovery of how to split the atom in more recent times.

It was not long before it became known that these rays, which could penetrate the human body and reveal its inner workings, could also harm the tissue through which they passed. Many persons who had been working extensively with X-rays began having serious ulcerations and cancers on their hands which had been held under the rays over extended periods. Many fingers had to be amputated, and many persons lost their lives because they had received too much radiation. Thus it became known that the damaging effects were cumulative — that is, small quantities received over a period of time could result in eventual death of the exposed tissue. In Hamburg, Germany, there is a monument commemorating over a hundred early physicians who lost their lives as a result of exposures received in their pioneer work with X-rays. About 30 years were to pass before it became known that these rays could also cause the more insidious changes, known as mutations, which could cause damage to children of future generations.

Discovery of Radioactive Elements

Only a few months after Roentgen announced his discovery of X-rays, a Frenchman, Henri Becquerel, found that radiation of a similar nature was emitted by uranium ore. He observed the same sort of fluorescence of certain minerals exposed to this radiation and found that they could expose photographic plates just as X-rays do. He placed some crystals of a uranium salt on a plate covered by black paper. After several hours the crystals were removed and the plate was developed. At each place where a crystal had stood over the plate there was a blackened outline, but the rest of the plate was clear. It was evident that the crystals were giving off invisible radiation which penetrated the paper and exposed the plate beneath.

Another Frenchman, Pierre Curie, and his wife Marie, isolated **radium** from uranium ore and learned of the value of radium in treating cancer. The cancer tissue appeared to be more susceptible to damage from the radium rays than normal body tissue and thus cancer could be destroyed by an exposure which would not destroy normal tissue. Today the number of radioactive elements (radioactive isotopes) which are known is very large. We have been able to isolate many of them from existing mixtures and many new ones have been created. The extensive scientific, medical, industrial, and military uses of these isotopes have increased the load of radiation to which man may today be exposed. It has, therefore, become of increasing importance to understand the somatic and genetic effects of this radiation.

Kinds of High Energy Radiation

High energy radiation may be classified into two general types. One, **electromagnetic radiation,** is in the form of short wave, high energy rays. These

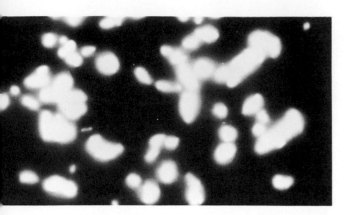

FIG. 20–4 *Effect of radiation from radioactive element upon photographic film. Crystals of uranium nitrate were placed on a film covered with black paper. This photograph is a print made from the film after it was developed. The film was exposed wherever it was beneath one of the crystals.*

are similar to ordinary light rays in that they are forms of radiant energy traveling in waves; the primary difference lies in the length of the waves and their energy. The other, **particulate radiation,** is in the form of actual subatomic particles emitted from atoms with rather high energy. The primary forms of high energy radiation are listed below.

Gamma rays. These are electromagnetic radiations including X-rays and rays of a similar wave length emitted from radioactive isotopes. X-rays emitted from tubes of relatively low voltages lie below the arbitrary designation of gamma rays as far as wave length is concerned. Practically all modern X-ray tubes, however, emit rays in the wave lengths known as gamma rays, although the longer wave, so-called soft, X-rays are used for particular purposes. Ultraviolet rays lie between X-rays and rays of visible light with respect to wave length. The penetration of electromagnetic radiation is inversely proportional to the wave length — the shorter the wave length the greater its penetrating power. Since germ cells generally lie beneath the surface of an organism, the shorter wave lengths are of great significance.

Alpha particles. These are positively charged particles given off from the nucleus of an atom. Alpha particles consist of two protons and two neutrons. They are of very low penetrating power because, being positively charged, they are easily deflected and slowed by the negative charges in matter. They have less genetic significance than some of the more penetrating forms of radiation since they do not generally reach the reproductive glands.

Beta particles. These are electrons, negatively charged particles, emitted from some unstable isotopes. They are much smaller in size than alpha particles, since an electron is only about $\frac{1}{1800}$ the size of either a proton or neutron. Beta particles vary considerably in their penetrating power because there is considerable variation in the energy with which they are ejected from the atom, but even the highly energetic beta particles are not very penetrating because they are deflected and slowed by positively charged particles in matter. Nevertheless, because of their very small size and relatively high energy, they are more penetrating than the alpha particles.

Neutrons. Free neutrons ejected from radioactive isotopes are extremely penetrating because, lacking any electrical charge, they are not deflected or slowed when they pass near charged particles in matter. They tend to move in a straight line until they collide with the nucleus of an atom. Since atoms are made mostly of empty space, this means that they may travel for quite a distance without such a collision. They can be very damaging to living tissues and, by affecting the structure of the atoms with which they collide, they can cause certain materials to become radioactive. A metal coin, for instance, when exposed to a neutron beam will for a time be radioactive and dangerous to carry.

Cosmic radiation. We should not close our list of the common types of radiation without mentioning the cosmic radiation which "rains" down upon us continually from outer space. Primary cosmic radiation comes to the earth in the form of stripped nuclei of such elements as carbon, nitrogen, and oxygen. Few of these particles ever reach the earth, for our atmosphere acts as a shield with a density equivalent of 3 feet of lead at sea level. Some of the particles, however, collide with nuclei of the atoms of the air and give rise to a shower of both particulate and electromagnetic radiations known as secondary cosmic radiation, which is the form to which all matter on earth is continually exposed.

THE MEASUREMENT OF RADIATION

High energy radiation cannot be detected by any of man's senses. A piece of radioactive cobalt (Co^{60}), the size of a pea, could be affixed underneath a chair bottom and any person sitting in the chair would receive a lethal dose of radiation within a few minutes without ever being aware that he was being exposed. Therefore, we must depend upon instruments which are sensitive to such radiation to tell us of its presence and intensity.

The Geiger-Müller tube is the best known and most widely used of the detecting instruments. This tube contains a gas which is ionized by radiation, and these ionizations can be detected with suitable amplifiers and counters. This set-up is efficient for the detection of gamma and beta radiation, but other instruments must be used for alpha particles because of their very low penetrating power; they will not even get through the thin mica window of the tube. Neutrons, on the other hand, are so penetrating that they go right through the tube with little ionization and also require other means of measurement.

The most widely used unit of measure is known as the roentgen (*r*). A roentgen is the amount of radiation which will produce 1 electrostatic unit of charge in 1 cc. of air at standard temperature and pressure. In some measurements, however, this unit is modified. Particulate radiation, for instance, was found to differ in its effect on living tissue as compared with the amount of ionization in a gas. When the occasion demands, scientists may use the *rad* (radiation-absorbed dose) as a unit. This is based upon the energy absorbed per gram of matter, either in living or non-living matter. In studies of the effects of radiation on man we also may use the *rem* (roentgen equivalent man).

IONIZATION INDUCED BY RADIATION

High energy radiation is often called ionizing radiation because it produces ions in matter through which it travels. Atoms carry positive charges in the protons of their nuclei, but these are balanced by an equal number of negative charges in their orbital electrons. Thus, atoms are electrically neutral in their total charge. The loss or the gain of an electron by an atom, however, upsets this balance and the atom becomes either a positive or a negative ion. Alpha and beta particles cause ionization because, as they pass through matter, they tend to pull orbital electrons out of the atoms. This loss of electrons creates positively charged ions. The free electrons are usually captured in the orbit of other atoms, thus converting them into negative ions. Neutrons do not have a charge and thus do not draw electrons out of orbit as do the charged particles. They do cause ionizations, however, by a more direct means. When a neutron strikes the nucleus of an atom it causes the atom to become excited. An excited atom is unstable and may eject an electron. Gamma rays cause ionization in a similar way. When the energy of the rays is absorbed by an atom, it becomes excited and throws off one of its orbital electrons. We sometimes speak of ion pairs which are created by radiation because, whenever one ion is created, the lost particle joins another atom and creates a second ion. One roentgen of gamma radiation creates about two billion pairs of ions in the matter, yet ionize only about one atom out of each ten billion in 1 gram of living tissue.

Both the somatic and genetic damages done by radiation appear to be caused by the ionizations produced in living cells. Free ions are produced which may combine with oxygen and produce highly reactive chemicals. These, in turn, react with genes, chromosomes, and possibly other vital cell parts. This concept is borne out by the fact that a cell held at a very low concentration of oxygen receives much less damage from radiation than a cell with normal oxygen concentration. A colony of bacteria, *Escherichia coli,* will all be killed by a dose of about 35,000 r if they are growing in oxygen. If they are placed in pure nitrogen, however, they can withstand dosages of up to about 100,000 r. Experiments with *Drosophila* and many other organisms show that the mutation rate is much lower if the organisms are kept in an oxygen-free environment during the time of radiation. Also, the somatic damage can be greatly reduced if the exposed organisms have been previously injected with certain chemicals which absorb the reactive radicals that combine with oxygen within the cells. Such a chemical is AET, 2-aminoethylthiouronium bromide hydrobromide. All rats exposed to a dosage of 900 r will normally die, but, if they are injected with AET shortly before the exposure, there is a high rate of survival. In an experiment conducted by the author, six out of eight rats injected with AET survived a dosage of 900 r of X-rays. The two deaths which occurred were probably due to the toxicity of the AET rather than to the radiation, because both of these rats died on the first day after radiation. Rats irradiated without the AET injections, on the other hand, did not begin dying until the third day after receiving the radiation, and the final survivor died on the eighth day. All of the

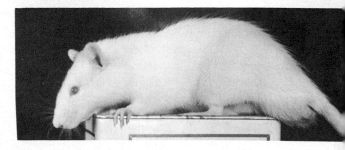

FIG. 20–5 *Protection from radiation damage by AET. These young male rats are all of the same age and were all of about the same weight (125 grams) when the experiment was begun 8 days before these pictures were made. The rat at the top received 900 r of X-rays and is the sole survivor of eight so treated; his weight is only 123 grams. The rat in the center also received 900 r, but was injected with AET just before the radiation was administered; his weight is 214 grams. The rat at the bottom was injected with AET, but was not given any radiation; his weight is 232 grams.*

rats irradiated without the AET showed greatly retarded growth, whereas those receiving the AET first showed almost normal growth as compared with the controls. It is unfortunate that the amount of AET required to give protection from the radiation is at a level which is rather toxic, since this fact seems to rule out the use of AET for radiation protection in man.

INDUCTION OF CHROMOSOME ABERRATIONS

Not long after Muller discovered that ionizing radiation caused mutations, it was found that the number of chromosome aberrations also was increased by such radiation. Deletions, inversions, duplications, translocations, and non-disjunction, as well as the various forms of ploidy, all show an increased incidence after radiation. Some types of aberration result from a single break of a chromosome in a cell, while others require at least two breaks. Single-break aberrations, if they persist, are usually lethal in their effects since an important block of genes may be missing from one of the chromosomes. Many cells survive single-break aberrations, however, because the portion of the chromosome which has been broken off becomes reattached where it broke loose and the chromosome is exactly as it was before the break occurred. When cells are exposed to the radiation and studied shortly thereafter, however, the single breaks can be seen and tabulated. When there is more than one break, the broken ends may become reattached to broken ends at other regions and thus establish a stable aberration which persists.

The chance of a single break in a chromosome seems to be in direct proportion to the amount of radiation which the cell receives and is not influenced by the intensity of the radiation. In other words, the radiation may come from a high voltage X-ray machine and be delivered in a few seconds, or it may be a much longer exposure to a much weaker source of radiation. As long as the total amounts of radiation received are the same, the number of single breaks will be the same. Two-break aberrations, however, increase exponentially as the amount of the radiation increases. This can be explained by the fact that two breaks in the same cell will occur as the square of the probability of a single break. At low dosages of radiation the number of cells with single breaks is higher, but, as the dosage increases, the proportion of cells with two or more breaks increases until they will exceed the number of cells with single breaks. To illustrate, Bender's study of chromosome breaks in human cells growing in tissue culture indicates that a dose of 2 r induces one chromosome break in about 1 out of every 100 cells. The chance of having two breaks in 1 cell would be only 1 in 10,000 ($\frac{1}{100} \times \frac{1}{100} = \frac{1}{10,000}$). If the dosage is doubled, (4 r), the number of single breaks in cells will be doubled to 1 in 50 cells, but the number of cells with two breaks will be quadrupled to one in each 2500 cells. Another doubling of the dosage (8 r) will result in four times the number of cells with single breaks, but eight times the number of cells with two breaks. Thus, we see that the number of cells with two breaks increases according to the square of the amount of the radiation received, whereas cells with single breaks increase only in direct proportion to the dose. Also, the intensity of the dose is an important factor in two-break aberrations. The figures given above have assumed

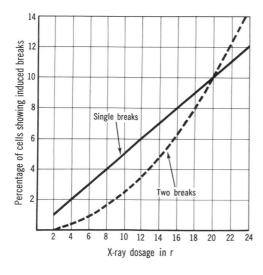

FIG. 20–6 *Relation between the number of single-break and two-break chromosome aberrations induced by different dosages of X-rays on human cells growing in tissue culture. Note that there are more single-break aberrations with low dosage, but there is an exponential increase of two-break aberrations with increased dosage and a linear increase of the single-break aberrations.* FROM DATA BY BENDER.

an intense dosage given within a short time. A long-continued exposure to radiation of low intensity will result in a smaller number of two-break aberrations. There will be just as many cases where two breaks occur in the chromosomes as when the dose is intense, but there will be many cases where one break occurs early, and the chromosome portions unite in the original position before the second break occurs. Hence, there is no chance for a two-break aberration to be formed.

SIGNIFICANCE OF INDUCED CHROMOSOME ABERRATIONS

Chromosome aberrations may result in phenotypic effects which are detectable in the first generation after their occurrence, whereas most gene mutations are recessive in nature and do not produce easily detectable effects when heterozygous. Chromosome aberrations induced in mature germ cells can be of a rather extreme nature and yet not cause the death of the cells, since the balance of the genes within the cells remains normal. If such a germ cell unites with a normal germ cell from the opposite sex, and mitosis of the zygote begins, there will be established an unbalanced condition and an abnormal embryo will be formed. The abnormality may be so extreme as to cause early death or, if less extreme, a viable but abnormal individual may be produced. Such an individual will be sterile in most cases because of the abnormal combinations of chromosomes and genes formed in meiosis, but in some cases viable gametes are produced and the abnormalities may be passed on to future generations. The Philadelphia chromosome, with its relation to leukemia, and the translocated chromosome 21, with its relation to Down's syndrome (see Chapter 16), are good examples of the transmissible type of aberrations occurring in man. In some cases the aberrations may not show any phenotypic effect in the first generation. In meiosis it is possible for a normal complement of chromosomes and genes to be passed to the germ cell, but abnormal complements may be formed in future generations.

That sterility is one of the effects of ionizing radiation became evident during the early work with X-rays. Many of the people who were exposed extensively to these rays had no children. Chromosome aberrations are the cause of such sterility to a great extent, although gene mutations are involved in some cases. In animals the sterility may come about because of the death or inactivation of the gamete-producing tissue, or embryos may form but will die during their early development. In plants the radiation may cause elimination of both male and female lines because a gametophyte generation must be formed before fertilization can occur, and the gametophyte is haploid. Chromosome aberrations or lethal gene mutations in either the microspore or the megaspore may prevent the formation of a viable male or female gametophyte. Some lethal genes, however, do not express themselves until the sporophyte is produced, so the life cycle may proceed to the sporophyte generation.

Studies on mice show that radiation-induced sterility may be temporary if the dosage does not greatly exceed the minimal amount needed to cause sterility. Observations in Hiroshima indicate that the same is true of man. Many of the inhabitants of this city who received rather extensive radiation had no children for several years even though they were married and made no attempts to prevent conception. In the male a few of the spermatogonial cells in the germinal epithelium may survive a dosage which renders the spermatocytes impotent. These few surviving cells may, in the course of time, gradually repopulate the germinal epithelium, and spermatogenesis will again begin producing viable sperms which can function in fertilization. These sperms, however, may carry induced mutations and small aberrations which can be very harmful to future generations. In a similar manner surviving oogonial cells or immature oocytes in the female may begin to produce functional eggs after the mature oocytes and eggs have been rendered incapable of producing a viable embryo.

DIFFERENTIAL EFFECTS OF RADIATION

When a person is exposed to a rather heavy, but sublethal, dose of radiation, certain parts of the body are affected much more severely than others. The skin may become reddened and develop many small lesions. The lining of the intestine is severely damaged and this causes nausea and diarrhea. As the days pass there is a drop in the number of blood platelets in the blood and as a result there will be many small areas of bleeding, some within the intestine and some, just under the skin, which will be apparent. Also, there will be excessive bleeding from small wounds which break the skin. Anemia develops as the red blood cell count drops. Leucopenia, that is, a reduced leucocyte count, develops and the person becomes very susceptible to infections. Sterility, as we have learned, may develop. Other body tissues, such as bone, muscle, and nerve, as well as the tissues of the kidney and liver, are much less sensitive and are damaged only when the radiation is much heavier. As we examine these differential effects, we can note that it is the cells which are most actively growing and dividing that are most sensitive. The epithelial cells lining the skin and gastrointestinal tract are very active in this respect and, therefore, very sensitive. The same is true of the blood cell-forming tissue in the bone marrow. This tissue is very active, as it is constantly providing new cells to the blood to replace those which are constantly being lost. The same activity is to be found in the cells of the gonads. Also, cancer cells, which grow and divide more rapidly than any of the normal body cells, are the most sensitive to radiation damage. We take advantage of this sensitivity and destroy localized cancers by applying radiation in a quantity which can kill the cancer without destroying the healthy tissue surrounding it.

Experiments on tadpoles indicate that the aberrations take place in cells which are not actively growing and dividing, but these cells are not damaged so easily because they still retain the normal complement of genes and chromosomes. Tadpoles are cold-blooded, and the growth and division of the cells are in direct relation to the temperature. At a warm temperature the tadpoles will be killed by a radiation dose of 700 *r,* but they can survive this dosage if they

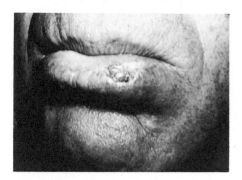 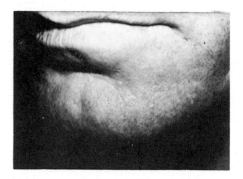

FIG. 20-7 *Cancer destruction by radiation treatment. Cancer cells are more sensitive to radiation than the normal tissues of the body, hence it is possible to give a carefully measured dosage and destroy cancer without destroying the healthy surrounding tissue. These photographs show the lip cancer before treatment and the same lips three months after treatment with X-rays.* COURTESY FRED J. MULLINS.

are refrigerated before it is applied. They will remain alive as long as they are kept at refrigerator temperatures, but, when they are allowed to warm up, they begin to die. Once growth and cell division are resumed, many of the cells receive abnormal combinations of chromosome complements, and a sufficient number of cells die to cause death of the entire embryo.

Cause of Differential Effect

In the light of what we have learned about gene mutations and chromosome aberrations it is easy to understand the differential sensitivity of various types of tissue. Such genetic changes in a single cell of the kidney of a man is not likely to cause any damage to the body. In a mature person the cells of the kidneys are stable, and the radiation will have little somatic effect unless the dosage is very high. In parts of the body where the cells are growing and dividing, however, a mutation or chromosome aberration will be propagated over and over again until a considerable island of tissue contains the altered genetic complement. Should the change be of a type which makes the tissue abnormal, an effect will become apparent. Most of the damage done to cells by radiation probably comes about because of chromosome aberrations, although mutations no doubt play some part.

Greater Sensitivity of the Embryo

In the light of the concept that the more actively growing and dividing cells are more sensitive to radiation, we would expect an animal embryo to be highly sensitive in all areas of its body since all of its cells are actively growing and dividing. Observations and experiments confirm this line of reasoning. An embryo can be destroyed within a female animal which is itself not seriously affected by the radiation. In Hiroshima many abortions resulted from radiation

TABLE 20–1

Variation in Sensitivity of Rat Embryos to Radiation (from data of James Wilson)

Age of embryo	Radiation dosage (r) required to cause 100 per cent mortality
8 days	200
9 days	400
10 days	400
11 days	600
6 weeks (birth)	750
Mature	900

received by pregnant women who were only mildly affected by radiation sickness. Many modern physicians will not use X-rays on expectant mothers if there is any alternative, because they feel that even the small amount of radiation required to make an X-ray photograph carries some small degree of risk to the highly sensitive embryo. Growing children are more sensitive to radiation damage than adults, although much less sensitive than the embryo. Such children have dividing cells in most areas of the body, but the rate of growth and division is much less than that in the embryo.

There is even a considerable variation in the sensitivity of different embryonic stages. The period of greatest growth, division, and differentiation of tissue occurs in the early embryo, and this is the time when it is most susceptible to radiation damage. Experiments on rat embryos carried out by James Wilson at the University of Florida showed that 200 r caused 100 per cent mortality when the embryos were in the eighth day of development (see Table 20–1). On the ninth and tenth days, however, 400 r were required for 100 per cent mortality. By the eleventh day 600 r were required, and the embryos became progressively more resistant up to the time of birth at about 6 weeks. Doses of less than the lethal amount resulted in abnormalities of the body parts which were being formed at the time the radiation was administered. Radiation on the ninth day, for instance, resulted in malformations of the eye, spinal cord, viscera, bladder, and blood vessels, since these are the body parts which are in a critical stage of development at this time. Radiation applied when the embryo was in its eighth or tenth day of development had much less effect on these particular body parts, but other organs were affected. In terms of human embryonic development, a 2-week embryo is at about the same state as the 9-day rat embryo. After 6 weeks the abnormalities of a human embryo induced by radiation are likely to be much less pronounced, since the body organs are in a more advanced state of development. Thus, the gross abnormalities are most likely to occur when a woman is perhaps not even sure that she is pregnant.

Variations in Sensitivity of Different Organisms

In addition to the variations in sensitivity of an organism at different stages of development, there is considerable variation in the sensitivity of individuals within any species. In man, for instance, whole-body radiation of about 250 r

will cause the death of some adults, about 450 *r* will kill about half of them, but about 600 *r* are required to kill all who are exposed. Because of this rather great range of tolerance within a species we often speak of dosages in terms of the LD-50, the amount of radiation required to kill 50 per cent of the exposed individuals. Why should some people be able to survive a dose of radiation more than twice as great as that which will kill some other people? Certainly general body condition plays a part. A healthy, well-functioning body can withstand the damage and repopulate the tissues of the body which are destroyed much better than a body which is already functioning poorly.

There is also good evidence that heredity plays a part in the resistance of the body to radiation damage. Experiments with four strains of mice showed inherited variation from an LD-50 of 500 *r* for one strain, known as BALB, to 630 *r* for another strain, known as C57BL. Figure 20–8 shows the effects of radiation on the four strains used in the experiments. Note that as the resistance increases the mice become more uniform in their response to the radiation. The inherited nature of this resistance is borne out by crosses between the different strains. When BALB was crossed with C57BL, the F_1 offspring showed a resistance which was intermediate between the two parent strains. A great variation was found in the following generations produced by inter se crosses. By selection these offspring could be separated into resistant and sensitive strains. This is the result which would be expected if the resistance was due to multiple genes.

The variation in sensitivity between different species is, as would be expected, much more pronounced than that within the species. Table 20–2 shows the LD-50 for a number of different forms of life. Note that, in general, the simpler the organism in terms of body complexity, the greater the resistance to radiation. With fewer genes and chromosomes there is less likelihood of mutation or chromosome aberration in any specific cell. Also, in one-celled forms of life, there is a purging of the damaged cells. Even though most of the cells are destroyed or so severely damaged that they cannot reproduce, the few normal cells which remain can grow and soon will have replaced the other cells. Higher plants as a whole are much more resistant than higher animals. The regions of

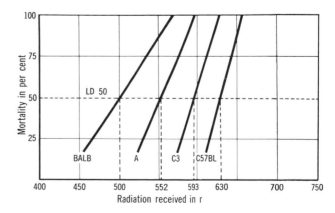

FIG. 20–8 *Inherited variation in susceptibility to radiation damage as illustrated by four strains of mice. The dosages required to kill 50 per cent (LD-50) ranged from 500 to 630 r. Note that the lines become steeper as the resistance increases.* FROM DATA BY GRAHU.

Organism	LD-50 in *rads*
Dog	325
Man	450
Mouse	530
Frog	700
Rat	850
Turtle	1,000
Triton (salamander)	3,000
Bacteria (*E. coli*)	25,000
Yeast	30,000
Drosophila (adult)	46,000
Paramecium	350,000

TABLE 20–2

Lethal Effects of Radiation on Different Organisms

growing tissue are localized and, while heavy radiation may cause some somatic deformities, there may remain enough growing tissue to repopulate the growing areas. Corn plants can withstand a dosage of 3000 *r* without being killed in most cases, but they will have a reduced yield and produce many deformed offspring. Experiments on pine trees and other gymnosperms, however, show that they can be killed by a dosage of as little as 500 *r*. Hardwood trees, on the other hand, can withstand about 1000 *r* before any deaths begin to occur. The onion can withstand the amazing dosage of up to 25,000 *r*. Plant seeds, as a whole, are highly resistant. Even though they contain an embryo, the latter is dormant. Many seeds will sprout after exposures as high as 25,000 *r* although the plants produced may show retarded growth and deformities. Thus, if man should ever succeed in elevating the radiation of the earth to such a high level that all human life would perish, he would leave the earth to the plants, insects, protozoans, and other invertebrates.

OTHER MUTAGENIC AGENTS

Although high energy radiation has received the greatest attention in studies of mutations, there are other agents which can also induce mutation.

Ultraviolet Rays

Ultraviolet rays can induce mutations, although, because they are much less penetrating than the shorter wave lengths of gamma rays, their effect is limited to nuclei which lie very near the irradiated surface. Edgar Altenburg was the first to show this effect. He obtained a small but significant increase in the mutation rate of *Drosophila* through the use of ultraviolet rays, but had to use extreme measures to permit the rays to reach the nuclei of the cells. Male flies, for instance, had to be used when newly hatched. At this time their bodies are so soft that they could be flattened between quartz cover glasses. This flattening brought the testes up near the surface where the rays could penetrate.

The most significant results have come from work with pollen grains of maize, the spores of molds, and bacteria, which are all sufficiently small to allow

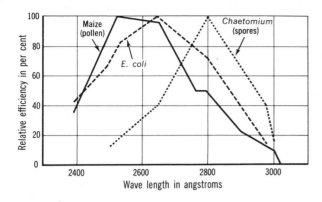

FIG. 20–9 *Variation in efficiency of ultraviolet light in producing mutations in different organisms. The most efficient wave length is expressed as 100 per cent, while the other wave lengths are given as a proportion of this.* FROM WAGNER.

penetration of the rays. L. J. Stadler found that ultraviolet rays can cause mutations in maize pollen grains, although, even in these very small bodies, the nuclei had to be on the upper surface which received the exposure if mutations were to be induced. He also found that some wave lengths of the rays were more mutagenic than others. Those with a wave length of about 2600 angstrom units were most effective, and these are the very wave lengths which are most readily absorbed by DNA. There was, however, some variation in the effectiveness of different wave lengths on different organisms. Other workers have found that the spores of the mold, *Chaetomium,* were more responsive to wave lengths of 2800 angstrom units and comparatively insensitive to those of 2600 angstroms. This is shown in Figure 20–9. Bacteria are so small that ultraviolet rays are frequently used instead of X-rays or other radiation to induce mutations.

Heat

Temperature can be a factor in the induction of mutations in at least some organisms. Muller and Altenburg found that one group of *Drosophila* raised at 27° C showed a mutation rate about 2.5 times greater than that in flies raised at 17°C. This is equivalent to about a fivefold increase in terms of the life of the flies, because flies raised at the warm temperatures lived at a faster rate and had a life span only half as long as those flies raised at a lower temperature. Temperature is equivalent to time in the life of the flies.

Even man may not be exempt from this possible effect of temperature. Although he is a warm-blooded animal and, therefore, has a constant internal body temperature, his testes are in a position where they are exposed to greater temperature fluctuations. A recent Washington newspaper carried the headline, "Tight underwear more dangerous than nuclear fallout." Beneath this sensational headline was a story of a talk by a Nobel prize-winning scientist which included a remark that the elevation of the temperature of the testes of a man wearing tight underwear which held these glands close to the body could, on the basis of results found in *Drosophila,* cause a greater increase in the mutation rate than would be expected from the exposure of these men to the radiation from fallout at its present rate. A report by Lars Ehrenberg and his associates showed that the temperature of the testes of men wearing trousers was over 3° C higher than that of men wearing kilts. This, Ehrenberg pointed out, is sufficient to more than double the natural mutation rate, if man reacts to temperature in the same manner as *Drosophila*. We have no evidence as yet to make such an assumption.

Chemical Agents

The production of mutations by chemical treatment of cells was first reported by Thom and Steinberg in 1939. They found an increase in the rate of mutation in the mold, *Aspergillus*, after it had been treated with nitrous acid. The importance of these results was largely overlooked by most geneticists until the report of Auerbach and Robson on chemical induction of mutations in *Drosophila*. This study, which was begun in 1940, showed an increase in sex-linked mutations from about 0.2 per cent in the controls to as high as 24 per cent from flies treated with mustard gas (allylisothiocyanate). This discovery was made during World War II, and, since mustard gas has been used in wartime, the results were considered classified information and were not published until 1956. Subsequent work by various workers has shown that many other chemicals also have mutagenic properties. These include formaldehyde, ethylurethane, nitrous acid, nitrogen mustard, phenol, manganous chloride, bromouracil, and even caffeine and theobromine which are so widely consumed by man.

In some experiments, after the eggs are laid, they are exposed to the chemical being used. In others, the testes or ovaries are removed from the body, treated by the chemicals, and then reintroduced into the body of experimental animals. In *Drosophila*, it is common to inject small portions of the chemicals into the body so that they come in contact with the testes or ovaries. In bacteria, yeasts, and molds it is necessary only to put the chemicals in the food upon which the organisms are growing. In higher plants the chemicals can be applied directly to the pollen, ovules, or the seeds.

As a rule, the chemicals which cause mutations also cause chromosome breaks. This property makes it possible to use the comparatively simple and quick cytological checks for chromosome aberrations, rather than the much slower techniques required for the detection of mutations, when chemicals are being tested for possible mutagenic effectiveness.

Chemical agents are not uniform in their effects. One may be very highly mutagenic in one species, but have no detectible effect on another. Hydrogen peroxide will cause mutations in *Neurospora*, but has no apparent effect on higher plants. Urethane and phenol produce mutations in *Drosophila* and higher plants, but no mutations in *Neurospora*. We have no verified explanation for this difference, but it could be related to the time of the life cycle when the chemical is applied. Formaldehyde has a mutagenic effect on *Drosophila* only during an early stage of the developing sperms. There is no effect on earlier or later stages of sperm development, nor any effect on the eggs at any stage of development. Urethane, on the other hand, is effective on mature sperms, but seems to have no effect on the spermatogonia. Since treatment usually is given at one particular stage in the life cycle of the organism, it is possible that some organisms seem insensitive simply because the treatment was not given when the reproductive cells were in the sensitive stage of development. Also, there is the possibility that some chemical compounds do not penetrate into the cells of some organisms as well as others. Still another explanation might be the fact that some cells, at certain stages at least, might have enzymes which break down the compounds before mutations can be induced.

Particular study has been made of purines and pyrimidines, since it seems logical that these compounds might have a good chance of causing mistakes in the replication of DNA since they are a vital part of it. Naturally occurring purines, such as adenine, caffeine, and theobromine, are mutagenic in some organisms at least, but the natural pyrimidines, like thymine, have not yet been found to have mutagenic effects. Synthetic analogues of pyrimidines, such as bromouracil, however, are mutagenic. Bromouracil has been of special interest because it is an analogue of thymine and has been shown to replace it in the DNA of bacteria, bacteriophages, and cells of mammals. Thus, thymine-adenine (T-A) in the DNA chain could become bromouracil-adenine (BU-A) after such a substitution. When a DNA chain containing this substitution opens out in replication, the single chain with the A would form a double helix with T-A and the normal condition would be restored. The single chain with the BU, on the other hand, might form hydrogen bonds with the purine guanine (G) to give BU-G. In the next replication the G would unite with the pyrimidine cytosine (C) to give G-C. Thus, a G-C would be substituted for a T-A in a DNA helix and a mutation would have occurred. Such a theoretical explanation for chemical mutagenesis has been called a transition by E. Freese who proposed the theory. There is a transition from one purine to another purine (A to G) or one pyrimidine to another pyrimidine (T to C). The term transversion has been suggested by Freese for the substitution of a purine for a pyrimidine (A to C) or vice versa.

This explanation could not account for all chemical mutagenesis because many mutagenic chemicals are not of the nature of purines or pyrimidines. It is postulated that these other chemicals act directly on the DNA to cause it to change its bases by chemical reaction, or by reacting with other chemicals in the cell which are thereby converted into purine or pyrimidine-type mutagens.

DIRECTED MUTATION

As the study of chemical mutagens progressed, it became apparent that different chemicals have different mutagenic effects. Studies of Kolmark and Westergaard have shown that bromoethyl methane sulfonate, when used on *Neurospora,* causes mutations of the gene ad⁻ to ad⁺, but causes practically no mutations of inos⁻ to inos⁺. On the other hand, ethyl methane sulfonate causes mutations of both genes. The mutations produced in *Drosophila* by mustard gas cause mutations of the same kind as are produced by radiation, but with different ratios. Other studies have also shown that some chemicals will produce mutations of only one particular kind. These discoveries have led to some rather wild speculation that man is on the verge of discovering a method of controlling heredity: that we may soon be able to cause genes with harmful effects to mutate to genes with normal effects and that we may even be able to produce a race of supermen by manipulating the genes to produce mutations which will be of maximum benefit. Such speculations are premature; no evidence today indicates any possibility of such control of mutations. All we can do is produce a whole group of mutations of related kinds with one chemical. No chemical is known which will produce mutations only at a specific locus, and the great majority of all mutations continue to be primarily harmful to the organism expressing them.

THE CAUSE OF NATURAL GENETIC CHANGES

With the discovery of methods of inducing mutations and chromosome aberrations, speculation turned to the possible causes of natural or spontaneous genetic changes. These changes occur in all forms of life without any special treatment. Since ionizing radiation has proved to be so effective in increasing the number of these changes, could it be that the high energy radiation which occurs in nature is the cause of natural genetic alterations? All forms of life on earth are constantly being exposed to a background of cosmic radiation as well as the ever-present radioactive elements. Radiation coming from the rocks and soil is about double that from cosmic radiation, and radiation in the food, water, and air taken in by animals or in the water, minerals, and carbon dioxide taken in by plants is about equal to that from cosmic radiation. Each breath you take, for instance, contains a very small number of atoms of carbon-14. Plants incorporate some of this carbon-14 into the food material which you eat. As a result, the carbon in your body contains the same proportion of carbon-14 atoms to stable carbon-12 atoms as occurs in air. Since carbon is such an important element in all protoplasm, all living things are constantly receiving beta radiation from the carbon-14 in their bodies. Other radioactive elements, especially potassium-40, also contribute to this internal radiation.

In all probability, these natural radiations do account for some of the naturally occurring genetic changes. Calculations of the amount of natural radiation reaching the reproductive cells of organisms before reproduction show, however, that it falls far short of the amount required to account for the total number of natural genetic changes. We must look for other forces to supplement this radiation. There are fluctuations of energy within molecules, and it is possible that a concentration of this energy on one bond of the DNA may cause breaks and rearrangements which could be mutations. Likewise, breaks of the entire chromosome could result in those rare cases when an unusually large concentration of energy occurred at a particular point on the chromosome.

Whatever the agents that induce the natural genetic changes, it is evident that their effects accumulate with the passage of time during the life of an

TABLE 20–3

Relation of Age of Mothers to Frequency of Mutant Children
(From Morch's records on chondrodystrophic dwarfed children from normal
parents in Denmark)

< Age groups of mothers >	< 25	25–29	30–34	35–39	40–44	> 45
Per cent total children	32	30	22	12	3.5	0.5
Per cent mutant children (chondro-dystrophic dwarfism)	19.4	19.3	24.8	21.3	13.2	2.0
Ratio, per cent mutant children to per cent normal children	0.606	0.643	1.13	1.78	3.77	4.0
Relative per cent mutant children in each group	5.0	5.4	9.4	14.8	31.4	33.3

individual. Older parents transmit a slightly greater number of mutations to their offspring than younger parents. Table 20–3 shows the results of a study on the frequency of a single mutation, chondrodystrophic dwarfism, in the children of mothers in different age groups in Denmark.

RADIATION HAZARDS TO MAN

The Dawn of the Atomic Age

After many years of intensive research, man's efforts to tap the great source of energy contained within the atom were finally successful. On December 2, 1942, a group of scientists gathered at the University of Chicago to witness a demonstration of the first controlled release of energy from the atom. Until that time, man was largely dependent upon the sunlight as a source of energy. Almost all of our heat, light, electricity, and other forms of power came from this source outside our planet. With the discovery of a means of inducing atomic fission, however, man received the promise of a vast new source of energy coming from materials on the earth. The first uses of this great power were destined to be turned toward destructive rather than constructive purposes, however. In 1945, on July 16 the first atom bomb was exploded and on August 6 the people of the world became aware of the great power of atomic energy as the first bomb destroyed the city of Hiroshima, Japan. What most people did not realize at the time, however, was that much of the energy of the atoms in the bomb was released in the form of high energy radiation which could have potential dangers to future generations as well as those of the present.

As the war came to an end, man turned his attention to peacetime uses of atomic energy. It was soon being applied to valuable purposes in medicine, industry, power generation, and scientific research. Also, there was continued production and testing of bigger and more powerful bombs, and soon the super hydrogen-fusion bombs were being tested as well as the atom-splitting, or fission, bombs. These tests continued to pour radioactive elements into the high atmosphere, and these gradually settled down all over the earth as a fine fallout which caused, and still is causing, a gradual increase in the radioactivity of the entire earth. Many of the radioactive isotopes generated in nuclear bombs are short-lived and their radiation is soon lost, but others have a half-life of thousands or even millions of years and so will continue to emit radiation for a long time. In addition, in many countries there has been an increase in the use of medical X-rays. Also, a few highly selected individuals are being shot out into space where they are exposed to the high intensity radiation belts which girdle the earth as well as to the more intense cosmic radiation from which they are not shielded by the earth's atmosphere. In the light of our knowledge of the genetic effects of radiation, all of these additional exposures have aroused grave questions as to the possibilities of increased genetic damage. Unfortunately, there have been conflicting opinions and even outright arguments about the degree of danger involved, even among persons in authoritative positions. This debate has served to confuse the issue in the public mind; yet the issue is so important to the future of mankind that it deserves clarification.

Popular Misconceptions

Common notions about the genetic effects of radiation are often very different from those held by geneticists. Many people seem to feel that nearly all of the children of parents who receive heavy radiation will be abnormal monsters, with perhaps a few supermen as a result of beneficial mutations. When the first reports of the studies of the children born to parents exposed to radiation from the bomb at Hiroshima showed no significant increase in abnormalities, many people concluded that there was no genetic danger from such radiation. Both viewpoints indicate a great misunderstanding of the principles of genetics. The effects of the radiation would not be expected to produce a significant increase in gross abnormalities in the first generation in the number of children studied, about 65,000. This conclusion does not mean that there was no genetic damage. Detrimental mutations, recessive lethal and visible mutations, and chromosome aberrations could have been induced which would not be expressed until future generations.

Hiroshima Studies

The Atomic Energy Commission of the United States, in collaboration with similar agencies in Japan, has built a laboratory on a hill overlooking Hiroshima and dedicated it to the express purpose of studying the effects of radiation from the atom bomb. James V. Neel and W. J. Schull of the University of Michigan, in conjunction with many Japanese physicians and geneticists, have been making an exhaustive study of the irradiated people and their children. Their first report, made before the International Congress of Human Genetics in Copenhagen in

FIG. 20–10 *This group of buildings overlooking the city of Hiroshima, Japan, has been built and equipped to permit extensive study of the people exposed to the atom bomb blast and the descendants of these people.* ATOMIC CASUALTY COMMISSION, HIROSHIMA, JAPAN.

1956, revealed no significant increase in the number of abnormalities in the children born to parents who were in Hiroshima at the time of the explosion. Some reporters seized upon this information as proof that there were no genetic effects from the radiation from nuclear bombs and that we need only to worry about damage to the individuals exposed. Those with a knowledge of genetics, however, withheld judgment until further study could be made; 65,000 children are too few to show the small first-generation effects which would be expected. Most of the damage done would be expected to be deferred to future generations.

Within a few years another report was made which indicated that some genetic damage from the radiation emitted from the bomb had occurred. This report was concerned with a study on the possible induction of sex-linked lethals. Since lethal mutations appear in a frequency much greater than visible mutations, it was expected that it might be easier to detect any possible genetic damage in this manner. The study concentrated on the sex ratio among the children born to exposed as compared to non-exposed parents. Recessive lethals induced on the X-chromosomes of exposed women would be expressed in the male embryos receiving these chromosomes, but the female embryos would be protected in most cases by a dominant normal allele. Thus, there would be fewer boys in proportion to girls in the children which were born alive to exposed women if the radiation increased the number of sex-linked lethals. This difference would be counterbalanced to some extent, however, in those cases where the fathers were also exposed because some lethals are dominant and would cause the death of female embryos, but not male embryos of exposed fathers. Also, any detrimentals induced on the X-chromosome of the fathers would cause some female deaths without a corresponding effect on males.

In six studies where the fathers were exposed and the mothers were not, the number of daughters were slightly fewer than in control groups. In six studies where the mothers were exposed and the fathers were not, there were fewer boys in five out of the six groups. Both of these results are indicative of genetic

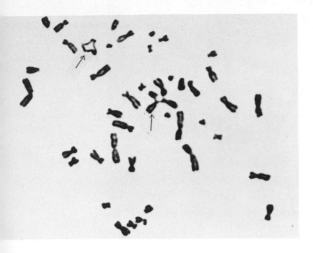

FIG. 20-11 *Damage to human chromosomes induced by radiation. This is a photograph of a smear of a human cell which received 50 r of X-rays. The arrow at the top of the photograph shows linkage of two chromosomes at two different points due to chromosome breakage and reattachment. The arrow in the center shows two small chromosomes attached to one large chromosome by bridges because of radiation-induced translocation. This cell could not have produced normal daughter cells.* COURTESY THEODORE T. PUCK.

FIG. 20-12 *Giant cells appearing in a tissue culture of human cancer following radiation. This tissue has received sufficient radiation (about 450 r) to destroy the reproductive capacity of some of the cells without destroying the enzyme systems required for growth. Hence, these cells continue to grow without division until they have become giants about one mm. in diameter. The smaller cells shown did not have their reproductive capacity destroyed and, therefore, continued division and are a more normal size.* COURTESY THEODORE T. PUCK.

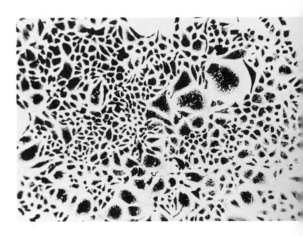

damage from the radiation, but the differences were small in both cases. This is understandable because the great majority of the parents included in the studies received an estimated dose of only 8 *rem,* with a very few receiving as high as about 200 *rem.* Many more children will have to be studied, perhaps as many as a million, to accumulate sufficient figures to give significant results on the basis of mutations expected based upon the results obtained from experimental animals. It should be remembered, however, that this study was directed only at differences on the basis of mutations on the *X*-chromosome and this is only one of 23 pairs of chromosomes.

The Doubling Dose in Man

Natural or spontaneous mutations occur in any population regardless of exposures to man-made radiations. The mutations induced by man-made factors are superimposed upon those which occur naturally. In an effort to evaluate the genetic damage done to man by extra radiation, geneticists have tried to determine the amount of radiation required to produce a doubling of the natural mutation rate. On the basis of careful evaluation of information from many sources, leading geneticists have agreed that the human doubling dose probably lies between 30 and 80 *rem,* with 50 *rem* as a good working estimate. Let us now see what a doubling of the mutation rate in man would mean.

Statistics show that about 4 per cent of all children born have, or will later develop, clearly defined defects which have a genetic origin. These include mental abnormalities, malformed body organs, blood or hormone defects, impaired sight or hearing, neuromuscular defects, etc. Throughout man's past history an equilibrium has been established between the rate of appearance of new mutations for these defects and their elimination through genetic deaths (see Chapter 24). Genetic death means that persons carrying the genes either die before their reproductive period or do not reproduce for one reason or another. In addition, about 2 per cent of the children born will be defective because of damage done to the child during its embryonic development. These defects have no genetic significance since they are not transmitted from one generation to another.

If all the people in the United States were exposed to an average mutation doubling dose of radiation for generation after generation, we would expect a gradual increase in these inherited abnormalities until they would occur at double their present rate — 8 per cent rather than 4 per cent. To be more specific, it is estimated that about 260 million children will be born to the people living in the United States today. Among these, we will expect about 10 million abnormalities resulting from natural mutations which occur in the present as well as in past generations. If all of these people were to receive the doubling dose of extra radiation, then there would be perhaps a 10 per cent increase in these abnormalities, an extra 1 million. Twice as many harmful mutations would be induced, but most of these would not show in the first generation; they would be recessives or detrimentals. This is a small increase in terms of percentage,

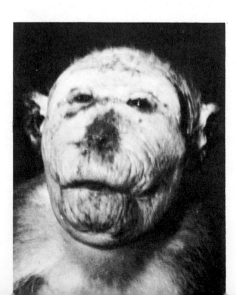

FIG. 20–13 *Physical effect of heavy exposure to radioactive substances. The head of this monkey was exposed three months previously to 2,000 r units of gamma radiation from radioactive cobalt* (Co^{60}). COURTESY A. J. RIOPELLE.

and it would be difficult to measure in terms of the total number of abnormalities whether hereditary or environmentally induced. Yet we could hardly say that 1 million abnormal children are not a cause for concern. Even one abnormal child is a tragedy, especially if the abnormality could have been prevented. If the children of the next generation also receive the doubling dose of radiation, and so on for generations to follow, eventually the average number of inherited abnormalities would be 20 million per generation rather than 10 million (on the basis of our present population and its rate of reproduction).

Effects of Low Dosage

Is a threshold of radiation necessary before any mutations are induced, and is the number of mutations induced proportional to the amount of radiation? Also, does the mutation rate vary according to the intensity of the dose, so that a very intense dose of short duration (a high dose rate) yields a mutation rate different from that obtained by the same dosage spread out over a longer period of time (a low dose rate)? We have already learned that the number of two-break chromosome aberrations does vary according to both the amount and the intensity of the radiation; can the same be true for gene mutations? These are important questions because there is a tendency for man to be exposed to an increased low-level, but long-continued, radiation. Studies on the frequency of recessive sex-linked lethals in *Drosophila* indicate that there is no threshold. In the range of 25 *r* to 3000 *r* the data from Spencer and Stern show the induced mutations to have a linear frequency (see Figure 20–14). H. Bentley Glass has shown that mutations for minute bristles in *Drosophila* show a linear relationship down to a dose as low as 10 *r*. Some older experiments of Muller showed that mutations are independent of the dose rate, but these experiments did not include the very low dose rates.

An extensive study on the induction of visible mutations in mice by W. L. Russel at Oak Ridge was the first to demonstrate a difference in the effects

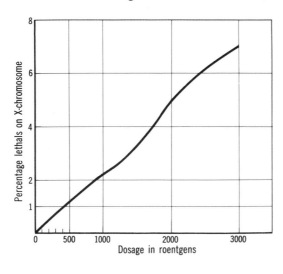

FIG. 20–14 *The percentage of lethal mutations induced on the X-chromosome of Drosophila by various dosages of radiation. The response is linear within the bounds of the standard error of the average for each dosage used.* DATA FROM SPENCER AND STERN.

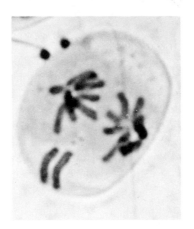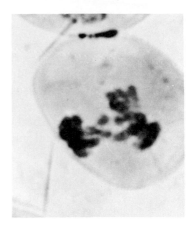

FIG. 20–15 *Radiation damage to pollen-mother cells of Trades-cantia, the spiderwort. The photo at left shows a normal anaphase with the chromosomes migrating to the poles. The center photograph shows a cell which has received 400 r of X-rays; the centromere of one chromosome has broken loose from the chromosome and the two chromatids can be seen in the cytoplasm separated from the dividing chromosomes. The photograph on the right shows a cell which received 750 r of X-rays. This is a very abnormal anaphase; chromosome bridges have formed between dividing chromosomes which will not separate because of these attachments.*

of high and low dose rates. He exposed large numbers of mice to radiation and bred these to other mice carrying a large number of recessive traits. Any visible mutations induced would show in the first generation if they happened to be the same as one of these recessive traits. At low intensities of radiation the number of such mutations appeared to be only about one-third to one-fourth as great, proportionally, as that produced by higher doses of radiation. It is quite possible, however, that many of the apparent mutations at high doses are actually small two-break deletions which allow the recessive genes to show. This idea is confirmed by the work of Grant Brewen who studied the effects of radiation dosages on chromosome breaks in the epithelial cells of the cornea from the eyes of hamsters. He found about one-third as many breaks from the low as from high dosages in proportion to the total *r* administered. Hence, while we cannot make accurate corrections for aberrations in the mice, it is possible that the proportionality idea still holds true for gene mutations.

At least we can still conclude that no dosage of radiation is too small to be ignored. For instance, a dose of 10 *r* is generally considered too small to be significant, but, if a large enough number of persons were to receive this dose to their gonads, it would cause just as many mutations as a much larger dose to a few persons. A heavy radioactive fallout that exposed 5 million people to 10 *r* would result in just as many mutations as if 500,000 survivors of a bombed city were to receive an average of 100 *r*. A dose of 10 *r* to the gonads of all the people of the United States would perhaps result in about 200,000 inherited defects in the first generation offspring and an eventual 2 million defects as the mutations were reproduced and appeared in the future generations.

These figures do not include all of the genetic damage. There are also lethals which would increase the number of abortions and stillbirths, and some of the detrimentals which could not be measured as clear-cut abnormalities, but which would lower the body efficiency in various ways. Six leading geneticists were asked by the United States National Academy of Sciences to give an estimate of the total number of mutations of all kinds which would be transmitted to the next generation if all the people of the United States received 10 *r* to their gonads. Their independent estimates were bunched around the figure of 5 million.

Long-Delayed Somatic Effects on Man

In addition to the genetic effects, there are long-delayed effects of radiation. For instance, the people who were exposed to radiation from the bomb in Hiroshima are still developing cataracts of the eye. Over 4 per cent of those who were within 3,000 feet of the blast center have developed cataracts, and there is evidence that another 4 per cent will develop it in the future years. The incidence of leukemia is also greatly elevated. The United Nations Scientific Committee on the Effects of Atomic Radiation reported in 1964 that leukemia is the predominant delayed malignancy triggered by the radiation, and that even very low doses can induce leukemia. A study of the number of cases of leukemia among the children in one region of England, which had received an unusually heavy fallout, showed a significant increase in this affliction. Also, experiments on radiation on mice and other animals show that there is a significant shortening of life in proportion to the amount of radiation received. Some even feel that the natural aging of animals is due to the background radiation to which they are exposed. It is estimated that 50 *r* of whole-body radiation in man, above the background, will result in an average reduction of life span by 1.4 years. More or less radiation would have a proportional effect. This can be estimated as a loss of about 9 days of life per roentgen. It has been known since the early days of X-ray work that cancer is one of the possible long-delayed effects of radiation. Cancer may appear at the site of exposure, sometimes years after the radiation was received. These long-delayed effects, as well as the more immediate effects, appear to be almost altogether due to chromosome breakage or gene mutation within the somatic cells.

Medical Uses of Radiation

Although X-rays are one of the most valuable tools used by the medical diagnostician, they do carry with them a certain hazard in view of the fact that there seems to be no threshold for gene and chromosome changes. In recent years, through the use of faster films and developers, improved techniques and machines delivering a more highly penetrating ray, the amount of radiation reaching the body during diagnosis has been greatly reduced. For instance, the amount delivered to the internal body tissues when an X-ray picture of the chest is made has been reduced from a former average of about 2.0 *r* to a present average of about 0.2 *r*. This can be directed so that only about 0.005 *r* reaches the gonads. Dental X-rays require about 2.0 *r* because of the density of the teeth, but the rays are highly localized. Extensive fluoroscopic examinations of the gastrointestinal tract may require as much as 100 times the exposure used in making an X-ray picture. The fluoroscope was once used in shoe stores to show customers how the feet fit inside a shoe. Today this practice has been discontinued as the dangers to both customers and salesmen became apparent.

In 1956 a report of the National Academy of Sciences backed by a British study indicated that the 30-year dose of medical radiation to the gonads of the population of the United States and Britain would be about 4.6 *r* if continued at the rate in use at that time. This is more than the background exposure of about 3.1 *r* and is 23 times the exposure from radioactive fallout produced to date. To the credit of the radiologists, it should be said that, since that time, many efforts have been made to reduce medical X-ray exposures. Other means of diagnosis are substituted wherever possible, and greater care is being taken to prevent the rays from reaching any body tissues except those requiring them. Also, there has been a reduction in the therapeutic use of X-rays for such conditions as skin infections (acne of the face, for example), warts, or fungus infections. In such treatments, rays of low penetrance (soft X-rays) are used, but there are still the dangers that go along with the cumulative dose of roentgens to any part of the body.

Radioactive isotopes also have valuable medical uses in diagnosis and treatment. Radioactive iron tends to be localized in the red blood cells and serves as a tracer which shows up occluded areas of blood circulation. Radioactive iodine is taken up primarily by the thyroid gland and may be used in treating thyroid cancer. Radioactive calcium becomes localized in the bones and may be used to treat bone tumors and to show bone defects. Gamma radiation from cobalt-60 is being used with greater frequency for treating localized cancers, because the beam can be directed more easily than in the case of radiation from the X-ray tube. Radioactive gold is sometimes injected into the body as a treatment for diffused body cancer. All of these contribute to the lifetime radiation load.

Tolerance Dosage

Workers who handle radioactive materials in atomic power plants, research laboratories, atomic ships and submarines, and industrial plants, as well as our space astronauts, are subjected to additional sources of radiation exposure. For such persons it is necessary to set a tolerance dosage. This was originally set at 0.1 *r* per day, but, if a person received this amount each working day for 2 years, he would have the estimated dose for doubling the mutation rate. Also, if the exposures were continued at this rate, his life would be shortened appreciably. Today this tolerance dosage has been reduced to an average of 0.03 *r* per day, and every effort is made to prevent even this small amount from reaching the body.

Exposure from Nuclear Weapons

In any war using nuclear weapons it is quite certain that vast numbers of people would receive large quantities of gamma rays from bomb explosions. A medium-sized atom bomb exploded over a large city would subject about 200,000 people to an average exposure of about 150 *r*. The radiation within about a quarter of a mile of the blast would be much greater than this, but those close enough to receive as much as 450 *r* would be killed by the blast and the heat unless well shielded. There would be many deaths from radiation effects among those receiving less than 450 *r*, but most such persons would survive.

Of greater genetic importance than the sudden release of gamma rays at the time of the explosion would be the radioactive fallout which would expose a much larger number of people to smaller amounts of radiation. A group of people in a small town in Nevada received an average of about 7 *r* when a sudden shift of the wind brought down some of the heavier particles from an atom bomb test 100 miles away. Some Japanese fishermen, 250 miles from the site of the first hydrogen bomb explosion, were showered with radioactive dust that caused severe radiation sickness and one death. Months later, tuna fish caught in the waters nearby were found to be dangerously radioactive.

In addition to the comparatively heavy fallout there are much smaller particles which rise higher and settle much more slowly. The mushrooming cloud from an atom bomb carries these particles up to a height of about 40,000 feet. There they are transported by high velocity winds and gradually settle. About half of the dust settles every 22 days. The H-bomb sends the radioactive particles to a still greater height, about 100,000 feet, and, as a result, they settle much more slowly. Since the suspension of atmospheric bomb tests in 1961 by all nuclear powers except France, the amount of fallout has gradually subsided, but some fallout will continue for many years. The beginning of such tests by China and other nations may raise the figure again.

The rate of decay of the radioactive elements, such as are found in the fallout, varies greatly. It is measured in terms of half-life — the period of time required for the elements to lose one-half of their radioactivity. For some isotopes this is only a few seconds, for others a few hours, days, or weeks, and in still others it is many thousands of years. It is this last type which is of greatest concern since the radioactivity generated today will be superimposed upon that which might be generated in future generations. Carbon-14, for instance, is generated by nuclear weapons and has a half-life of 5600 years. It emits only beta radiation, but carbon becomes a part of the protoplasm in all parts of the body. Even the gene itself contains much carbon and the incorporation of carbon-14 could lead to an internal explosion which would greatly alter the gene. Strontium-90 is an isotope, with a half-life of about 28 years, which may be concentrated in the body. It is similar to calcium in its chemical properties, and living things tend to substitute it for calcium. When it finds its way into the soil, plants absorb it and use it as they would use calcium, thus concentrating it. People then may get this isotope by eating the plants or by drinking milk produced by cows which ate the plants, and it will be absorbed mainly by the "calcium-hungry" bones. Bone cancer and leukemia might result from such concentrations. The beta radiation from strontium-90 is not highly penetrating and very little reaches the gonads, so its genetic effect is not great. Cesium-137, however, is a fallout isotope which gives off a powerful gamma ray that can have genetic significance. It is not incorporated into the protoplasm, but may remain in the body for a time after it is taken in.

Although the total radiation received from fallout of the bombs exploded in the atmosphere up to date is very small for any one person, we must consider the fact that everyone on earth is exposed to this radiation, and those yet unborn will be exposed to the lingering long-lived isotopes. Some have expressed the viewpoint that this radiation from fallout is so low that it is not significant; but others, in the light of the no-threshold concept, feel that some genetic changes will be induced by the fallout from bomb tests and that any change at all is significant. The suspension of atmospheric bomb tests by some of the world powers indicates that some are in agreement with the latter concept.

Recommendations to Reduce Radiation Hazards

Some years ago the United States National Academy of Sciences Committee on the Genetics Effects of Atomic Radiation made a careful study of the problem and made certain recommendations which included the following:

1. Records should be kept for each person, showing his lifetime exposure to ionizing radiation in each part of his body. Each person should keep such a record and each time he receives any form of radiation, the number of *rem* received and the part of the body receiving them should be entered. This would have many values, such as the prevention of over-exposure of a particular body part because of lack of knowledge of previous exposures of that body part.

2. The medical use of X-rays should be reduced as much as is consistent with medical necessity, and particular care should be taken to shield the reproductive cells from radiation. Each X-ray technician should use the smallest exposure necessary to obtain the desired results.

3. The average exposure of the population's reproductive cells above the natural background should be limited to 10 *rem* from conception to age 30. Some persons would certainly need more than this for therapeutic reasons, while most persons would need less, but this average should not be exceeded for the whole population.

4. No one person should receive a total accumulated dose to the reproductive cells of more than 50 *rem* up to age 30 and not more than an additional 50 *rem* up to age 40. (About one-half of the children in the United States are born to parents under 30 and nine-tenths are born to parents under 40.)

The people of the world today face a responsibility greater than mankind has ever faced before. We now own a great source of energy which can not only greatly harm those now alive, but which can also cause untold damage to persons of future generations. Nuclear energy can be a great blessing, if used for peaceful purposes, but an understanding of its dangers is absolutely necessary if the damages done are not to outweigh its values.

PROBLEMS

1. In the ClB cross why does the female not die when she receives an X-chromosome carrying an induced lethal from the male parent? There is a lethal on the X-chromosome received from the female parent, thus making a lethal on each of her two X-chromosomes.

2. What did the early workers with X-rays not know about the nature of these rays which caused most of them to suffer severe radiation damage?

3. How does electromagnetic radiation differ from particulate radiation? Which of the radiation particles has the greatest genetic significance and why?

4. Why is the *rad* or the *rem* sometimes used as a designation of a unit of radiation exposure rather than the roentgen?

5. Explain how ionization may be produced in living tissue by each of the four major forms of radiation to which living matter may be exposed?

6. Why is radiation damage less when tissue is in an oxygen-free environment?

7. All of the evidence points to a linear proportionality between number of gene mutations induced and the total amount of radiation administered. Such a linear relationship does not hold for the chromosome aberrations which persist in the cells. Explain.

8. Some tissues of an animal body are much more sensitive to radiation damage than other tissues, yet these sensitive tissues also show the greatest potentiality for recovery from radiation damage. Explain.

9. Many women in Japan had abortions shortly after they were exposed to radiation from the bomb explosions even though they did not show any severe symptoms of radiation sickness. Explain.

10. One study showed that the age of the mothers was correlated with the incidence of dominant mutations appearing in their children. Explain.

11. A radiation dosage of 50 *rem* to a human being in the adult period of his life will produce no effects which are easily detected by gross examination. Does this mean that a person need not be concerned about radiation damage as long as he receives no more than 50 *rem* at one time? Explain.

12. Radiation from X-rays and radioactive isotopes are very effective agents for the destruction of cancer, yet we know that this same radiation is a potential cancer-inducing agent. How can you correlate these statements?

13. Various forms of radiation are considered to be very valuable for the induction of mutations which may have commercial value in domestic animals and cultivated plants, yet the genetic effect of these same radiations on man is considered to be very harmful. Explain.

14. A man and a woman work in an industrial plant where they handle radioactive isotopes. Both receive small amounts of radiation from the isotopes, yet within the tolerance dosage. They marry and their first child is an albino. There are no cases of albinism in their ancestry as far as they can trace it and they feel that this abnormality has resulted from the radiation which reached their reproductive cells during the course of their work. How would you evaluate the possibilities of such an occurrence?

15. A manufacturer in Chicago once installed an X-ray machine together with a large fluoroscopic screen near the time clock. Here, unknown to his employees, he could see through them and determine whether any of them were carrying out any of his products in their pockets. When his scheme was discovered he was forced to discontinue it, but not until after his employees had received considerable radiation. A few years later one of the men who was subjected to this

radiation bore a son with hemophilia. He sued the manufacturer, claiming that the radiation had caused a mutation which resulted in the hemophilia in his son. As a genetic expert in court, what would your testimony be?

16. The first report of the study of the children born to parents exposed to radiation from the atom bomb in Hiroshima showed no significant increase in the number of abnormalities as compared to non-exposed controls. Evaluate the conclusion that radiation from atom bombs has no genetic significance.

17. If an entire population of a region receives an average doubling dosage of radiation (estimated at 50 r), there will be only about a 10 per cent increase in inherited abnormalities which appear in their children. Why are not the abnormalities doubled?

18. Which would do the greater genetic damage, a bomb which exposed the surviving population of a city of 500,000 people to an average of 100 r, or a radioactive fallout which exposed an entire nation of 100 million people to an average of 0.5 r? Explain your answer.

19. The radioactive fallout from hydrogen bombs is so great that efforts are being made to produce a "clean" bomb that will yield less fallout. Since the object of bombs is to destroy the enemy, why should the nuclear powers be concerned with the production of a so-called "clean" bomb?

20. A great amount of research in recent years has been directed toward finding whether there is a threshold amount of radiation required to produce genetic damage from radiation. Why is it considered of such importance for us to know whether a threshold exists?

21

Modification of Gene Expression

AFTER THE REDISCOVERY OF MENDELISM near the beginning of the present century there was a tendency to think of the genes as all-powerful forces destined to exert a particular effect on the individual who carried them, and an inclination to believe that they would be expressed with unvarying precision regardless of other conditions. Arguments arose as to whether certain characteristics were the result of "heredity or environment," with the simple assumption that they must be due to one or the other. The more we study gene action, however, the more we realize that an individual is the sum total of his heredity and his environment. We are coming more and more to realize that the genes cannot react in a vacuum and that the various environmental influences which come to bear on the developing organism generally have their bearing on the phenotypic expression of those genes. In some instances the environmental influences may be **intracellular,** and may have their origin in other genes in the same cell. Other influences may be **extracellular** in origin: temperature, sunlight, and food, for example, are a few of the agents originating from outside the cells which may influence gene action.

In a multicellular organism, such as a human being, the immediate and most important part of the environment around each cell is made up of other cells and fluids, such as blood and tissue fluid, that bathe the cells, supply oxygen and nutrients, and remove carbon dioxide. and other wastes. As the great French physiologist Claude Bernard emphasized a century ago, maintenance of the constancy of the internal environment is of paramount importance and a criterion of living organization. Many genes are concerned with the production of the substances that maintain the constancy of the internal environment, with the transport of oxygen or carbon dioxide, and with the production of hormones, the potent chemical messengers that promote or inhibit the activity of cells and organs. Any agents, either external or internal, which bring about an alteration of this cellular environment can have a profound effect on the growth and development of these cells.

Environmental Duplication of Genic Effects

In some cases certain environmental conditions may cause such alterations in the embryo as to produce an effect which is indistinguishable from similar effects that are influenced by genes. Harelip in mice serves as an illustration of this phenomenon, which is known as a **phenocopy.** There are certain strains of mice in which harelip is found in about one-fourth of all the offspring no matter how carefully the mothers are reared. In other strains this abnormality is almost unknown in a normal environment. Hence, we can have little doubt of the relation of heredity to harelip. Still, a very high incidence of harelip will be found in the strains which normally do not produce it if certain treatments are given the pregnant females at a critical stage of the development of the embryos. These treatments include injections of the hormone, cortisone, a diet deficient in certain vitamins, an atmosphere of low oxygen concentration, and removal of some of the amniotic fluid around the embryos. That such a variety of treatments can cause the same results can be explained as follows. During embryonic development there is just one opening at the anterior end of the body, and this becomes divided into mouth and nostrils by a growth of the upper lips inwards from either side. Any factor, be it genes or a variety of environmental agents, which causes an interference with growth at the critical stage of development could cause an incomplete fusion of the upper lips, with the resulting abnormality known as harelip. There is some evidence of a similar relationship in human beings, and some physicians are known to advise women in the early weeks of pregnancy to avoid extended exposure to conditions of low oxygen concentration such as might be encountered when they go from a low altitude to a high altitude. An interesting recent study by D. Grahn showed a slightly greater percentage of births with congenital defects in the Rocky Mountain area of the United States than in areas near sea level. Apparently, despite the adjustments of a woman's body to compensate for the lower oxygen concentration in the air, cases may occur in which the oxygen supply to a growing embryo is slightly below the optimum quantity. Also, there are various drugs which might affect the general metabolic rate of an embryo and bring about some serious deviation from normal development, even though the drug seems to be entirely harmless to the woman taking it. People were shocked in the summer of 1962 to learn of the birth of many babies without arms or legs to women who had taken the drug thalidomide during their early pregnancy as a means of relieving the nausea which they experienced during this time. For many years there had been rare occurrences of such abnormalities, and a gene

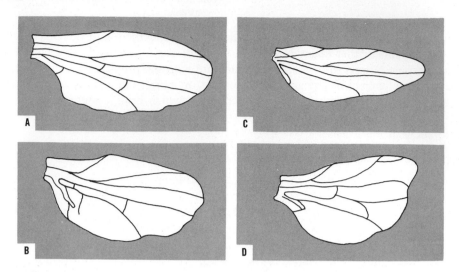

FIG. 21–1 *Phenocopies of wing traits in Drosophila induced by heat treatment of the pupae. A. Wild-type wing. B. Nicked wing, a phenocopy identical with the trait caused by the gene vg^{ni}. C. Lanceolate wing, a phenocopy of the trait caused by the gene vg^{la}. D. Truncate wing, phenocopy of the gene vg^{tr}.* AFTER RICHARD GOLDSCHMIDT.

was found to be involved in most cases. The sudden great increase in abnormalities of this particular type was traced to the thalidomide. Adequate tests had been made which showed that the drug did not permanently harm adults, although some persons did report a numbness in their fingers and toes that persisted for some time. The embryo, however, may be much more sensitive to certain drugs that depress growth; consequently, growth of limbs can stop at such critical times. Many physicians and radiologists now advise women not to have pelvic X-irradiation during pregnancy, if it can be avoided or postponed, since the embryo is much more susceptible to radiation damage than an adult, and development of the embryo may be affected by a dose too low to affect an adult.

The term "phenocopy" was first proposed by Richard Goldschmidt while working at the University of California. He subjected pupae of *Drosophila* to a relatively high temperature (35° C) for a short time at different periods in their development. Several phenotypes appeared which were indistinguishable from the phenotypes produced by certain mutant genes. In fact, in order to be sure that the effects of heat treatment were phenocopies rather than gene-induced effects, Goldschmidt had to conduct breeding tests. He found that the genes had not been changed by the heat treatment, and that the descendants of the phenocopies were normal in their phenotypes when they were grown at a normal temperature. Figure 21–1 shows some of the phenocopies which were induced in wings, all of them copies of effects produced by various other alleles of the mutant gene which produces vestigial wings. Goldschmidt found that he could actually predict what type of phenocopy could be obtained by giving the heat treatment at specific times during the period of pupal development. The meta-morphosis from larvae into adults seems to have a timetable for development and responds in a predictable manner to many kinds of mistreatment.

PENETRANCE AND EXPRESSIVITY

Environmental factors may so influence gene action in some cases as to suppress completely the expression of the gene. The sclerotic coat of the human eye is normally white, and the part of it we see at the front is sometimes known as the "white" of the eye. A certain dominant gene can cause this coat to be blue. About 10 per cent of the people who carry this gene, however, do not express the **blue sclera** but have a normal white sclera. These people can transmit the gene to their children and the children may express this dominant gene. We say that the gene has a reduced **penetrance** — a penetrance of 90 per cent. Genes with a reduced penetrance can cause problems in the analysis of pedigrees, for dominant traits can appear in children when neither parent expresses the traits.

The gene for blue sclera also illustrates variation in **expressivity.** Among the 90 per cent of those who carry the gene and express it, there will be a considerable variation in the shade of blue of the sclera. In some persons the sclera will be a blue so pale that it must be examined carefully to distinguish it from white, but in others the blue may be so dark that the sclera appears almost black. Some of the people with the very dark blue sclera, however, may have children with the very light blue sclera, so this is not a case of multiple alleles producing degrees of blueness. Exactly what factors influence the penetrance and expressivity of this gene we cannot say — there might be various modifying genes within the cells, or some factors in the external environment, such as the type of food eaten.

Unfortunately, the gene which causes a blue sclera does more than affect the color of the outer coat of the eye; it also may affect the bones. About 75 per cent of those persons who have blue sclera also have fragile bones which break easily. In some cases a bone will break while a person is turning over in bed. In other cases, breaks require greater strain, but the bones of these people still break more readily than those of unaffected persons. Some persons outgrow the affliction as their bones naturally harden with maturity, but in other cases the trouble persists throughout their lifetime. Thus, this one gene shows considerable variation in both penetrance and expressivity in its effects on the bones as well as the eyes.

In analyzing the cases of reduced penetrance and variable expressivity it appears that most, if not all, genes have complete penetrance and a constant expressivity on the primary level of gene action within the cell. The recessive gene for phenylketonuria, for instance, fails to produce the enzyme for phenylalanine breakdown to the same degree in all cases where it is homozygous within the cell. When we come to the secondary level of expression of traits as a result of this phenylalanine accumulation, however, we find considerable variation in expressivity. Some will have an I.Q. on the low idiot level while others with the same genes will be on the moron level. Some other genes or factors in the environment affect the degree of brain development which will take place in the presence of the excess phenylalanine. By reducing the intake of phenylalanine we can bring about an even greater variation in the expressivity of this gene, almost to the point of lack of expression, or a reduced penetrance. We find an even greater variation in expressivity and penetrance in the effect of the gene on melanin deposits in the hair and skin. Some persons with PKU have very fair skin and blonde hair, more so than would be expected on the basis of their inheritance of other genes affecting these traits. In other persons, even those who show the greatest mental retardation, there seems to be no noticeable reduction of the melanin deposits. Thus, we see that genes within the cells operating at the primary level have a consistent action, but as we go into the secondary consequences of the gene actions we find that there can be a considerable variation in the phenotypic expression. Many environmental factors can come into play to influence the final expression of the genes.

Hutt and Child have reported an interesting case of reduced penetrance and variable expressivity in chickens. In these birds there is a recessive gene for tremor which causes the afflicted individuals to shake almost continuously. But, out of 112 chickens homozygous for this gene, only 39 showed any detectable tremor. Among those birds that expressed the tremor there was great variation in degree. Some of the chickens shook so violently that they had to be fed by hand. Others shook so slightly that they had to be observed carefully in order to detect the tremor at all.

The gene for lobe eye in *Drosophila* is considered to be dominant, but it has a penetrance of only about 75 per cent when heterozygous. This penetrance may be varied by altering the conditions of food and other factors of the environment in which the flies live. When the gene is homozygous, however, penetrance is complete.

In some persons the infolded edge of the outer portion of the ear bears a point or tubercle. This was noticed by Charles Darwin, who regarded it as a remnant of the erect, pointed ear of some ancient ancestor, and it has come to be called **Darwin's ear point.** The trait appears to be inherited as a dominant with incomplete penetrance and variable expressivity. Some persons who transmit the trait do not show it, and hence we can say that the gene has incomplete penetrance. Among those who show the trait, some show it in only one ear while others show it in both. Thus, there is a variation in expressivity. A pedigree, shown in Figure 21–2, illustrates these points.

The recessive gene for vestigial wings in *Drosophila* seems to have complete penetrance, but varies considerably in expressivity. When raised at a normal room temperature of about 72° F, most of the flies homozygous for the gene will have only tiny stumps of wings, but, if the room temperature is warmer during the larval development, flies will hatch with longer wings. At a temperature of about 88° F (which is about as warm as they can stand for any considerable length of time) many of the flies will have wings which extend

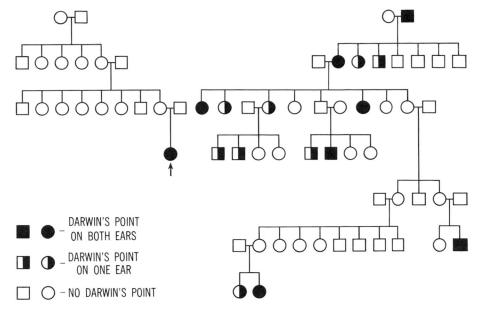

FIG. 21–2   *Pedigree of the inheritance of Darwin's ear point.*

as far as the posterior tip of the abdomen. These are variations in expressivity, in this case induced by temperature variation. Some of the gradations between the extremes of expressivity of the vestigial gene are shown in Figure 21–3.

ENVIRONMENTAL ALTERATION OF GENE EXPRESSION

When we say that certain genes have certain phenotypic effects it is usually on the assumption that the environment is kept within the bounds of what we think of as normal. By experimental alteration of the environment we can sometimes thwart the influence of specific genes. A certain recessive gene in corn produces the phenotypic effect of dwarfism when it is homozygous. It was found

FIG. 21–3   *Effect of temperature on gene expressivity. These fruit flies are all homozygous for the vestigial wings. The typical expression of this gene is shown in the first photograph; this fly was raised at a normal room temperature which averaged about 72° F. The second fly is taken from a vial kept at 80° F and shows the type of wing growth which may occur at this temperature. The third fly is from a vial kept at 88° F.*

FIG. 21–4 *Alteration of inherited dwarfism in corn. The two plants on the left have genes for normal growth; the two on the right are homozygous for a recessive gene for dwarfism. Gibberellic acid was applied to the second and fourth plants. This did not affect the plant with genes for normal growth, but caused normal growth in the plant homozygous for dwarfism.* COURTESY BERNARD O. PHINNEY.

that the dwarfism is due to the deficiency of a growth hormone, gibberellic acid. Somehow, the genes interfere with the normal production of this hormone. Plants homozygous for this gene can grow to normal size, however, if the gibberellic acid is added to the soil in which they are growing. Man can furnish that which the genes fail to produce. Figure 21–4 shows that the hormone has no effect on the normal plants since they produce sufficient quantities of it in their own metabolism.

Alteration by Temperature

Temperature is something from which we cannot escape. It is a universal environmental factor and one that has an important bearing on the expression of many genes. This is very nicely illustrated by a fur pattern known as **Himalayan** in domestic rabbits. It is caused by a simple recessive gene. Rabbits homozygous for this gene have white fur over most of the body, but black fur on their tails, ears, and tips of their legs and noses. This pattern is produced through the differences in temperature in the different parts of the body. The temperature of the skin is higher on the back or abdomen than on the extremities. The skin of a man, for instance, is often several degrees colder on the tips of his fingers than on his abdomen. This is due to the fact that there is a greater loss of heat by radiation at the extremities than there is in the trunk of the body. Hence, the Himalayan gene in rabbits does not produce the characteristic coat pattern directly but merely causes black pigment to be produced in the fur on all regions of the skin which remain below a certain temperature (about 92° F). Whenever the temperature exceeds this limit, no pigment is produced and the hair is white. This can be proved by a very interesting experiment which is illustrated in Figure 21–5. The white hair may be plucked from the back of a rabbit of the Himalayan variety, and then the plucked area may be packed in ice to reduce skin temperature in this region. As new hair grows and replaces that plucked out, it will develop black pigment, so that the rabbit will have a black spot in the middle of its back in addition to the black on its extremities. If the rabbit is kept at normal temperatures the black spot will gradually disappear, for these hairs will be shed and replaced with new ones which will lack the black pigment.

The reverse experiment may also be performed — black hair may be plucked from a region on one of the extremities and the bare region covered by a hot pad, which will diminish the loss of heat and raise the temperature of the skin which is covered. When replacement hairs grow out in this area, they will now be white. One theory which has been advanced to explain this condition holds that the gene for the Himalayan coat pattern causes the production of an enzyme which is necessary for the formation of the black pigment. According to this theory, the enzyme is not formed at temperatures above 92° F. If naked new-born rabbits homozygous for this gene are placed in a temperature of no more than 52° F for a short time and are then returned to a warm room, the hair, as it grows out, will be black all over the body. Only a few minutes in the cold is sufficient to chill the skin and cause the genes to produce the enzyme. Later, when the hair grows out, the enzyme is present and causes pigment to be produced even though the skin temperature may then be above 92° F. A more

Himalayan rabbit

FIG. 21–5 *Temperature and gene expression in the Himalayan rabbit. The fur is normally black on the ears, nose, feet, and tail. If the white fur is plucked from the back and this region is packed in ice for a time, the fur which grows back will be black. This shows that these rabbits inherit a gene which produces black hair at a low temperature and white hair at a higher temperature (above 92° F).*

Fur plucked from back
and ice pack applied

Black fur grown out
in plucked region

recent and more probable theory holds that the enzyme is present in all areas of the skin, but that the gene produces an inhibitor of the enzyme. According to this theory the gene would become inactive at temperatures below 92° and the enzyme would then be free to produce the black pigment.

This experiment with rabbits serves very well to illustrate the fact that, although we usually think of inheritance in terms of specific characteristics, actually only genes are inherited and the characteristics are brought out through environmental action on the gene-controlled processes of development. The rabbits inherit genes which can produce a pigment-forming enzyme, or an inhibitor of a pigment-forming enzyme, within a certain temperature range, but the actual coat pattern which results is due to the temperature of the skin. The Siamese coat pattern of cats is due to a similar temperature relationship.

A similar condition has been found to exist in some plants. One variety of the **Chinese primrose,** *Primula sinesis,* is homozygous for genes which produce white flowers at temperatures above 86° F, but red ones at temperatures below that point. The final flower color depends upon the temperature prevailing during a critical period in the early formation of the flower, so that, by changing a plant back and forth from hot to cold rooms, it is possible to get both red and white flowers on a single plant. We could not say that such a plant inherits red flowers; but only that it inherits a gene which can lead to the production of red pigment in the petals of the flowers under certain conditions of temperature.

There is also a condition of the feathers in domestic chickens which shows how far-reaching may be the effects of a single gene through the effects of temperature. There is a gene intermediate in dominance for **frizzled feathers** which causes the feathers to be wavy when heterozygous, but when homozygous causes the feathers to be very narrow, curled, and fragile, so that they break off quite easily. This frizzled condition of the feathers is apparently the only direct effect of the gene, but chickens which are homozygous for this gene are different from normal chickens in many other ways. One of the primary functions of the feathers of birds is to insulate the body and prevent an excessive loss of body heat through radiation from the skin. Feathers are very efficient in providing this protection — a light down comforter will keep a person as warm and cozy on a cold winter night as several heavy quilts or blankets. Birds have a rather high constant body temperature, running up to about 105° F, and depend rather heavily on their feathers to maintain this body heat. With frizzled feathers, however, most of this insulating property is lost, since the feathers are too thin and sparse to do much good and tend to break off easily, leaving large areas of the body almost naked and resulting in excessive loss of body heat. To compensate for this loss, certain changes occur within the body. The thyroid gland becomes more active and its hormones cause a rise in the rate of body metabolism. The extra metabolism requires more oxygen, which leads to an increase in the rate of respiration and a more rapid heart beat. Extra metabolism also requires more food, and the chickens eat more than the normal amount. All these physiological changes are reflected in physical changes in the body. The thyroid gland becomes larger as an adaptation to its increased duties, the heart also becomes larger, the digestive organs (crop, gizzard, pancreas) enlarge, and the kidneys also enlarge because of their greater task of removing the extra wastes generated in metabolism.

Thus extensive changes in the development of the internal organs are brought about by temperature variations induced indirectly by the effects of genes which influence the development of body feathers. The same effect on the internal organs may be induced by shearing the feathers from normal chickens, for this results in the same loss of body heat and the resulting internal adaptations. Also, the extent of the influence of the gene for frizzled feathers may be modified by the temperature of the external environment. If the chickens are kept in heated rooms constantly, where the temperature is at the optimum point, these internal effects of the frizzled condition will not be so pronounced. On the other hand, if the chickens are raised at a temperature which is considerably below normal, there will be such a drain on the thyroid glands that these eventually become atrophied, and the birds will die.

Effect of Sunlight

Sunlight plays a necessary role in the expression of many genes. In corn there are a number of genes which result in **albinism.** When homozygous for any of these genes, the young seedling will grow several inches in height, using the stored food in the seed, and then it will die because it fails to develop chlorophyll. Thus in plants the genes for albinism are lethal because chlorophyll is necessary for the manufacture of food. The character continues to crop up in many plants, however, since the genes may be carried in the heterozygous condition by normal plants. Normal plants have genes which cause the production of chlorophyll in the presence of sunlight. They do not inherit the chlorophyll itself, as can easily be proved by raising corn seedlings in the dark, where chlorophyll never develops because of the absence of sunlight. Suppose we plant seed from heterozygous parents and for a time keep the seedlings in a dark place. According to the laws of probability about one-fourth of them should be albinos and three-fourths should have the ability to produce chlorophyll. Grown in the dark, however, they all look the same — they all appear to be albinos. If we then place the plants in the light, within a period of a few minutes the green color of chlorophyll will begin to show in the plants which have the ability to produce this vital substance. Thus an environmental factor, darkness, can duplicate the condition which may be caused by genes in a more natural environment. This is another case of a phenocopy.

Another excellent illustration of the environmental effects of sunlight is also found in corn. There are some genes which cause the grains of corn to be red, while other genes cause them to be without color, or white. At the old-fashioned corn-husking contests, the red ears were highly prized, for the young man who shucked such an ear was rewarded by a kiss from the young lady of his choice. There is another gene, however, which produces white grains if no light strikes them, but red grains if they are exposed to the light. As long as the husk remains intact the ear will be white, but, if some of the shucks peel back from the tip, the exposed portion of the ear will be red. If all of the shucks are peeled back and light is allowed to reach the entire ear, it will be red throughout. This is known as **sun-red corn.** Figure 21–6 shows an ear that was exposed to the sun but partially covered with a black cloth with the word "sun" cut in it. The results clearly show the effects of sunlight.

In man the effects of sunlight on the skin are clearly evident. Some persons of the fair-skinned Nordic races inherit genes which permit them to obtain a deep coat of tan when exposed to abundant sunlight. Others, such as members of the Negro race, develop a rather large amount of pigmentation in the skin regardless of whether they are ever in the sunlight or not. A fair-skinned life-guard on a beach may easily develop a pigmentation as deep as that of many members of the Negro race. Thus in one case the effect is due to heredity and sunlight and in the other to heredity alone. Other members of fair-skinned races may inherit genes which do not permit the formation of heavy deposits of pigment in their skin in response to sunlight. Such persons are severely sunburned even by very short exposures and cannot develop a tolerance to sunlight by a deep coat of tan. These three groups of persons might be compared to the condition described in corn. Some develop the dark pigmentation regardless of sunlight; some have the power to develop it, but can do so only in the presence of sunlight; and others remain fair-skinned even though exposed to sunlight.

Some persons shun the sunlight because the pigment which is formed in response to the rays of the sun tends to accumulate in embarrassing little spots, better known as **freckles.** We often say that freckles are inherited as a result of a dominant gene. Still the gene for freckles lacks complete penetrance, for some persons who carry the gene avoid exposure to the direct rays of the sun long enough to escape the phenotypic expression of a dominant gene which they carry.

The disease of **rickets** often develops in children who receive insufficient exposure to sunlight. There are precursor substances in the plant foods which an animal eats that can be converted into vitamin D when sunlight strikes these substances in the skin. It is possible for man to stay out of the sunlight and obtain sufficient vitamin D which has been manufactured in other animal bodies

FIG. 21–6 *Sun-red corn. The corn plant which produced this ear was homozygous for a gene which results in the production of red pigment in the aleurone cells of the corn when exposed to sunlight. The shucks of this ear were peeled back and the ear was covered with a black cloth with the word "sun" cut in it.* COURTESY R. A. EMERSON, CORNELL UNIVERSITY.

or through artificial radiation of certain types of foods, such as milk. In an institution, however, there will quite often be a group of children who receive the same kind of food and the same exposure to sunlight and yet some will develop rickets while others do not. The reason for this seems to be a matter of heredity. It appears that some children inherit a constitution which requires greater amounts of vitamin D than others. Streeter and others confirmed this hypothesis through carefully controlled experiments on rats. Following 14 generations of careful selection, these workers obtained two strains of rats which looked identical but differed greatly in the amount of vitamin D required in their diets. When both groups were fed on diets deficient in this vitamin, one group developed severe rickets, while the other group was much less sharply affected. Other studies show that the same situation holds true for other vitamins. We must not be too arbitrary when we attempt to set mininum daily requirements of vitamins for human nutritional needs.

Alteration by Food

Food is such a vital and important environmental factor that it is bound to have an extensive effect on the expression of genes. A person may inherit genes for a strong husky body, yet it is quite evident that such a body cannot develop unless the proper food elements are furnished. In *Drosophila* there is a gene for **giant body size.** Such a gene will have a very low penetrance if the vial in which the fruit flies are raised is crowded and there is a rather scanty food supply. Under these circumstances very few of the flies will show the giant body size even though all may be homozygous for the gene. If the food is very sparse, none will show the character. On the other hand, if the food supply is abundant, the majority of the flies will be giants.

When rabbits are dressed for human consumption, it is found that some have **white fat** and some have **yellow fat.** A study of this characteristic shows that yellow fat is inherited as a simple recessive character but one which does not have complete penetrance, for on some occasions a group of rabbits homozygous for the gene for yellow fat will all have white fat. This condition results from the food which the animals have eaten. If the diet includes green plants the

FIG. 21–7 *Effect of environment altered by genes. These two rats are of the same age and sex and feed on the same diet which is slightly deficient in Vitamin D. The rat on the left thrives on it, but the one on the right has rickets. Genes make the difference — the rat on the left is from a strain selected for resistance to rickets while the rat on the right comes from a strain selected for susceptibility to rickets.*

fat will be yellow, but, if no green food is consumed, the fat will be white. This is explained by the fact that green plants also contain a yellow pigment, xanthophyll, in their leaves. The dominant gene seems to function by causing oxidation of this pigment. In the absence of this gene the pigment is not oxidized and, since it is fat-soluble, it tends to accumulate in the fatty tissues of the body. It cannot accumulate, however, unless it is present in the food which is consumed.

HORMONES AND GENE EXPRESSION

Hormones play such an important part in the growth and activities of animals that we would surely expect them to be a vital factor related to the expression of certain genes, and indeed many of the effects of genes on body characteristics are brought about through their influence on hormone production. In mice a recessive gene produces a **dwarf body size** when homozygous. Dissections of dwarf mice reveal that the pituitary gland is abnormally developed. The anterior lobe of this gland secretes a hormone, the growth hormone, which stimulates body growth. The dwarf mice do not receive a sufficient quantity of this hormone and are consequently stunted in growth. The other genes for normal body growth are present in such animals, but they cannot express themselves in the absence of sufficient quantities of the growth hormone. This can be proved, for if the gland from a normal mouse is transplanted into a young dwarf mouse, its growth will soon pick up, and a normal-sized mouse will result even though all of its body tissue is homozygous for the gene for dwarf size except the cells of the transplanted pituitary gland.

Alteration by Hormones

Landauer has made some very significant discoveries of the effects of insulin on the development of domestic fowls. In chickens a recessive gene produces a condition called **rumplessness.** Birds showing this character are peculiar looking creatures that look as if a near-sighted slaughterer had attacked his victims at the wrong end. The caudal vertebrae, tail feathers, and oil glands are absent. This throws the chickens off balance, so that they must bring their heads up and

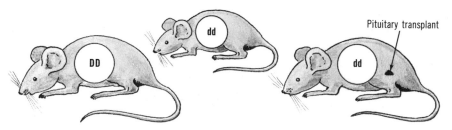

FIG. 21–8 *Artificial alteration of inherited dwarfism in mice. The mouse at the left is of normal size, but the mouse in the center is dwarfed because he has a pair of recessive genes which reduce the output of the growth hormone from the pituitary gland. The mouse at the right also has inherited the genes for dwarfism, but has received a transplant of a pituitary gland from a normal mouse; he is of normal size.*

back to produce a new center of gravity for the body. This results in a bird with an upright rather than a horizontal posture. A non-inherited phenocopy of this condition may be produced by shaking the eggs at certain stages of incubation. Landauer also found that injection of insulin into the air chamber of incubating eggs up to 72 hours after incubation started resulted in rumpless phenocopies. Insulin speeds up carbohydrate metabolism, which in turn affects other vital body reactions. These reactions, coming at a critical time during the development of the embryo, result in the rumpless condition.

If the insulin is injected later than 72 hours after the beginning of incubation, the rumplessness is not produced. Apparently the embryonic rudiments for the posterior end of the body are already formed by this age and no longer respond to the insulin treatment. Other body structures may be affected by later injections, however. Injections on the fifth or sixth day produce a shortening of the legs and wings and abnormalities of the skull and beak which are similar to conditions produced by other mutant genes. Through such studies it may be possible to learn more about the time and method of action of certain genes.

Diabetes

Another excellent illustration of the effect of hormones on gene expression is found in man. *Diabetes mellitus* is a disease caused by the inability of the body to supply sufficient insulin to take care of normal carbohydrate metabolism. All of the carbohydrates are converted into simple sugars in the process of digestion. In a normal individual the pancreas secretes insulin which regulates the metabolism of these sugars in all parts of the body. If the supply of insulin is insufficient, the unused sugar will accumulate in the blood. The kidneys remove this extra sugar, and the only outstanding symptom of mild diabetes may be sugar in the urine. As the quantity of sugar in the blood becomes greater, however, the kidneys cannot remove it fast enough and other serious symptoms appear, the most severe of which is diabetic coma and possibly even death unless proper treatment is administered. Diabetes is definitely influenced by heredity, yet environment does play a part. A person may inherit the tendency for diabetes, but if he does not overtax the pancreas through excessive intake of carbohydrates over a prolonged period he may never show the disease. This is well illustrated by a pair of one-egg twins, one of whom had diabetes while the other did not. This difference in penetrance is explained by the different modes of life of the two. One had owned a tavern and had subjected his body, which was inclined to diabetes through heredity, to the strain of heavy beer-drinking. The strain on the pancreas in producing sufficient insulin for the large amount of carbohydrate in the beer brought on the diabetes. The other twin did not subject his body to such a strain and did not develop the disease, but clinical tests showed that his body had a very low tolerance for a high carbohydrate diet.

A person may inherit diabetic tendencies and develop a very serious manifestation of the disease, but if he takes proper doses of insulin he will not show the symptoms of the disease. Thus an environmental modification of the phenotype is achieved. Such is possible with many inherited abnormalities. A person born with a thyroid gland which does not secrete sufficient thyroxine will become a cretin, with dwarfed body and serious mental deficiency. Administration of extra quantities of thyroxine, however, will cause such a child to develop into a normal human being if the treatment is begun soon enough.

MODIFYING GENES

There are instances in which a group of organisms with the same genotype for a particular gene and raised under similar environmental conditions will yet show some variation in the expressivity of the character involved. This may be due to the effect of other genes which influence the same character. It is customary to call such genes **modifying genes,** because they modify the expression of a certain character. It is easy to understand such a situation — almost all characters result from the interaction of a number of genes and, although mutation of one vital gene of the group may produce a definite phenotype effect, there still may be minor variations in other genes which will influence the expression of this effect. A good example of this is found in Holstein cattle. The black and white spotted condition which characterizes this breed is due to a single pair of recessive genes. There is a great variation in the degree of spotting, however, and some animals are almost pure white with a little black on the head and legs, while others are almost pure black. These variations are due to modifying genes. The spotting itself is inherited as a simple recessive, but the modifying genes determine the degree of spotting which will show. These modifying genes seem to have no influence upon coat pattern unless the cattle are homozygous for the gene for spotting.

Such variations due to modifying genes would not come under the heading of environmentally induced phenotypic variations, for we commonly think of the word environment as referring to factors outside the cell (extracellular environment). These variations, however, might be considered as influenced by intracellular environmental agents.

HEREDITY AND ENVIRONMENT IN HUMAN TWINS

Most studies in human genetics dealing with the relative effects of heredity and environment have been made on twins. Such studies make an excellent substitute for the kind of experimental work which is carried out on plants and other animals but cannot be employed on human beings. As we learned in Chapter 3, there are two types of twins — the so-called "identical," monozygotic, or one-egg twins, which arise from a single zygote; and the fraternal, dizygotic, or two-egg twins, which arise from two zygotes. Monozygotic twins have identical genes and, as a result, make ideal subjects for studies of the effects of environment upon development, since any differences which they show must be environmentally induced (barring somatic mutations). Fraternal twins serve very nicely to illustrate the effects of differences in heredity in a constant en-

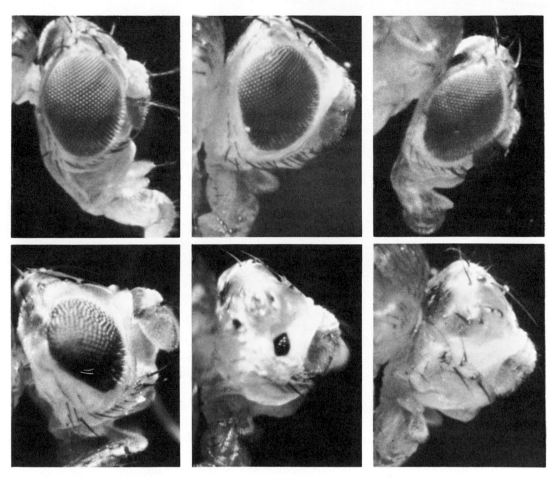

FIG. 21-9 *Effect of modifying genes on gene expressivity in Drosophila. The fly in the upper left photograph has a normal-sized eye; all the others are homozygous for the recessive gene for eyelessness, but they vary greatly in the degree of expression of this gene, ranging all the way from an eye almost as large as the normal one down to no eye at all. The flies were all raised in the same vial, and modifying genes no doubt play a part in this variation in expressivity.*

vironment. They will have differences in many of their genes, yet are most likely to have a very similar environment. Comparisons of monozygotic and dizygotic twins raised under rather constant environmental conditions show very significant effects of heredity. For instance, monozygotic twins are always of the same sex, have the same blood types, hair and eyes of the same color, and the same type of hair growth, while dizygotic twins may differ in all these characteristics just as children born to the same parents at different times may differ. This indicates that such physical characteristics as these result primarily from gene action and are little influenced by environment. Other physical characteristics such as body height and weight show variation among monozygotic twins, thus indicating that environment plays a part in these characteristics.

Concordance

On the other hand, studies of a large number of twins show that there is a closer agreement between monozygotic twins with respect to height and weight than there is between dizygotic twins. This indicates that heredity plays a part also. Thus, by extensive studies of both types of twins it is possible to establish the relative part played by heredity and environment in different human characteristics. This is often expressed in terms of **concordance,** which is the percentage of agreement. For instance, suppose we find fifty surgeons who have identical twin brothers. Suppose we now find that forty of these twin brothers are also surgeons; this is a concordance of 80 per cent.

Of course, those cases of multiple births where more than two children arrive at once are quite valuable for studies of heredity and environment, but the comparative scarcity of such cases limits their usefulness. The most widely studied case of human multiple births was that of the Dionne quintuplets. Exhaustive tests established the fact that these began life as a single cell; in fact, there is good evidence that there were six of them for a time, but that one was crowded out of existence very early in the course of embryonic development. These girls had identical environment as near as this can be provided in childhood, but many important bits of information have been obtained by studies of their differences as adults. The untimely death of one of these girls was regrettable from scientific as well as personal viewpoints. Another noteworthy case is that of the Keyes quadruplets of Texas, four girls who were produced from three eggs. Apparently what happened was that one of the eggs produced two embryos while the other two eggs produced one embryo each. This unusual situation makes it possible to study both fraternal and identical relationships among children of the same parents and of the same sex and age. In spite of the excellence of such rare cases for genetic study, however, twins still remain the best source of material for studies of heredity and environment because of their greater frequency.

Monozygotic Twins Reared Apart

The most valuable twin studies, from the standpoint of genetics, are those rather rare cases of monozygotic twins reared apart. Dr. H. H. Newman of the University of Chicago succeeded in finding nineteen cases of monozygotic twins who had been separated in infancy and reared by different families. Living in different homes, often in different parts of the country, receiving different educational opportunities, having different social contacts, these pairs of twins had a considerable opportunity to develop environmentally-induced differences. One case revealing some of the most striking differences was that of Gladys and Helen, which we will review briefly as an example of Dr. Newman's work.

These monozygotic twins were born in Ohio, were separated at the age of 18 months, and did not see each other again until they were 28 years of age. Gladys was reared in a medium-sized town in Ontario, Canada. She left school after finishing the second grade and remained at home doing housework for her foster father, who was in poor health. She went to work in a knitting mill at 17 and 2 years later moved to Detroit and took a job as a saleswoman. At 26 she changed positions, moving to a small printing establishment. She held this position at the time of the study, when she was 35. Helen's foster-parents lived in a rural area in southern Michigan. They were prosperous enough to send Helen to a good college in Michigan after she finished high school. After graduation from college, she obtained a teaching position and for 8 years preceding the tests had been a teacher in a Detroit public school. The physical traits of the two were found to be very similar when they were brought to Chicago for an examination. Helen was slightly more than 1 inch taller than her sister, but their weights were almost identical. The study of physical traits showed that they were monozygotic twins. The greatest differences shown were in personality and intelligence. Helen had a distinct charm and grace, and was an interesting and animated conversationalist, while Gladys seemed ill at ease and diffident, and lacked the other attractive personality traits of her sister. On intelligence tests, Helen scored 24 points higher than Gladys. No greater argument for the value of education and social contacts could be made than a survey of the contrast between these two girls who started their lives as a single individual.

Twins Reared Together

Twins raised in the same family group also have their value in studies of heredity and environment. When twins of the same sex are reared in the same family, the environment is about as nearly uniform as it is possible for human beings to have. Comparisons of monozygotic and dizygotic twins who have not been separated give some indication of the effect of the difference of genes which is bound to occur in the dizygotic group. Of course it must be remembered that the dizygotic twins have many genes in common because of their common parentage, but these genes are no more alike than would be expected in brothers and sisters born at different times (sibs). The first two columns of Table 21–1 show the results of a comparison between fifty pairs of each type of twins as made by Newman and others for a number of different characteristics. It can be seen from this table that differences do exist between monozygotic twins, even when they are raised in the same family, but these are much less pronounced in every case than the corresponding differences between dizygotic twins. This observation applies particularly to the physical traits which are less subject to those small variations in environment that are bound to occur. The third column of the table shows the results of a study of the differences between monozygotic twins and their sibs of the same sex in the same family group, with due allowance for age differences. It can be seen that these differences are very much like those found for fraternal twins. The fourth column shows the comparable results for identical twins reared apart. It is apparent from these figures that identical twins reared apart are more alike, on the average, than fraternal twins reared together.

	Identical	Fraternal	Sibs	Identical (reared apart)
Height (diff. in cm.)	1.7	4.4	4.5	1.8
Weight (diff. in lbs.)	4.1	10.0	10.4	9.9
Head length (diff. in mm.)	2.9	6.2	—	2.2
Head width (diff. in mm.)	2.8	4.2	—	2.8
Total finger ridges	5.9	22.3	—	—
Age of first menstruation (diff. in mo.)	2.8	12.0	12.9	—
I.Q. (Binet) (diff. in points)	5.9	9.9	9.8	8.2
Concordance in criminal record (diff. in per cent)	32.0	72.0	—	—

TABLE 21–1

Differences between Twins and Non-Twin Sibs in a Number of Characteristics

Criminal Tendencies

The tendency toward criminality is often thought of as an acquired trait brought about by pathological social conditions, yet studies of twins indicate that this trait may be influenced by heredity. Five different surveys have been made during the past 20 years, each by a different person, on the possibility of hereditary influence in criminality. Each investigator concluded from the results of his survey of twins that there is a much greater likelihood of criminal careers among both members of a pair of identical twins than among both members of a pair of fraternal twins of the same sex. One study found 37 cases in the United States where at least one member of a pair of identical twins had been convicted of a major crime. In 25 of these cases the other twin had also been convicted of a major crime. In other words, about 68 per cent of the cases included both members of a pair of one-egg twins. The same study found 28 cases where at least one member of a pair of fraternal twins of the same sex had been convicted of such a crime. In only 5 of these cases, however, had the other twin been convicted of a major crime, that is, in about 18 per cent of the total number of cases. These results strongly support the idea that heredity plays a part in criminality.

Studies in Germany point to the same conclusion. Of 66 pairs of identical twins of which one member had been convicted of a crime, there were 45 cases in which the other member of the pair had also been convicted. From 84 pairs of fraternal twins of the same sex, there were 32 cases where both had been convicted. This gives percentages of 68 and 38 respectively, values in close agreement with the results of the study in the United States. In analyzing these results, we should keep in mind the fact previously mentioned, that similar heredity may play a considerable part in the concordance among fraternal twins also, for they possess many genes in common because of their common parentage, but they also show hereditary differences because of the principle of independent segregation of genes. If it were possible to study non-related persons of the same age and the same sex who were raised by the same parents, it is very likely that the concordance would be lower than with fraternal twins. There are very few such cases to be found, however.

These studies do not by any means minimize the importance of environment in the production of criminality. But they do strongly suggest that there are inherited differences which would lead one person into a life of crime under certain social conditions while another person under the same circumstances would not be so influenced.

One fact which stands out in the studies is that so-called "identical" twins are never exactly identical. Even though they are reared together, it is evident that environmental differences can cause slight differences between them. Some of these differences may be prenatal. At birth one twin is nearly always more robust than the other, perhaps because a more favorable position in the uterus has enabled it to obtain a better food supply or to have more room to exercise. Such a prenatal advantage may be sufficient to have its effect on potential physical development. Then after birth, in the struggle for self-expression, one twin may gain the advantage and develop a somewhat different personality as a result.

Disease

The effect of heredity on disease is another of the many aspects of the studies of twins. Table 21–2 shows the concordance for a number of different diseases and other body abnormalities for the two kinds of twins. It can be

FIG. 21–10 *Variable expressivity of a dominant gene in man. These men are brothers and both carry a dominant gene affecting the growth of the arms. In one brother the arms are shortened and the hands are deformed. In the other brother, the arms are reduced to mere flipper-like stumps.* COURTESY KARL STILES.

TABLE 21–2

Concordance (Percentage of Similarity) between the Two Different Kinds of Twins with Respect to Disease and Other Body Abnormalities

	Identical	Fraternal
Harelip	33	5
Mongolism	89	6
Mental retardation (feeble-mindedness)	97	37
Schizophrenia (adjusted to allow for age)	86	15
Cancer	61	44
Site of cancer the same (for those cases where both have cancer)	95	58
Measles	95	87
Tuberculosis	65	25
Diabetes mellitus	84	37
Rickets	88	22

seen that in every case the concordance is greater for the monozygotic twins, even in the case of the diseases caused by germs, a fact which indicates that heredity plays a part in susceptibility to germ diseases. This is particularly noticeable in tuberculosis, where the concordance for identical twins is 65 and for fraternal twins is only 25. The non-infectious diseases of the group — diabetes and rickets — show a greater difference in concordance than the infectious diseases as a whole. This seems to indicate that heredity is relatively more important in such diseases than in the infectious ones. The relationship between heredity and mental disease is also indicated by the rather great difference in concordance between the two groups for schizophrenia.

The further we go in studies of twins, the more apparent it becomes that heredity places its imprint upon almost every aspect of human development and behavior. The studies of twins serve to re-emphasize the facts brought out by the experimental studies on environmental modification of gene effects discussed in the first part of this chapter. The individual is a product of his environment as it acts upon his hereditary potentialities.

PROBLEMS

1. In a mating between a normal hen and a normal rooster, four chicks out of a total of forty-eight produced showed tremor. Show the probable genotype of the parents and the chicks. Explain the unusual ratio in the offspring.

2. If you wanted to produce a Himalayan rabbit with white ears, tell how you would proceed to do so.

3. Is it correct to say that a green plant has inherited chlorophyll through its genes?

4. A group of rats of mixed hereditary background are fed on a diet which is low in riboflavin (vitamin B_2). Some of the rats develop characteristic abnormalities of the skin, but others remain normal in appearance even though all of them receive the same quantity and type of food. Explain.

5. A farmer kills several rabbits on his farm in June, and on dressing them, he finds that some have white fat and some have yellow fat. The next January he kills several more and finds that they all have white fat. How would you explain this to him?

6. Why are monozygotic twins reared apart of such great genetic interest?

7. Identical twins reared apart show greater deviation in I.Q. scores than in body height. What would you conclude from this?

8. When a characteristic is known to be inherited, does that mean that environment may play no part in its expression? Give an illustration to support your viewpoint.

9. When a character is proved to be produced by external environmental agents, does that mean that genes may have nothing to do with the expression of the character? Give an illustration to support your viewpoint.

10. Is the relative role of genes and environment the same for all characteristics? If not, list some human characteristics which you think would be due primarily to gene action and some which you think would be due primarily to environmental agents.

22

Cytoplasmic Inheritance

THE FACT THAT THE NUCLEUS OF A CELL must be the location of the units of heredity was recognized by Oscar Hertwig and others in the 1870's. The equality of inheritance from both male and female parents implied that it was the nucleus and not the cytoplasm which was of primary importance in heredity. Nevertheless, in the course of investigations on the mode of inheritance of thousands of characteristics among hundreds of different kinds of plants and animals, a few cases have been found which seem to indicate cytoplasmic inheritance. We shall survey a few of these in the present chapter.

Plastid Transmission in Plants

The plastids which develop in plant cells are in general **self-perpetuating bodies** within the cytoplasm which develop and function in conjunction with nuclear genes. The cytoplasm of the ovule, which is incorporated into a seed, carries primordia of the plastids which in certain respects determine what type of plastid will develop in the plant growing from the seed. The sperm nucleus from the pollen grain, which carries very little cytoplasm, often does not influence the type of plastids which are formed. For example, in the four o'clock, *Mirabilis jalapa,* there is a variety which has variegated leaves and the normal green is spotted with patches of white or pale green. In some cases the spots may form small areas on the leaves, but in other cases entire leaves or branches may be affected. Seeds taken from a branch which is entirely green will produce plants which are entirely green, and seeds taken from a branch which is wholly white produce plants without chlorophyll, and these of course die. Moreover, seeds taken from variegated branches produce green, white, and variegated plants. In this case, it makes no difference what type of pollen is used to fertilize the plants — pollen from a wholly green plant placed upon the pistil of a white branch still produces seeds which grow into white plants. This is a true case of cytoplasmic inheritance. It is not inheritance in the same sense as that which occurs with nuclear genes, but it still is a generation-after-generation transmission of a characteristic through self-perpetuating bodies in the cytoplasm. Such a method of transmission has been found in twenty genera of plants. We know, however, that nuclear genes have their influence on plastids, for there are many genes for albinism or for modifications in the amount of chlorophyll in corn and other plants and these genes are inherited in a simple Mendelian manner. Also, some types of striping and even of variegation are inherited in the same manner.

Male Sterility in Plants

Another classic case of cytoplasmic inheritance is male sterility in plants. In certain species of plants aberrant forms have been found in which there is degeneration of the sperm-forming pollen cells within the anthers. This results in male sterility, although the ovules may function normally with full female fertility. Marcus Rhoades has made a thorough study of this characteristic in corn. Starting with a strain having male sterility, he replaced, through extensive breeding experiments, each chromosome with a corresponding chromosome from a strain in which the males were fertile. In the resulting plants the males remained sterile, however, in spite of this complete replacement of the chromosomes. This would seem to exclude the influence of nuclear genes and, since the trait was transmitted through the female line, the evidence is very strongly in favor of some form of cytoplasmic inheritance. This hypothesis was established rather conclusively by a **reciprocal cross.** This type of cross, using male-sterile plants as the male parent and male-fertile plants as the female parent, may seem impossible. It is possible, however, because a very few fertile pollen grains are sometimes produced on male-sterile plants. Using these exceptional pollen grains, Rhoades found that the male sterility was never transmitted through the male parent.

Maternal Inheritance in the Ephestia Moth

Another type of maternal inheritance is illustrated by the flour moth, *Ephestia.* There is a dominant gene, *A,* which causes a darkening of the skin in both larvae and adults. In a cross of a recessive female, *aa,* with a heterozygous male, *Aa,* one-half of the offspring are dark and one-half light in color, as would be expected by simple Mendelian inheritance. In the reciprocal cross, however, where the female is heterozygous and the male homozygous, all the caterpillars are dark when they are first hatched. As they grow, however, one-half of them gradually lose their pigment and become light while the other half remain dark. Such a condition can be explained if we assume that the gene *A* produces a substance which is necessary for the production of pigment. This substance may be stored in the cytoplasm so that the offspring of a dark female will be dark, regardless of their own genotype, because of the transmission of this substance through the cytoplasm of the egg. As the caterpillars grow and molt, however, they will become light if they do not have the gene to produce more of this substance, for that received from the mother loses its effect. This, then, is a case in which a product formed by a gene may be transmitted through the cytoplasm and may produce a temporary effect on the offspring. However, it is not true cytoplasmic inheritance of self-perpetuating bodies such as was found in the case of the plastids in plants.

Cytoplasmic Inheritance in Drosophila

There is one rather clear-cut case of cytoplasmic inheritance in *Drosophila*. One strain of these flies has been discovered which is much more sensitive to carbon dioxide than others. When insects are exposed to air with a rather high concentration of carbon dioxide, they soon become anesthetized. In fact, this is a very easy way to immobilize *Drosophila* for laboratory examination — it is much cleaner and less likely to cause death than ether, though unfortunately most laboratories are not equipped with the facilities to provide carbon dioxide for such purposes. This particular strain of *Drosophila* is affected by much smaller concentrations than those required to anesthetize normal strains, and may be killed by stronger concentrations. This characteristic was found to be inherited, and a true-breeding strain was established. When females from this strain are mated with normal males, all the offspring show the characteristic of the female parent and are highly sensitive to carbon dioxide. On the other hand, when males from this strain are mated to normal females, only very rarely are any of the offspring so highly sensitive. We expect some variation in reciprocal crosses when we are dealing with sex-linked genes, but nothing like this can result from sex-linked, sex-influenced, or sex-limited genes. Each of the chromosomes of the sensitive strain has been replaced with other chromosomes from non-sensitive stock without influencing the characteristic. Hence it may safely be concluded that the transmitting factor is located outside the chromosomes and in the cytoplasm, since it is transmitted regularly by the eggs and only rarely by the sperms. This self-perpetuating unit has been called **sigma.**

Further experiments show that sigma can be transferred from a sensitive to a non-sensitive strain. This is done by transplanting ovaries from normal females into the bodies of sensitive females. Some of the sigma from the body cells reaches the cells of the transplanted ovaries, and all offspring from such females will be sensitive even though the germ cells may have come from a normal strain. This is actually a case where the somatoplasm has an influence on the heredity of the offspring.

Cytoplasmic Inheritance in Paramecium

A great amount of significant work on cytoplasmic inheritance has been done on *Paramecium aurelia* by T. M. Sonneborn of Indiana University. *Paramecium* is a small, one-celled protozoan which is abundant in ponds of stagnant water. It reproduces asexually by fission, but on occasion there is a

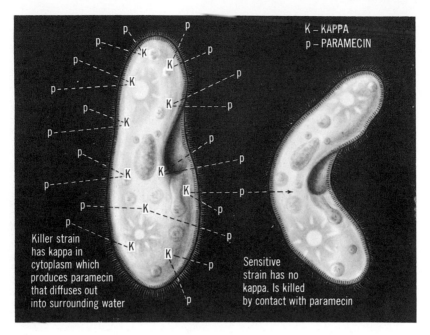

FIG. 22-1 *The relationship between killer and sensitive strains of Paramecium aurelia.*

type of sexual union known as **conjugation.** In this process two organisms come together, form a temporary bridge of cytoplasm between them, and exchange nuclear material over this bridge. This allows recombination of the genes with all the evolutionary advantages which accompany sexual reproduction in higher forms of life. It is possible to cross different types of *Paramecia* and study the heredity of distinguishable characteristics. Such studies show that *Paramecium* has nuclear genes which follow the same principles of heredity as those which have been demonstrated in multicellular organisms.

In the course of mixing different races of *P. aurelia* in a common culture dish to allow conjugation to occur, it was found that sometimes all the members of one race would die while those of the other race survived. It would seem from such results that one race must produce a substance which diffuses out into the surrounding water and causes the death of all individuals of certain other races sensitive to this substance. This was found to be true. One race was allowed to live in a culture dish for a time and was then removed. When members of a sensitive race were placed in the water in this dish they soon died. The two races were subsequently designated as **killers** and **sensitives.** The killing substance generated by the first race was called **paramecin,** a sort of protozoan antibiotic.

It is possible to cross these two races by placing them together in fresh water free of paramecin. After conjugation, however, each individual gives rise to organisms which are like itself — sensitives produce sensitives and killers produce killers. Now if these characteristics were inherited through nuclear

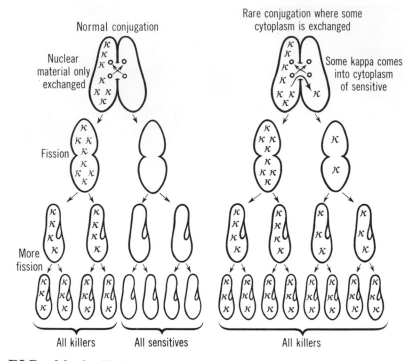

FIG. 22−2 *Evidence of cytoplasmic inheritance of kappa in Paramecium. The offspring of sensitives become killers only in rare cases of
conjugation when some of the cytoplasm of a killer strain passes into a
sensitive.*

genes, we would expect a mixing of the traits when the nuclear material is exchanged during conjugation. Since cytoplasm is not ordinarily transferred from
one organism to the other in conjugation, we would suspect this to be a case of
cytoplasmic inheritance.

Fortunately, a simple test of this conjecture is possible (see Figure 22–2).
Under certain conditions the conjugants form a cytoplasmic bridge between them
which is larger than usual, and there is a visible exchange of cytoplasm. If the
killer trait is determined by some factor in the cytoplasm, then this characteristic
should under such circumstances be transferred to a sensitive mate whenever the
latter receives cytoplasm from a killer. This was found to be the case, and
sensitives become killers and produce killer offspring after such a cytoplasmic
transfer from a killer.

The substance in the cytoplasm which generates paramecin has been designated as **kappa.** There may be hundreds of kappa particles in the cytoplasm of
killers, but none in the cytoplasm of sensitives. The kappa particles are self-
duplicating, they control a hereditary trait, they never arise *de novo,* and they
may mutate. Thus they bear many of the characteristics of genes, but they are
not on a chromosome and they are in the cytoplasm rather than in the nucleus.
Like viruses, they may duplicate themselves more than once or less than once

for each cell division. Because of this fact it is possible to convert killers into sensitives, as illustrated in Figure 22–3. If the *Paramecia* are given so much food that they multiply faster than the kappa particles in the cytoplasm, the numbers of these particles in each organism will become smaller and smaller. Finally, there may be only two or three particles in one cell. Then when division comes, it is quite possible that all these particles will be included in one cell, while the other cell will receive none. The latter will be a sensitive *Paramecium* and will never produce killers in its offspring. The same thing may be accomplished more quickly by the use of agents which can destroy kappa within the cell. This can be done with X-rays, nitrogen mustard, or unusually high temperatures.

It is also possible to accomplish the reverse procedure — to produce killers from sensitives. This is done by exposing sensitive animals to the crushed bodies of killers. The kappa particles may be taken into the bodies of the sensitive *Paramecia,* and, if they receive as much as one such particle under favorable conditions, that particle will multiply and the protozoan becomes a killer. Sensitive *Paramecia* can thus be exposed to the kappa particles, for not these particles, but the paramecin they generate, causes the death of the sensitives.

These studies raised the question of the exact nature of these cytoplasmic particles. Are the kappa particles of much the same nature as genes existing free in the cytoplasm and multiplying independently of cell division, but nevertheless influencing heredity in a manner similar to the nuclear genes? The evidence of recent research indicates that the kappa particle shows more of the nature of a parasite which has invaded certain *Paramecia* and is transmitted as described. For instance, it has been found that kappa contains DNA. The fact that kappa can act as an agent of infection when it is taken in by sensitives and converts them into killers indicates a phenomenon similar to the infection of other cells by a virus. This concept is supported by studies on the tick which transmits Rocky Mountain spotted fever. It was found that the offspring of the infected ticks are also infected. This disease is caused by a type of organism

FIG. 22–3 *How a sensitive strain of Paramecium may be produced from a killer strain. When fission takes place more rapidly than the multiplication of the kappa particles, some organisms will not receive any of the kappa particles when fission occurs. These become sensitives, and give rise to more sensitives by fission.*

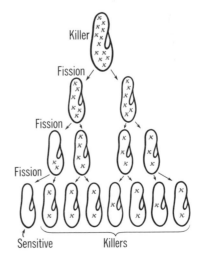

known as a **rickettsia,** which is similar to a virus in being able to develop and multiply only in living cells, although it is somewhat larger than a virus. The cytoplasm of the egg carries the rickettsia, and an indefinite succession of offspring through the female line will also carry the infection. In this case, however, the particles may be transmitted to higher animals by a tick bite, and it may cause a serious disease in man.

We have interesting evidence that viruses can be carried without damaging their hosts in the case of the healthy potato virus. Potatoes in the United States are infected with this virus and, since potatoes are propagated from the "eyes" of potatoes of the previous crop, the virus is carried in the cytoplasm from one generation to the next. It appears to do no harm to the crop in the United States, but when European potatoes are exposed to this virus they become diseased and die. This is not widely different from the case of kappa in *Paramecium.* We have self-perpetuating particles in the cytoplasm which do no apparent injury to the host, but which can kill other organisms that do not carry these particles.

One other fact about kappa should be mentioned. In the course of studies of this substance, it was found that a nuclear gene is necessary for kappa to exist and multiply in the cytoplasm, regardless of other conditions. Both the kappa particles and at least one of the dominant nuclear genes are necessary for the multiplication of kappa.

Other studies have been made by Sonneborn on the method of inheritance of those complex chemical substances known as **antigens.** We learned something about the inheritance of human blood antigens in Chapter 12. The antigens which have been studied in *Paramecium aurelia* are located on the cilia, the tiny hair-like processes which cover the entire body. The presence of these antigens can be demonstrated only by their effect on the blood of higher animals. In these experiments, *Paramecia* from a pure culture were injected into the blood stream of living rabbits. After several days blood was taken from the rabbits and a bit of the serum added to other *Paramecia* from the same culture. This caused the cilia to stick together and, since cilia are the locomotor organs, the organisms were paralyzed. The rabbits had developed antibodies in their blood which reacted with the antigens in the cilia to cause this effect. The same serum, when mixed with a different culture of *Paramecia,* failed to elicit this response. It was evident that the cilia of the animals in the second culture contained a different antigen. Eight different antigenic types of *Paramecia* were found as a result of these experiments. It was then found that when two *Paramecia* of different antigenic types conjugate, each of the two conjugants produces a culture like itself.

Further study on the cilia showed that each is formed by a small self-replicating body, a **kinetoplast,** which contains DNA and its own protein-synthesizing system. This was demonstrated by delicate grafting operations in which a row of cilia was transferred from one organism to another. The grafted row was placed in an upside down position with respect to the cilia of the recipient. The grafted cilia were found to replicate in this upside down position;

thus after more than 1000 generations the paramecia all had one row of cilia in an upside down position.

Investigations in Mammals

Many other investigations have been made of cases which appear to indicate cytoplasmic inheritance. Mammals are poor subjects for such studies because of the close connection between the mother and her offspring before and after birth. Some apparent cases of cytoplasmic inheritance have proved to be due to other factors after thorough investigation. As an example, J. J. Bittner found what appeared to be a case of cytoplasmic inheritance in mice. He had a strain of mice in which nearly all developed breast tumor. From these he took females and bred them to males that did not inherit this tendency. Ninety per cent of the offspring of such crosses developed breast tumors. On the other hand, when he took males from the tumor-susceptible strain and mated them to normal females, he found no cases of breast tumor among the offspring. These results suggested cytoplasmic inheritance. Later studies, however, showed this not to be the case. When Bittner took the newborn offspring from a tumor-strain female and put them with a foster mother from a normal strain for nursing, there were no breast tumors in the offspring. This established the fact rather definitely that the milk was the major element responsible for the transmission of breast tumor. This was identified as a virus-like "milk factor," and it was also found that hormones and chromosomal genes played a part by making the mice susceptible to this "milk factor."

FIG. 22–4 *Maternal transmission of breast tumor in mice as worked out by Bittner of Jackson Memorial Laboratory. Experiments indicate that the primary cause of this unusual transmission is a "milk factor" which is passed from females with breast tumor to their offspring.*

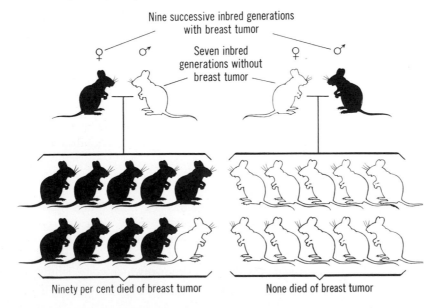

Nine successive inbred generations
with breast tumor

♀   ♂   Seven inbred   ♀   ♂
                  generations without
                  breast tumor

Ninety per cent died of breast tumor   None died of breast tumor

CYTOPLASMIC INHERITANCE IN HAPLOID ORGANISMS

Yeasts

Some of the best evidence for cytoplasmic inheritance has come in recent years through studies of haploid organisms. B. Ephrussi in France and several investigators in the United States have shown clearly that certain inherited characteristics of yeasts are transmitted through the cytoplasm. In one investigation a red strain was obtained from the common white strain after treatment with mustard gas, a mutagenic agent. At first it was assumed that this change was brought about by a mutation of a nuclear gene, but when the red mutants were propagated vegetatively under conditions which stimulated rapid growth, some white cells were found. It was then postulated that a cytoplasmic factor must be involved and that, when the cells grow and reproduce rapidly, some offspring may be produced which do not happen to include any of these cytoplasmic factors. Yeasts reproduce by budding, and a small bud might easily be formed without any of the red pigment-producing factors, if these factors were not very abundant in the cytoplasm.

In another investigation it was found that many populations of yeasts yield dwarf (petite) colonies at a frequency of about 1 per cent of the total number of colonies. These colonies were made of cells which lacked the proper enzyme system to utilize glucose efficiently. Since most of this enzyme system is located in the mitochondria, it was assumed that the petite colonies had lost their mitochondria. Mitochondria are self-reproducing bodies within the cytoplasm like the plastids of higher plants and, like the plastids, they have been shown to contain minute amounts of DNA. When normal cells are plated out on a medium containing certain acridine dyes (euflavine or acriflavine), the number of petite colonies was much greater. The dyes seemed to destroy the mitochondria. It was apparent that the frequency of these petite colonies was much greater, either with or without the dye treatment, than could possibly be accounted for by mutations of one or more nuclear genes.

Later a strain of yeast was found in France which gave petite colonies as a result of a mutation of a nuclear gene. This gene seemed to inhibit the activities of the mitochondria. When cells from this mutant petite were crossed with cells from the vegetative petite, it was possible to obtain some cells with normal growth. The gene for normal activity of the mitochondria was provided by the vegetative petite and the mitochondria were provided by the mutant petite. It was found that this combination gave some cells with mitochondria which were normally active.

Neurospora

A similar case of cytoplasmic inheritance has also been found by M. B. and H. K. Mitchell. They found a strain of *Neurospora* which was called poky (*po*) because of the slow growth of its filaments. This trait was found to have maternal inheritance and was due to an altered system of respiratory enzymes. At the same time mutant chromosomal genes are also known which cause similar alterations in respiratory enzymes and also result in slow growth of the filaments.

Chlamydomonas

Some of the most extensive investigations into cytoplasmic inheritance have been done on the one-celled green alga, *Chlamydomonas,* by Ruth Sanger and her colleagues at Columbia University. The cells reproduce asexually by fission, but also have a means of sexual reproduction. Two cells of opposite sex shed their cell walls and unite to form a diploid zygote. Later meiosis takes place within the outer zygote wall and four haploid cells break out. Sex is determined by a single chromosomal gene, so two of the four cells give rise to female progeny and the other two give rise to male progeny.

Strange to say, even though both male and female contribute equal quantities of cytoplasm to the zygote, there is primarily maternal inheritance of characteristics resulting from cytoplasmic factors. The explanation of this is not known, but it is clear that something more than the mere presence or absence of certain cell constituents is involved in such cytoplasmic inheritance. (Perhaps there is a casting off of the cytoplasmic factors from the male during divisions of the cells within the zygote wall; such a casting off is done by *Escherichia coli* when it becomes diploid during conjugation.) In about 1 per cent of the progeny of *Chlamydomonas,* however, there is an expression of the cytoplasmic factors from the male line as well as from the female line. In such cases we might assume that the cytoplasmic factors from the male are not cast off during the meiotic divisions.

If the characteristics inherited through the cytoplasm are due to extranuclear genes which are like the genes in the nucleus, then it should be possible to obtain mutation of these extranuclear genes with methods similar to those used to induce mutations of nuclear genes. Radiation and certain chemical compounds which have been effective in producing mutations of nuclear genes were tried, but no noticeable increase in cytoplasmic mutations was found. It was discovered somewhat by accident, however, that streptomycin could cause cytoplasmic mutations. Cells susceptible to streptomycin were placed on a culture containing this antibiotic and about one in a million cells survived. Each cell multiplied to form a colony on the medium. It was found that, in most cases, the resistance was due to a mutation of a nuclear gene from streptomycin-sensitive (*ss*) to streptomycin-resistant (*sr*). In about 10 per cent of the cases, however, the mutation showed maternal inheritance — it was transmitted through the female line and only very rarely through the male line. The chromosomal mutations arose at random and would arise even though the organisms were not exposed to streptomycin. The nonchromosomal mutations, however, arose only after sensitive cells were grown in the presence of streptomycin in sublethal concentrations — streptomycin was inducing the mutations.

At this point, the question arose as to whether the streptomycin was acting as a specific mutagen, inducing mutations only to streptomycin resistance, or whether it was non-specific and was inducing mutations in other nonchromosomal genes as well. A study of the progeny of individual cells from the colonies exposed to streptomycin showed a great increase in mutations of all kinds. The number of mutations found was much greater than anything that had ever

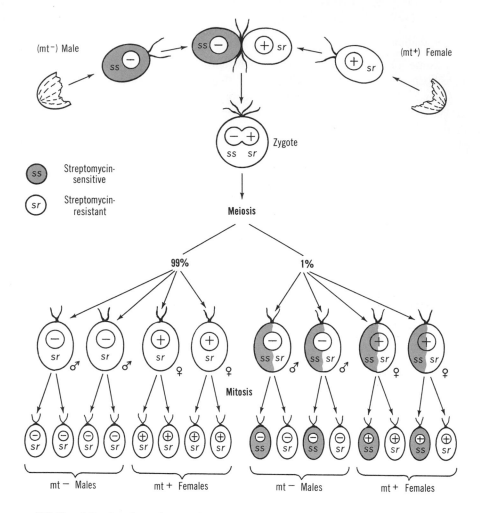

FIG. 22–5 *Sexual reproduction and maternal inheritance in Chlamydomonas. Even though both male and female gametes contribute equal amounts of cytoplasm to the zygote, the cells which come from the zygote by meiosis show maternal inheritance only of cytoplasmic genes in about 99 per cent of the cases. In about one per cent, however, diploid cells are formed with the cytoplasmic genes from both male and female. These segregate out and form haploid cells at the first fission.*

been found in chromosomal genes by any kind of treatment. About thirty different nonchromosomal mutations were studied and they were found to affect all aspects of the cells, just as is found in genes on the chromosomes.

Dr. Sanger and her colleagues were even able to study segregation and independent assortment among the nonchromosomal genes. Chromosomal genes behave in a typical manner. If a streptomycin-sensitive cell is crossed with a streptomycin-dependent cell, the four offspring segregate out into two of each type when the genes involved are chromosomal. When nonchromosomal genes are involved, however, all four progeny normally are like the maternal parent. In the exceptional 1 per cent of the cases the four progeny receive the genes from both parents and are both streptomycin-sensitive and streptomycin-dependent. Such an unusual combination can be grown only on a medium which contains low levels of streptomycin, not strong enough to kill a sensitive organism, yet sufficiently strong to supply the streptomycin needed because of the dependent trait. When one of these cells reproduces by fission, it gives two cells, one of which will be streptomycin-sensitive and one of which will be streptomycin-dependent. It appears as if two sets of nonchromosomal genes were received by the original four progeny, but these segregate out and only one set is transmitted to the offspring produced by the first fission after meiosis. Thus, there must be something like a pairing and segregation of the cytoplasmic genes like that so well known for chromosomal genes in meiosis.

These studies make it evident that nonchromosomal genes do exist in *Chlamydomonas* and that they carry genetic information much like chromosomal genes. The nonchromosomal genes seem normally to be haploid, but on occasion

FIG. 22–6 *Differences in inheritance of traits from chromosomal and nonchromosomal genes in Chlamydomonas. Chromosomal genes segregate out to yield a 1 : 1 ratio of traits from the male and female line. Nonchromosomal genes, however, are transmitted only from the female line in about 99 per cent of the cases. The nonchromosomal genes from the male line are lost somewhere in the process of meiosis.*

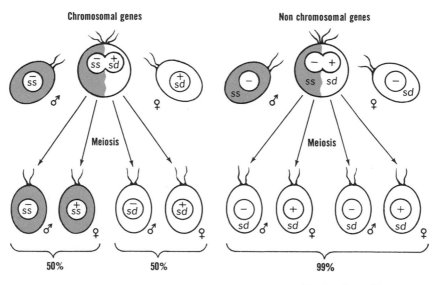

ss—Streptomycin- sensitive
sd—Streptomycin-dependent

they may become temporarily diploid when genes from both parents are included in the cytoplasm of the four offspring of the zygote. The haploid state is restored at the first division after meiosis when there seems to be a division of the cell without any duplication of the diploid nonchromosomal genes.

In this chapter we have seen that there is verified cytoplasmic inheritance in a number of very widely separated species. Why should there be two separate genetic systems yet the bulk of inheritance dependent upon chromosomal genes? Perhaps extrachromosomal genes were the only units of heredity in primitive organisms during the early days of the emergence of life on earth. In the course of evolution these genes might have coalesced into the more easily handled chromosomes of the nucleus. In certain instances, however, there may have been circumstances which made it more advantageous for the developing forms of life to retain some hereditary units in the cytoplasm. The replication of chromosomes and the genes which they bear is closely related to the division of the cell, but certain cytoplasmic bodies are better not so restricted. The replication of chloroplasts and mitochondria is better geared to the environmental needs and this is possible if they have a genetic system independent of the chromosomal genes which usually replicate only once for each cell division. Thus a greater flexibility is provided for the growth of certain organelles of the cell.

PROBLEMS

1. A nasturtium plant is found which has no chlorophyll in the leaves of one branch. Flowers appear on this branch. Pollen from one of these flowers is transferred to a flower on a normal green plant. What types of plants would you expect from seeds which are produced by this flower? Explain.

2. If you made the reciprocal cross, transferring pollen from a green plant to a flower on the branch without chlorophyll, what type of plants would you expect to grow from the seeds?

3. List the ways in which cytoplasmic genes differ from true nuclear genes.

4. Suppose a certain character in rabbits was suspected of being inherited through the cytoplasm. Explain the type of crosses you would make in order to determine if this is true.

5. *Drosophila* females of certain strains yield only daughters — about half of the eggs (presumably male) fail to hatch. Ooplasm from eggs of these females was removed and injected into the abdomen of normal females. After about 2 weeks some of these began producing only female offspring. Give possible explanations for these results.

6. One theory holds that all of the genetic material in the cytoplasm of cells got there originally as an invasion of small particles such as viruses. When these proved advantageous to the cell, they were established as a normal part of the cell. Give arguments for and against this theory.

7. It has been found that a *Drosophila* resistant to carbon dioxide can be made sensitive through the injection of a minute amount of the body fluid from a sensitive fly. Would you expect the offspring of this altered fly to be sensitive or resistant? Explain.

8. Certain strains of yeast produce an enzyme for digestion of a sugar, galactose. When these yeasts are raised on media without this sugar for several genera-

tions they cease the production of this enzyme. When transferred back to media with this sugar they produce the enzyme again after about seven generations. Could this be cytoplasmic inheritance? If not, how does it differ from cytoplasmic inheritance?

9. What experiments prove that the bodies which produce the cilia in *Paramecium* replicate themselves independently of the nuclear genes?

10. Suppose you find a characteristic of light sensitivity in cats that appears to be transmitted only from females to their offspring. Tell how you would go about determining if this characteristic was due to cytoplasmic inheritance rather than nuclear genes or virus infection.

11. Suppose you find a trait in *Chlamydomonas* which causes the alga to require streptomycin in its food medium before it can grow; it is streptomycin-dependent. Explain how you would determine if this dependence were due to a chromosomal gene or a nonchromosomal gene.

12. When two petite colonies of yeast are mixed together and conjugation is allowed to take place, some of the yeasts which result produce colonies of normal size. Explain how this might have come about.

23

Evolution and Population Genetics

THERE IS NO ROOM IN A BOOK OF THIS SCOPE to discuss all of the many-faceted aspects of biological evolution. Such a topic includes many phases of biology and related sciences. The basic mechanics of evolution, however, lie in the field of genetics — there can be no evolution without hereditary variations in the populations. Through selection there is a tendency to smooth out the variations which arise in any population living in a stabilized environment. Some characters will be weeded out; some will become more widespread. In the course of time a population will tend to become homogeneous throughout, and there would no longer be any variety to serve as a basis for further evolutionary change were it not for the fact that the hereditary material is capable of change. Such adaptive changes are not frequent, as we have learned, but they do occur at a steady rate throughout the ages, providing new material which becomes the basis for continuing evolution. Most of the types of changes have been studied in previous chapters; we will review them here and point out their relationship to evolution.

MUTATION AND SELECTION

The mutation of the gene lies behind all evolutionary changes. True, through changes in gene position or changes in chromosome number and arrangement, new phenotypic effects may be produced, but without gene mutation these changes are necessarily limited in scope. Mutation affords virtually unlimited scope for selection. The fact that over 99 per cent of the mutations which have been studied in various forms of life are harmful to some degree

404

may seem to rule out the importance of mutation as a factor in adaptive evolution. Yet it is just that fraction of 1 per cent which happens to be beneficial that forms the basis for most evolutionary developments. It is because of mutation that life has been able to attain the stupendously complicated organization which many forms now possess. Out of the chaotic mass of random mutations which have occurred through the ages, the phenomena of selection exert their influence and bring order out of chaos. Through selection there is a differential multiplication of the few mutations which contribute to the welfare of the race. Should this selection be relaxed, either through natural or artificial influences, the accumulating harmful mutations which are ordinarily not perpetuated would soon bring about disorder and degeneration in place of adaptation. Consider almost any domesticated animal or cultivated plant. Through artificial selection men have concentrated on establishing in them those genes which have a practical value to mankind, without much consideration for the welfare of the organisms themselves. As a result, few of these forms are able to live or reproduce without some aid from man. There can be no substitute for the rigor of natural selection. Mutation combined with natural selection forms a powerful force for the strengthening and continued evolution of species. Mutation without natural selection exerts a deteriorating influence.

The combination of mutation and selection is necessary, not only for improvement, but for the continued existence of the race. There is no such thing as a *status quo*. In man there are many genes which bring about the death of the individual receiving them, namely the lethal genes. Each such death purges the population of one such gene. It cannot eliminate them, for genes of the same nature arise anew, but an equilibrium becomes established between the rate of mutation and the rate of elimination. Suppose through medical science we could find a way to prevent the lethal effect of a certain mutation. Suppose the gene causes a heart deformity which results in death shortly after birth, and through surgery the condition can be corrected so that the person can live and reproduce. The delicate equilibrium is thus upset by counteracting the selective forces of nature, the gene becomes more widespread in the population, and more and more often will it be necessary to perform this operation. Extend this principle to many genes and we can see how a race of man might develop that would be completely dependent upon medical skill for survival. This is no argument for the abandonment of the practice of medical science, but it does point up some of the genetic problems which are to be expected as a result of the upset of the equilibrium between gene mutation and natural selection. The frequency of genes at equilibrium has such an important bearing on evolution that we shall consider it more fully in the following discussion.

DIFFERENTIAL SELECTION

Variation in frequency of distribution of specific genes may occur in populations living in different environments because of differential selection. As an example in man, let us consider the sickle-cell trait, which is found in about 8 per cent of American Negroes, but is almost completely lacking in Americans of northern European descent. See Chapter 18 for details of the nature of this trait and its method of inheritance. Why should this gene be so prevalent among Negroes and so very rare among whites? Certainly, there is no reason to expect a differential mutation rate of this gene in the two races. And, even if such could be the case, the rate of mutation would have to be much greater than any rate known in order to compensate for the selective pressure against it. We will have to seek another explanation. A study of the geographical distribution of the gene offers a clue to a solution of the problem. Its frequency is greatest in certain areas of equatorial Africa, in a small region of India, and in parts of Greece. Is it possible that the heterozygotes have some selective advantage in these areas and not in others? There is strong evidence that this is indeed the case. These are the areas of the world where malaria has been very prevalent. Allison and others have found a very high correlation between high frequency of the sickle-cell trait and the prevalence of malaria. This suggests a possible greater resistance to malaria by these individuals. Experiments with human volunteers show that this is the case. The advantage possessed by those carrying one gene for the abnormal hemoglobin counterbalances the disadvantage possessed by those with two of these genes, and keeps the gene at a high level in the population. Something about this hemoglobin is not compatible with the needs for the growth of the malarial parasite in the red blood cells. In Northern Europe, where malaria has been almost non-existent, the selection has kept the gene at a very low equilibrium.

This shows how a gene which may be entirely disadvantageous in one environment can have some advantage in another environment and selection can take place on a different basis. As a test of this theory, there should be a gradual decrease of this gene in a population removed from a malarial region and the advantage of the heterozygote is no longer present. A study of American Negroes in comparison with those in Africa shows that there has been a reduction of the gene in American Negroes. Of course, in this study due allowance must be made for the introduction of genes from the white race into the germ plasm of the American Negro; but the frequency of the gene in the American Negro is well below what would be expected had there been no selection against the gene since the Negro has been in America, and had the entire reduction in the gene frequency resulted simply from racial intermixture.

The prevalence of the gene for sickling varies in the different regions of Africa somewhat in proportion to the severity of malaria in those regions. Even in the areas of heaviest malarial infection, however, the frequency of the gene does not rise above 30 per cent. This is the point at which the selective advantage of the heterozygote becomes counterbalanced by the selective disadvantage of the homozygote; for, when the frequency of a recessive gene rises in a population, there is an increase in the proportion of homozygous individuals in comparison to the proportion of heterozygous individuals. Thus an equilibrium is reached. Table 23–1 shows how the ratio of heterozygotes to homozygotes decreases as the gene frequency increases. Figure 23–1 shows this relationship

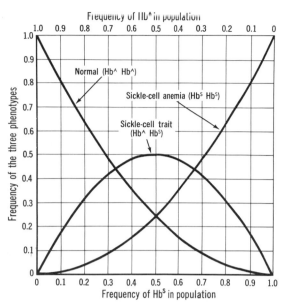

FIG. 23-1 *Graph to show how the percentage of homozygous and heterozygous individuals varies as the allele frequency varies in a population. Note that the number of persons with the sickle-cell trait becomes the same as those with sickle-cell anemia if the gene for sickling reaches about 67 per cent.*

in graphical form. You will note from this figure that the number of homozygous persons becomes equal to the number of heterozygous persons at about 67 per cent frequency of the allele in the population. This figure can be applied to any gene in a population.

TABLE 23-1

Ratio of Sickle-Cell Trait to Sickle-Cell Anemia in Populations

Frequency of gene for sickling in population (percentage)	Percentage of persons with sickle-cell trait (heterozygous)	Percentage of persons with sickle-cell anemia (homozygous)	Ratio (heterozygous to homozygous)
2	3.9	0.04	97.5 : 1
5	9.5	0.25	34.0 : 1
10	18.0	1.00	18.0 : 1
20	32.0	4.00	8.0 : 1
30	42.0	9.00	4.7 : 1

THE HARDY-WEINBERG PRINCIPLE

In 1908 two men made a very important contribution to the study of population genetics. Independently of each other, an English mathematician, G. H. Hardy, and a German physician, Wilhelm Weinberg, developed a simple mathematical method of analyzing the frequencies of alleles in populations. Although their work remained relatively unnoticed while geneticists were working out the principles of Mendelian inheritance, in recent times it has become an important part of the studies of gene distribution in populations.

The Hardy-Weinberg principle holds that in a population alleles tend to establish an equilibrium with reference to each other. As an example, if two alleles occur in equal proportions in a large, isolated population and neither allele has a selective advantage over the other and neither mutates more frequently than the other, they will be expected to remain in equal proportion generation after generation. It would make no difference if one allele was dominant and the other recessive or if there was an intermediate expression. Since this is true, it is possible to calculate the approximate frequencies of the alleles from a sample which is representative of the population as a whole. Of course, we will never find a large population where there is no selective advantage of one allele over another and no variation in rate of mutation, but this principle is of great value since there are many populations which approximate these conditions.

Determining Frequencies of Allelic Genes

By means of the Hardy-Weinberg principle it is possible to determine the frequency of a particular recessive allele in a population and the number of heterozygous carriers of the allele, as well as the number of homozygous dominant individuals. To do this calculation we must first take a representative sample of the population and determine the percentage of the people who show the recessive phenotype. The square root of this percentage will give the approximate percentage of the recessive gene in the population, assuming that there are only two variant alleles. The frequency of the dominant allele then is easily determined since it represents the balance of the genes at this locus in the population. From the percentages, it is easy to separate the people who show the dominant trait into heterozygous and homozygous individuals. The square of the percentage of the dominant allele in the population represents the homozygous persons who show the dominant trait, and the balance of those showing the dominant trait are, therefore, heterozygous. This can be understood better with an example.

Some persons have the ability to roll the tongue into a distinct U-shape when the tongue is extended from the mouth. See Figure 23–2 for an example of this ability. A dominant gene seems to be responsible for the tongue-rolling ability, while its recessive allele brings about an inability to roll the tongue. Suppose you wished to determine the frequency of these two alleles in your school population and what percentage of the rollers were homozygous. You should choose a rather large sample of the school population in such a way that it will be representative of the population as a whole. That is, your sample should be balanced as to sex, race, and other characteristics in the same proportions as those in the entire school population. In your selected sample suppose you find that $\frac{1}{16}$ of them do not have the ability to roll their tongues. These, you know, are homozygous for the recessive allele. Your first problem is to find what proportion of the genes at this locus in your sample are the recessive allele for inability to roll the tongue. You can do this by applying some of the principles of probability learned in Chapter 8.

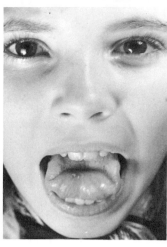

FIG. 23–2 *Tongue rolling and tongue folding. Both of these abilities appear to be present because of dominant genes. The gene for tongue rolling, however, is much more prevalent in the human population than the gene for tongue folding.*

Let $a =$ the frequency of the dominant allele (R) and $b =$ the frequency of the recessive allele (r). Since each person has two alleles for each locus of his genotype, we can represent the distribution of the two alleles in the population as follows:

$$(a + b)^2 \text{ or } a^2 + 2ab + b^2$$

We have already determined the value of b^2; this is the fraction of the sample who cannot roll their tongues, so we can determine the value of b as follows:

$$b = \sqrt{b^2} \text{ or } \sqrt{\tfrac{1}{16}} \text{ or } \tfrac{1}{4}$$

Since the alleles of the gene in the population are either a or b, then

$$a = 1 - b \text{ or } 1 - \tfrac{1}{4} \text{ or } \tfrac{3}{4}$$

This means that ¼ of the gene pool in the school consists of the r allele and ¾ are the R allele.

With these results at hand we can now determine the frequency of the heterozygous (Rr) persons in the sample as follows:

$$2ab = 2 \times \tfrac{3}{4} \times \tfrac{1}{4} = \tfrac{6}{16}$$

The frequency of the persons homozygous for the dominant allele (RR) will be:

$$a^2 = (\tfrac{3}{4})^2 = \tfrac{9}{16}$$

Thus, beginning with only the value of b^2, we have obtained the proportion of each of the two alleles and the proportion of each of the genotypes in the group sampled. If your sample has been representative of the entire school population, then we can conclude that the frequencies of the alleles and the proportions of genotypes are approximately the same for all of the students in the school.

In many cases it is more convenient to use percentages or decimals rather than fractions in working problems by the Hardy-Weinberg principle. The same problem with the tongue-rolling ability is shown in tabular form, using decimals, in Table 23–2. The results, when translated into the actual number of persons, will come out the same as when the fractions were used.

TABLE 23–2

Use of Hardy-Weinberg Principle for Analysis of Frequencies of Alleles and Genotypes in a Population. In This Example $^{15}\!/_{16}$ of the Sample Could Roll the Tongue While $^{1}\!/_{16}$ Could Not.

Eggs→ Sperms ↓	$R\ (a) = 0.75$	$r\ (b) = 0.25$
$R\ (a) = 0.75$	$RR\ (a^2) = 0.5625$	$Rr\ (ab) = 0.1875$
$r\ (b) = 0.25$	$rR\ (ab) = 0.1875$	$rr\ (b^2) = 0.0625$ (known)

R = gene for ability to roll tongue; r = gene for inability to roll tongue; a = frequency of R; b = frequency of r.

$b^2 = \frac{1}{16} = 6.25\% = 0.0625$ cannot roll tongue, known rr
$b = \sqrt{0.0625} = 0.25$, the proportion of r allele in population
$a = 1 - b = 0.75$, the proportion of R allele in population
$a^2 = 0.5625$, the proportion of homozygous (RR) tongue-rollers in population
$2ab = 0.3750$, the proportion of heterozygous (Rr) tongue-rollers in population

Applications to Poultry Breeding

There are many practical applications of the Hardy-Weinberg principle to breeding practices. We learned in Chapter 5 that rose comb is dominant over single comb in domestic poultry. Since certain breeds are required to have the rose comb, it helps a poultry breeder to know what proportion of the rose-combed birds in his flock carry the gene for single comb. He can obtain this information by first noting the proportion of chicks with single comb which are produced. Let us assume that this is ¹⁄₆₄. By use of the Hardy-Weinberg principle we can calculate that about ⅞ of the genes in his chicken population are alleles for the dominant rose comb and ⅛ are alleles for the recessive single comb. We also calculate that about fourteen out of every sixty-four chickens are heterozygous rose-combed. This information may be of value to the poultryman if he wishes to try to eliminate the recessive gene from his flock.

Albinism in the United States

When a recessive gene is rather rare, the number of heterozygous carriers of the gene far outnumber the homozygous individuals who show the recessive trait. We can illustrate this with albinism in the United States. About one person out of 20,000 is an albino (homozygous for the recessive gene for albinism). By means of the Hardy-Weinberg principle, we find that at the locus affecting pigmentation about one allele out of every 141 in the population is for albinism and about one person in 70 carries the allele. This means that far more albino children are born of normal parents than are born when either or both parents are albinos. Although heterozygous parents have only one-fourth albino children as compared to 100 per cent albino children from two albino parents, the heterozygous persons so far outnumber the albinos that most of the albino children come from two heterozygous parents. Should some dictatorial regime decide that albinism was not a desirable trait and try to eliminate it by preventing reproduction of all albinos, it would find that even after many generations of such a practice there would be hardly any perceptible reduction in the number of albino children being born.

Efficiency of Selection

When a completely recessive gene has a rather high frequency in a population, the efficiency of selection, either natural or artificial, is relatively high. As the frequency of the gene decreases, however, the efficiency of the selection for or against it drops. If half of the alleles at a certain locus in a population produce an undesirable recessive trait when homozygous, it will require only 2 generations to reduce the frequency of the allele to one-fourth if all individuals showing the trait caused by the allele are prevented from reproducing. If only one-twentieth of the alleles in the population are of the harmful recessive type, however, it will require 20 generations of complete selection to reduce the frequency of the allele in the population 50 per cent. The efficiency of selection continues to decrease rapidly as the frequency of the allele diminishes, and 100 generations of selection are required to reduce a gene frequency of 1 per cent to one-half of 1 per cent. Table 23–3 shows how this decrease in efficiency operates. With this information in mind, it is easy to see how complete elimination of a recessive gene by selection is almost impossible in a large population, even if no new alleles of this sort were entering the gene pool through mutation.

TABLE 23–3

Decreasing Efficiency of Selection with Decreasing Frequency of a Recessive Allele in a Population

Generation number	Percentage of a recessive allele in population
1	50.0
2	33.3
3	25.0
4	20.0
5	16.7
10	9.1
20	4.8
30	3.2
40	2.4
50	2.0
100	1.0
200	0.5
1000	0.1

The above figures show the efficiency of total selection — the elimination of all who show the recessive trait. Note that it takes only 2 generations to halve the number of recessive genes when the frequency starts with 50 per cent. It requires 20 generations when the frequency is about 5 per cent and 100 generations when the frequency is 1 per cent.

Selection in Cases of Intermediate Traits

The Hardy-Weinberg principle also has value in studies of alleles that show intermediate inheritance. Even though we can recognize all three genotypes in such instances, the calculations can often show whether there is some degree of selection that favors one genotype over another. Studies of sickle-cell anemia show this to be the case. The actual number of persons with the sickle-cell trait in a population where malaria is prevalent is greater than the expected number calculated on the basis of the Hb^S allele's frequency in the population. This is true because so many of the homozygous persons with sickle-cell anemia never live to be included in a tabulation of the population. Also, many persons homozygous for Hb^A die of malaria while very young and are not included. Hence, the percentage of persons who are heterozygous for the two alleles will be considerably higher than would be calculated on the basis of the frequency of either of the homozygotes. As has been pointed out previously, many so-called recessive genes actually show some degree of expression when heterozygous, and selection for or against the heterozygote must be taken into account. Because of the greater numbers of heterozygotes in most cases, this selection can be much more effective than selection solely against (or for) the homozygous recessive.

Another case, one which does not show such a selective advantage, can be illustrated by a study of the **M** and **N** blood antigens made by William C. Boyd. This is a case of co-dominance since each of the alleles is fully expressed in the heterozygote, in comparison with intermediate inheritance where each allele is partially expressed. Boyd found that about 20 per cent of the Italians living in Sicily were of type **N**. These persons were all homozygous $L^N L^N$. Taking the square root of 20 per cent, we obtain the frequency of the allele, L^N, in the population — about 44 per cent. Calculations by means of the Hardy-Weinberg principle show that 31.36 per cent of the population should be type **M**, that is, $L^M L^M$. Boyd actually found 32 per cent. The agreement of his percentage with that calculated by the Hardy-Weinberg principle indicates that **M** and **N** individuals survive equally well. The complete figures are given below:

M and **N** blood antigens of Italians (Sicily)

	Phenotypes			Alleles	
	M	**MN**	**N**	L^M	L^N
Observed	0.32	0.48	0.20	0.56	0.44
Expected	0.3136	0.4528			

It is also possible in the cases where the heterozygote can be recognized to calculate the allele frequencies directly from the phenotypes. Using **MN** as an example, take half of the frequency of the phenotype **MN** and add it to the frequency of **M** and obtain the frequency of the L^M allele in the population.

$$0.32 + \tfrac{1}{2}(0.48) = 0.56$$

Similarly, for L^N

$$0.20 + \tfrac{1}{2}(0.48) = 0.44$$

The actual frequencies of the L^M and L^N alleles in the population sampled are thus in agreement with the frequencies calculated from the Hardy-Weinberg principle by either method. Such agreement is not found when selection is acting differentially on the alleles.

ABO Blood Groups

Some of the most widespread applications of the Hardy-Weinberg principle have been made in studies of distributions of the **ABO** blood groups in the various races. In the case of multiple alleles, however, we cannot assume that all persons who do not carry one allele must carry a particular other allele. We must expand the principle to include three or more alleles. In the case of the blood groups we begin with persons of type **O** blood who are homozygous for the allele, I^O, which behaves as a recessive gene, and we determine the frequency of this allele in the population by taking the square root of the frequency of type **O** individuals. The remainder of the alleles are I^A or I^B, so we must find a way to estimate the frequencies of these alleles from the phenotypes observed. The method of doing this, shown in Table 23–4, is based upon the distribution of the four basic **ABO** blood types among people living in Brooklyn, New York.

TABLE 23–4

Method of Calculating Frequency of Genes for Blood Groups

	O	**A**	**B**	**AB**
No. of persons in each group	808	699	259	83
Percentage in each group	43.7	37.8	14.0	4.5

Let a equal frequency of allele I^A, b equal frequency of I^B, and c equal frequency of I^O.

	I^A (a)	I^B (b)	I^O (c)
I^A (a)	I^AI^A (a^2) type **A**	I^AI^B (ab) type **AB**	I^AI^O (ac) type **A**
I^B (b)	I^BI^A (ab) type **AB**	I^BI^B (b^2) type **B**	I^BI^O (bc) type **B**
I^O (c)	I^OI^A (ac) type **A**	I^OI^B (bc) type **B**	I^OI^O (c^2) type **O** 0.437

$c = \sqrt{c^2} = \sqrt{0.437} = 0.66$ (gene frequency of I^O)
B + **O** $= b^2 + 2bc + c^2 = (b + c)^2$
$\sqrt{\mathbf{B} + \mathbf{O}} = b + c$
$\sqrt{0.14 + 0.437} = b + c$, or $0.76 = b + c$
$c = 0.66$, so $b = 0.76 - 0.66$, or 0.10 (gene frequency of I^B)
$a = 1 - (b + c)$, or $1 - 0.76$, or 0.24 (gene frequency of I^A)
Summary of gene frequencies: $I^A = 24\%$; $I^B = 10\%$; $I^O = 66\%$

The distribution of the genes for blood group antigens have been used extensively in studies of human populations and their racial relationships. While most races have been found to have all of the **ABO** phenotypes, the proportion of the different antigens varies among them. The distribution of the **ABO** antigens in certain ethnic groups in various regions of the world is shown in Table 23–5. Such studies can show routes of past invasions, migrations, and intermixing of the different races. As an example, at one time all of the inhabitants of Europe probably had very few **B** antigens. Europe, however, was subject to repeated invasions by peoples from the east. These people had a relatively high proportion of **B** antigens in their blood, and some of their genes were scattered among the conquered peoples. Today we find that the proportion of **B** antigens in Europe varies from a relatively high frequency in the eastern parts to a relatively low one in the west. In addition to the invasions, as methods of transportation improved, there was more intermarriage among the people in the contact zones which tended to spread the antigens from east to west. An interesting group of people, the Basques, live on the northern coast of Spain and the southeastern coast of France. The people in other regions of Spain have about 9.2 per cent type **B** blood, but the Basques have only 1.1 per cent. This striking difference is explained by the fact that the Basques live in an area separated from the rest of Spain and southern France by high mountains. Armies which tried to invade these regions found the high mountains a considerable bulwark against invasion. The mountains also served as a barrier to intermarriage between the Basques and the rest of the Spanish population. Hence, we have this small isolated group of people who are probably more like the original European population with respect to their **B** antigens than any other present-day Europeans.

T A B L E   2 3 – 5

Per Cent Frequencies of ABO **Blood Groups in Some of the World Populations***

Ethnic Groups	**O**	**A**	**B**	**AB**
American Indian (Utes)	97.4	2.6	0	0
Australian (Aborigines)	42.6	57.4	0	0
Basque (Spain)	57.2	41.7	1.1	0
Eskimo (Alaska)	41.1	53.8	3.5	1.4
English (London)	47.9	42.4	8.3	1.4
Spanish (Spain)	41.5	46.5	9.2	2.2
French (Paris)	39.8	42.3	11.8	6.1
German (Berlin)	36.5	42.5	14.5	6.5
Italian (Sicily)	45.9	33.4	17.3	3.4
Japanese (Tokyo)	30.1	38.4	21.9	9.7
Russian (Moscow)	31.9	34.4	24.9	8.8
Egyptian (Cairo)	27.3	38.5	25.5	8.8
Chinese (Huang-Ho)	34.2	30.8	27.7	7.3

* After William C. Boyd.

H. Bentley Glass and C. C. Li have made an extensive study of the blood groups of North American Negroes in comparison with the North American Whites and the West African Negroes. The proportions of the different antigens indicate that the gene flow from the White race into the Negro population of the United States during the 275 to 300 years in which the two races have been living together has amounted to a total of about 30 per cent. With an average generation length of 27.5 years, this much influx of genes has come about in some 10 generations. The average gene flow required to bring about this degree of intermixture is 3 per cent per generation. If the flow continues at this rate, there will be no physical line of distinction between the two races in 60.7 generations, or about 1,699 years. In actuality, the time required will probably be much less. In other parts of the world where one racial group has moved into a region occupied by another, population studies show that, at first, the races tend to remain distinct, with marriage mostly within each racial group. In most cases, however, there is a gradual breakdown of the barriers and some mixing occurs. Then, as the distinctions between the two become less marked, the gene interflow becomes accelerated.

SELECTION IN HAPLOID ORGANISMS

In those organisms which have only one set of genes, selection takes place much more rapidly and without any of the complications growing out of the presence of dominant and recessive genes, heterozygotes, etc. Selection for resistance to drugs and antibiotics in bacteria is a good example. Streptomycin is an antibiotic which inhibits the growth of many kinds of bacteria. In those bacteria which have never been exposed to streptomycin, any genes which might make them resistant to the antibiotic have no advantage over genes for sensitivity to the antibiotic. In the absence of any selective advantage, the genes for resistance, which in general slow down the growth rate, will remain at a very low level in the population. If the bacteria are placed on a medium containing streptomycin, however, there is an immediate selection for any genes which may be present or any mutations which may appear that make the bacteria resistant. Those not resistant either die or grow very poorly and, as a result, the bacteria growing will be primarily those which contain the gene or genes for resistance. Figure 23–3 shows how a strain of resistant bacteria, *Sarcina lutea,* can be developed by this technique.

We sometimes unwittingly exercise a similar selection on certain disease germs. During the mid-1930's, the sulfa drugs were found to be excellent agents for the treatment of gonorrhea. In most cases the response in infected persons was dramatically rapid. It appeared that we had discovered drugs which would control this widespread venereal disease. Today, however, the sulfa drugs are effective in only about 15 per cent of the cases. When the drugs were first administered, the great majority of the gonorrhea-causing bacteria were sensitive and killed by them. In a few cases, however, there were mutant genes which

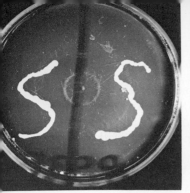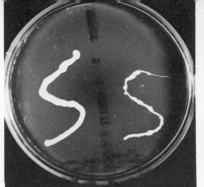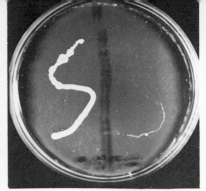

FIG. 23–3 *Establishment of a streptomycin-resistant strain of Sarcina lutea through selection. The strain growing on the left of each culture dish has previously been exposed to streptomycin. The dish at the left contains nutrient agar without any streptomycin. The dish in the center contains 0.0002 mg. of streptomycin per ml. (2 parts per million). The dish at the right contains 0.002 mg. of streptomycin per ml. (20 parts per million). There is practically no growth of the unselected strain in the stronger concentration.*

provided the bacteria with enzymes which would break down the drugs and render them harmless. These were the bacteria which survived the widespread usage of the sulfa drugs in treating gonorrhea. In a few years most of the cases of gonorrhea were being caused by the selected resistant strains. Fortunately, penicillin was discovered about the time the sulfa drugs lost their effectiveness, but there is always the chance that a similar mutation and selection may establish penicillin-resistant strains. In the laboratory we have established penicillin-resistant strains of many bacteria.

It should be made clear that the antibiotics or drugs in such cases act merely as agents of selection and do not induce the mutations to resistance. This fact has been verified by a number of experiments. In one such experiment, small bits of a colony of pneumococcus were transferred to test tubes and allowed to grow. By chance, some of these tubes should have more mutants than others. Samples from each tube were then plated out on media containing streptomycin, and it was determined which tube had the most mutants by noting which sample yielded the most colonies. The original tube, never having been exposed to streptomycin, was then selected and small samples from it were transferred to other tubes. These tubes were likewise tested, the one with the most mutants was selected, and the process was repeated. After this had been done several times, it was found that some of the test tubes contained pure cultures of resistant cells even though they had never been exposed to streptomycin. Hence, since it is clear that mutations for resistance take place in cultures not exposed, we can conclude that the antibiotic is merely an agent of selection which allows mutants for resistance to develop.

GENETIC DRIFT

Another factor which may be of evolutionary significance is known as genetic drift. This is the term applied to the fluctuation of gene frequency which is due merely to the chance assortment of genes in meiosis and fertilization and is unrelated to the benefits or detriments of the genes involved. The effects of such genetic drift are, of course, much more pronounced in a small isolated population group than in large groups where the magnitude of the numbers would level off random variations in gene distribution.

The Dunkers

An excellent case of genetic drift in human populations is shown in a study made by H. Bentley Glass of a group of people in Franklin County, Pennsylvania, known as the Dunkers. These people are descendants of a religious sect known as the Baptist Brethren, who lived in the Rhineland region of Germany near Krefeld. In 1719 a group of twenty-eight of these people migrated to Pennsylvania. Later others of the same sect came from Germany to join them. They have remained relatively isolated from others in America, with their own customs and manner of dress distinct from those around them. Among a number of characteristics of the Dunkers which were studied, the **ABO** blood groups will serve to illustrate the possible influence of genetic drift in this rather small isolate. Comparisons were made with the people in the Rhineland of West Germany today and with people in the eastern United States. The proportions of the different blood types is shown below.

Group	No. people	Percentage of **ABO** blood groups			
		O	**A**	**B**	**AB**
Dunkers	228	35.5	59.3	3.1	2.2
Rhineland Germans	3,036	40.7	44.6	10.0	4.7
U. S. A.	30,000	45.2	39.5	11.2	4.2

The distinctly higher percentage of those with type **A** blood among the Dunkers appears to give good evidence of the influence of genetic drift. Somewhere along the line, probably in the early generations after migration, the ancestors of today's Dunkers produced children with a greater abundance of the gene for the **A** antigen than the exact proportions of probability would indicate. There is no reason to believe that selection or anything other than pure chance was involved. This has come down through the generations with the higher proportion of type **A** blood which the tabulations show.

Drift and Selection

Other studies indicate that genetic drift may be a factor in determining gene frequency in a population, even when selection is also operating. In a study of certain African tribes where selection should be about the same, Foy and his coworkers in England found that the sickle-cell blood trait showed considerable variability among the different tribes as well as among isolates within the same tribe. This study indicates that genetic drift as well as selection may operate to determine gene frequency in a population.

Evidence that genetic drift may exert an influence on adaptative evolution is also furnished in recent work on *Drosophila* resistance to DDT. King has found that each of two selected strains of *Drosophila* developed a resistance after many generations of exposure to increasing quantities of DDT. Each strain, however, developed its own integrated multiple gene system which differed from that developed by the other strain. This indicates that there may be more than one, and possibly many, ways of adaptation to the same environmental conditions. Genetic drift could play a part in determining which line is followed among two isolated groups of the same species.

Within each species there probably lie more potentialities for adaptation to a specific environment than can ever be realized. As long as the species remains a freely interbreeding unit, there will tend to be a somewhat uniform direction of evolutionary adaptations along the lines of one of the potentialities. When a small sample is isolated from the others, however, some of the potentialities are unavailable in the pool of genes contained within the isolate. Adaptation, therefore, may proceed along the line of a potentiality which is different from that taken by the balance of the species. This can result in genetic diversity between the two. In certain cases it may be found that, with a change in environmental conditions, the small isolate has developed a superior pool of genes which would give it a distinct advantage over its ancestors. Thus we see that the chance variation of gene quantity brought about through drift can have a bearing on the evolutionary development of a species.

EQUILIBRIUM FREQUENCY OF GENE MUTATIONS

Any mutation which occurs time and again in a population will eventually reach an equilibrium in frequency. If it happens to be one of those rare beneficial mutations, it may not reach an equilibrium until nearly all the members of the population carry the gene and express its phenotypic effects. If, on the other hand, it is one of the more common harmful mutations, it will reach an equilibrium at a frequency where the rate of mutation to it balances the rate of its elimination through genetic death. We might compare the condition to the equilibrium which is established in a fleet of trucks owned by a certain hauling concern. Suppose the concern starts with five trucks and adds five new trucks each year. Then, let us assume that each truck has an average life of 5 years. For some it will be shorter — one may have a wreck and be eliminated the first year, while others may last as long as 10 years. The number of trucks in the fleet will increase each year until the number reaches about twenty-five. Then, with a constant rate of average addition and a constant rate of average elimination, an equilibrium will be reached and the number of trucks will remain nearly constant. It is possible that some change of conditions may alter this frequency. The company may double the number of trucks which it adds to its fleet each year, there may be a change in the servicing of the trucks to increase their life, or the company may begin using the trucks for heavier hauling, which will shorten their life. Under any one, or any combination, of such changes there will be either a gradual increase or a gradual decrease in the number of trucks until a new equilibrium frequency is reached.

In the case of a newly mutated gene the condition will be similar. If the gene is a completely recessive lethal, it must multiply in the population until it becomes homozygous before there can be an elimination. Thus, as with the trucks, the number of genes of this type will increase in the population through multiplication of previously mutated genes and through the appearance of new mutations of the same kind. When homozygous individuals begin to appear, however, the equilibrium will become established at a level where the elimination will balance the rate of new mutation.

Lethal Genes

As an example, suppose such a lethal gene appears as a mutation of a normal allele once in every 100,000 germ cells. When equilibrium is established, the gene will be homozygous in 1 individual out of every 100,000, and that person will consequently die before reproducing. The frequency of the gene in the population, however, must be much higher than this, for if two germ cells carrying this gene come together once in each 100,000 fertilizations, then 1 germ cell in about 316 must carry the gene. This figure is obtained by taking the square root of 100,000, since the chance of independent events of like frequency occurring individually is equal to the square root of the chance that both will occur at once.

Detrimental Genes

Many mutations are lethal, but, as we have learned, there is a greater number which are detrimentals. Let us now see how the equilibrium frequency will be established for such a gene. Suppose the gene produces only a slight handicap, so that only 1 individual in 100 who is homozygous for the gene will die as a result of its effects. Now, if this gene has the same rate of mutation as taken above, namely, 1 in 100,000 germ cells, the gene frequency will be much greater in the population. When equilibrium is established 1 individual in each 100,000 will die, as in the case of the lethal genes, but that one is only 1 in each 100 who are homozygous for the gene. Hence 1 individual in each 1,000 in the population must be homozygous for the gene. When we take the square root of this figure, we find that about 1 germ cell in each 32 carries the gene.

Changes in Equilibrium

Once the equilibrium is established, it may become altered through changes in any of the factors involved. A change in the environment, for example, might lessen the detrimental effect of the gene and result in an increase in its frequency in the population. If the chance of survival were doubled, the genes would multiply until 1 individual in each 500 would be homozygous, which means that about 1 germ cell in each 22 would carry the gene at this new equilibrium. There would still be 1 death in each 100,000 and 1 mutation in each 100,000, however. Conversely, a change which would double the detrimental effects would at first double the death rate, but as the equilibrium became reestablished the death rate would return to a frequency equal to the mutation rate. There would then be only 1 individual in each 2,000, however, who would be homozygous for the gene, and only 1 germ cell in each 45 would carry the gene.

Also, the equilibrium can be changed by an alteration in the rate of mutation of the gene. Suppose we develop a civilization within which each person receives an average dosage of high-energy radiation of 50 *r* on the gonads during his pre-reproductive life. We have estimated that this would double the natural mutation rate. If a lethal gene has a spontaneous mutation rate of 1 in each 100,000 germ cells, one such mutation would now appear in each 50,000 germ cells. The frequency of this gene in the population would then increase until 1 person in each 50,000 would die from it and about 1 germ cell in each 222 would carry it. In the case of a slightly detrimental gene which resulted in only 1 genetic death out of each 100 persons who are homozygous, a doubling of the mutation rate would again cause 1 death in each 50,000, but 1 person in each 500 would be homozygous and 1 germ cell in about 22 would carry the gene.

Heterozygous Expression and Equilibrium

We have assumed these two genes to be completely recessive in effect, but recessive genes are seldom completely recessive in their total effects on the body. Even a slight departure from recessiveness will have a pronounced effect on the equilibrium frequency which is established. Assume that the lethal gene lowers life expectancy by about 1 per cent when heterozygous (a rather conservative estimate on the basis of the work of Muller, Stern, and Novitski). In such a case about one individual in each 100 who is heterozygous is eliminated as well as the one in each 100,000 who is homozygous. Thus each newly mutated gene would survive only about 100 generations before its existence is terminated in a genetic death. This would allow it to attain only about 100 times its mutation frequency among the germ cells of the population, which would be one in each 1,000 rather than one in each 316 as calculated for the completely recessive gene. The same calculations would also apply to those detrimental genes which cause some lowering of vitality or fertility when heterozygous, but the effects would be less.

THE EVOLUTION OF DOMINANCE

The great majority of mutations which appear in any group of living organisms are recessive or have only a slight effect when heterozygous.. Since it appears very likely that all the genes which any one organism possesses have arisen through mutation at some time or other, how can we explain the fact that the mutations which we observe today are mostly recessive or almost so? Why weren't some of those ancient mutations which became established in the race ages ago recessive? As a matter of fact, many of them probably were recessive when they first arose, but have become dominant since that time, through the evolution of dominance itself. Let us see how this might occur.

The Theory of Dominance Modifiers

Assume that a new mutation arises for the first time in a given race of animals. Like most mutations it might be harmful and it might be dominant. Hence, every individual that receives this gene is going to suffer some loss of viability or fertility. We have already learned, however, that genes vary in expressivity, and hence some individuals carrying this gene will be more strongly affected than others. Some of these variations are due to environmental influences and have no evolutionary significance, but others are due to the genotypes of the individuals. Some individuals will have genes which reduce the intensity of the effect of the newly mutated, harmful, dominant gene. Those with such a reduced intensity will survive and reproduce at a rate greater than those that show a more severe effect of the gene. Hence, there is a selection for those modifying genes which reduce the intensity of the newly mutated gene. In the course of time the dominant gene may be eliminated through the process of natural selection, but the modifying genes may remain. Hence, when the mutation appears a second time, it is not likely to be as severe in its effects because of the presence of these selected modifying genes in the population. Over long periods of time, as the mutation appears again and again, it will tend to lose its dominance because of the establishment of these modifying genes which reduce the effects of the mutation. Mutations among the modifying genes may further reduce the harmful effects, and if so they too will become established. Eventually, the harmful dominant gene becomes intermediate and then recessive in its expression, because of the inexorable accumulation of these modifying genes. Conversely, the normal allele of this gene, which at first was recessive, gradually acquires a dominance over the mutation.

The reverse situation would prevail in the case of a newly mutated gene which is beneficial to the race. It might arise as a recessive, yet in some individuals there may be a slight penetrance in the heterozygous condition. Such a penetrance may be due to the presence of other genes which reduce the dominance of the normal allele. The advantage gained by individuals with this heterozygous expression would tend to perpetuate the modifying genes and to select those with the greatest effect. As mutations of the modifying genes increased the penetrance and expressivity of the original gene they would be selectively accumulated in the course of time and the beneficial gene would become dominant. If the benefits were great enough, it might become established along with its modifiers as a normal part of the genotype of the race.

The Theory of Increasing Potency

It is equally possible that selection based on variations in the potency of recurring mutations has played a part in the evolution of dominance. Studies of multiple alleles have shown that mutations appearing at the same locus on a chromosome will have the same general effect on the organism, but that the effect may vary in potency. Assume that a certain type of mutation is beneficial to a certain species in its particular environment. Through a period of time this mutation appears time and again, but not always with the same potency, since some of the mutations are entirely recessive, some have a slight effect when heterozygous, others when heterozygous have a more pronounced effect, and a few appear which are entirely dominant. It is easy to understand how, through selection, the dominant form of the mutant would become established throughout the race in preference to the mutant alleles with a lower potency.

Suppressor or Enhancer Genes

Suppressor or **enhancer genes** in a population may also influence the expression of other genes. Their effect is closely related to the evolution of dominance. Suppressor genes are common in all species that have been investigated genetically. One of the commonest types of reversions of mutants to the wild type, or original type, can, by genetic analysis, be shown to result from a mutation, not at the chromosomal locus of the original mutation, but at one elsewhere in the same or a different chromosome. The suppressor gene for erupt eye in *Drosophila melanogaster* has already been mentioned. Many suppressor genes are known in *E. coli* and *Neurospora* that suppress one or another of the biochemical mutants commonly studied. For example, there is a mutation in *Neurospora* which acts as a suppressor of the gene for pyrimidineless, and the mold can synthesize pyrimidine even though it has the gene for a lack of the enzyme needed to synthesize this substance. Also, there are suppressors of the gene which blocks the formation of tryptophan synthetase (the enzyme which synthesizes tryptophan).

Enhancer genes have the opposite effect and make a certain gene expression more extreme than it would normally be. Through natural selection, suppressor genes could become established in a species and thus keep harmful gene mutations from being expressed. Similarly, enhancer genes could be selected which would bring a fuller expression of beneficial genes, either dominant or recessive.

It is probable that most of the mutations that arise today in any organism have arisen many times in the past, when we consider the eons of time during which life has existed on the earth. Thus there has been ample opportunity for those genes which have become established as the normal, or wild-type genes, to evolve dominance. By such a line of reasoning we might speculate that mutations which are both harmful and dominant are fairly new in a race or species, or that they occur with such a low frequency that there has been no opportunity for the evolution of dominance on the part of their normal alleles.

CHROMOSOMAL ABERRATIONS AS A FACTOR IN EVOLUTION

Whenever a homogeneous population is separated into two isolated groups by some environmental factor, such as a geographical barrier which is difficult to cross, they will tend to develop along different lines and eventually may form two distinct species. Let us suppose that a race of animals lived in a hot, humid forest area near the seacoast, and a group of them migrated over a mountain range to a dry, desert-like area. In this different environment selection would operate somewhat differently than on the coast. Mutations which formerly were of no benefit, or even were harmful, might be beneficial in the new environment and would become established in the race through selection. At the same time chromosomal aberrations would be occurring in the two groups and phenotypic effects resulting therefrom might be beneficial in one environment and harmful in the other. Through the effects of selection, new arrangements in the chromosomes would ultimately become established in the group now living in the new environment. Still other aberrations would have no particular survival value, but might become established through sheer chance. In the course of time the two groups would thus become quite distinct from one another, and their chromosomal differences might even become so extensive that the two could not be crossed together to yield a fertile hybrid. We would then be justified in calling the two groups different species. This is an over-simplification, of course, but it serves to illustrate the role of chromosomal aberrations in the production of new species and, in the course of time, new genera, new families, and so on.

Nature is so slow in its evolutionary processes that we cannot ordinarily observe these events as they occur. There is an interesting case, however, which clearly illustrates the results of the process even if not the actual process visibly taking place. During the fifteenth century a group of European rabbits were released on the island of Porto Santo in the Madeiras. Today, their offspring are smaller, darker in color, and more timid than the European species. Furthermore, the two groups cannot be mated successfully and have been classified as two different species. Here we have a verifiable case of the evolution of a species within the recent history of man. And it is possible to duplicate the process in the laboratory. Using X-rays to speed greatly the rate of mutation and chromosomal aberration, plus a substitution of artificial selection for natural selection, it is possible to produce two groups which look different and which do not yield highly fertile hybrids. For the most part, however, we must study the existing species and judge the course of evolution in the past accordingly.

Fragmentation and Translocation

Extensive studies have been made on some species of *Drosophila* in an effort to determine the possible relationships between them. Figure 23–4 shows some diploid sets of chromosomes of eleven of these species. It can be seen from a study of these diagrams how fragmentation and translocation may have played a part in the possible derivation of these from a common ancestor. When the chromosomes of an intraspecific hybrid are studied in the salivary glands, however, it becomes apparent that inversions play an even greater part in the differentiation of closely related species. For instance, *D. pseudoobscura* and *D. miranda* look very much alike and have the same number of chromosomes, yet they yield a completely sterile hybrid. Salivary gland studies of the hybrid larvae show that there are fifty or more breakage points with subsequent rear-

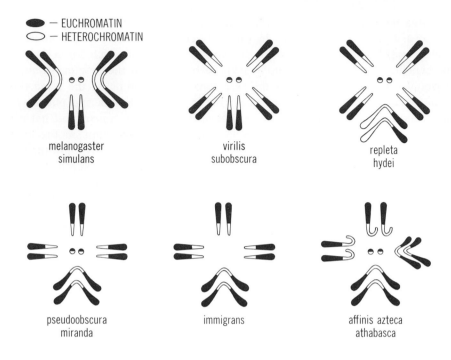

FIG. 23–4 *The chromosomes of some of the species of Drosophila. By comparison of these chromosomes it is possible to see how different species might have arisen through chromosomal aberrations of various sorts. These diagrams also show that two or three species may share the same chromosome pattern. There are differences, however, in the arrangement of genes within the chromosomes.*

rangements which apparently have occurred since these two species diverged from a common ancestor. Of these, only about five involve deletion or translocation of small pieces of a chromosome; all the others are inversions within the chromosomes.

Polyploidy and Aneuploidy

Various types of polyploidy and aneuploidy have no doubt also played a part in species formation. Much evidence exists that in plants new species have arisen through polyploidy. For instance the native wild woodland strawberry, *Fragaria vesca,* is diploid; a wild strawberry of central Europe, *F. moschata,* is hexaploid; and our wild meadow strawberry, *F. virginiana,* is octoploid. The same condition was found to exist among many other species of fruits.

Hybridization

Hybridization between species is another possible way in which new species may be produced, although it is probably not an important one. Normally, such a hybrid is sterile, but occasionally diploid gametes are formed as a result of the doubling of the chromosomes without cell division (such as is induced by colchicine). If one parent species had twenty-four chromosomes and the other had eighteen, then the gametes would carry twelve and nine, respectively. These would unite to give a hybrid with twenty-one, which might have normal division of its somatic cells since the chromosomes do not normally pair in somatic cell division. When the time for meiosis comes, however, proper synapsis is impossible, and a non-viable gamete is usually produced. If, however, these twenty-one chromosomes become duplicated and produce tetraploid tissue with forty-two chromosomes, then there can be twenty-one matched pairs in meiosis and consequently viable diploid gametes can be produced. Some of our species have arisen by just such a method; for example, our common tobacco plant, *Nicotiana tabacum,* arose from a cross of *N. sylvestris* and *N. tomentosa,* in which there was a doubling of the chromosome number. Figure 23–5 illustrates such hybridization in tobacco by colchicine-induced tetraploidy.

A similar cross has been made by man between the radish and the cabbage. Both these plants have eighteen chromosomes as their diploid number, but there are so many differences between them that it is difficult to get them to cross at all. A few hybrids may be obtained, but they are almost completely sterile. In one such cross, however, a few tetraploid plants were obtained with thirty-six chromosomes. This type of plant proved highly fertile, and from it a new species has been established. It is called *Raphanobrassica* from the radish, *Raphanus,* and the cabbage, *Brassica.* The meiotic behavior of its chromosomes is entirely normal, for eighteen pairs are formed and the gametes carry nine radish and nine cabbage chromosomes each. When crossed back to either radish or cabbage, this synthetic species is almost completely sterile. This shows how new species may be produced through hybridization combined with tetraploidy.

FIG. 23–5 *Hybridization in tobacco by colchicine-induced tetraploidy. The blossom on the left is from an untreated hybrid of Nicotiana glutinosa and N. sylvestris. It is sterile with aborted pollen in the anthers. On the right is a blossom from a growing point of the same type of hybrid which was treated with colchicine to induce tetraploidy. This blossom is fully fertile.* FROM WARMKE AND BLAKESLEE IN JOURNAL OF HEREDITY.

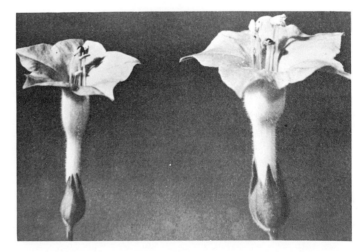

PROBLEMS

1. Let us assume that a certain recessive lethal mutation appears once in every 400,000 human germ cells which are produced. When the equilibrium has been reached, what proportion of the gametes will carry this gene?

2. Suppose that radiation from all sources doubles the mutation rate in man. How many germ cells will then carry this gene after an equilibrium has been established?

3. Suppose that in a species a certain harmful and recessive mutation has been recurring over a long period of time. Some members of this species are introduced into new surroundings where the mutation produces a beneficial effect. What would probably happen to the gene, both as to its frequency in the race, and as to its recessiveness?

4. Among the white population of the United States, approximately 70 per cent detect a very bitter taste when a small amount of phenyl thiocarbamide (PTC) is placed in their mouth. The other 30 per cent do not detect this bitter taste. The tasting ability in this case is due to a dominant gene, and the inability to taste is due to its recessive allele. By means of the Hardy-Weinberg principle, determine how many of the tasters are heterozygous.

5. A breeder of Jersey cattle finds that the recessive mutation for brindling has become established throughout his herd, and he obtains about one brindle calf out of about every twenty-five births. He wishes to know how many of his non-brindle cattle carry this gene. Use the Hardy-Weinberg principle to determine this.

6. A group of animals living on an island is similar to another group living on a nearby island. Cytological studies show the same number and size of chromosomes, yet when bred together, members of these two groups yield sterile hybrids. Explain the most likely cause of this.

7. Would genetic drift be more likely to be greater in a small or in a large isolate of a population? Explain.

8. Explain the part which may be played by chance in the direction of evolution.

24

Applications of Genetics to Plant and Animal Breeding

THROUGH APPLYING THE PRINCIPLES OF GENETICS to the breeding of domestic animals and cultivated plants, man has already achieved remarkable improvement in these commercial types, in spite of the fact that for most of his history he had proceeded by trial and error only. With the great increase of knowledge in genetics which has been acquired during recent years, however, there is every reason to expect an acceleration in the rate of improvement in the years to come. Unfortunately, the practical breeder of plants and animals often does not have a thorough foundation in genetics, and as a result sometimes uses methods which are slower and less reliable than would be possible through the application of our present knowledge of genetics. On the other hand, the highly trained geneticist who has worked primarily with those species used experimentally in the laboratory or in the field often fails to realize all the problems which confront the practical breeder who must produce to meet market requirements. A combination of the knowledge and skills which have been developed in the two fields is highly desirable. In this chapter we shall present some of the problems of the commercial breeder and some ways in which they may be solved on sound genetic principles.

PROBLEMS OF THE COMMERCIAL BREEDER

From the beginning, it should be realized that the problems of the plant or animal breeder are almost always rather complex. True, there are problems comparatively easy to solve, but these are few. It is easy to tell an animal breeder how to eliminate single comb from his flock of Wyandottes, red coat from his Angus cattle, or black from his Wensleydale sheep. These are characters which are caused by variations in one or two genes and, once he learns the method of inheritance of the character, he can eliminate it. It is not so easy, however, to tell him how to eliminate broodiness, susceptibility to coccidiosis, or low egg production from his poultry. These, and most other characteristics which have commercial application are dependent upon the interaction of multiple genes. Hence, most programs for the improvement of commercial types involve the application of the principles of selection based on quantitative variations. Also, the commercial breeder is faced with changing fancies about what is desirable and what is not, and such standards vary as much by whim and as suddenly as fashions in clothes. In Ayrshire cattle, for instance, fancy may favor a certain distribution of the colored areas on the body in a pattern which has nothing to do with the utility of the animals. And as soon as a breeder has established such a pattern in a good proportion of his breeding stock, the fashion may change and require an alteration of breeding techniques in order to establish the new type without sacrifice of the useful features of the breed. Again, colored eggs are just as nutritious as white ones, but in certain sections of this country the demand favors white eggs, so that the poultryman must work not only for a flock with high egg yield, resistance to disease, and so on, but he must also consider the color of the eggs his hens are laying! Quite often the insistence upon fancy show points in a breed leads to the use of comparatively inferior animals for breeding purposes and the rejection of animals superior in utility. Ideally, of course, the standards of a breed should not be rigidly fixed unless they bear a definite relationship to utility.

SELECTION FOR DESIRABLE QUALITIES

Selection is the most widely practiced method of maintaining and improving domestic animals and cultivated plants. Long before the basic principles of genetics were discovered, man had learned that the use of desirable animals as breeding stock tended to produce desirable offspring, and that seed selected from desirable plants tended to grow into desirable plants. Thus **artificial selection** began. But the wise application of this type of selection is not as easy as it may seem. For instance, one might select and breed hens on the basis of the size of eggs they lay, and in the course of several generations might secure a flock which lays eggs distinctly larger than the average of the original flock, only to find that fewer eggs are laid and more hens die of disease. In other words, selection on the basis of one factor alone often defeats its purpose by producing organisms which are deficient in other desirable qualities. Selection, to be of the greatest value, must take all desirable qualities into account. It may be of interest if we tabulate some of them.

Body Form

Body form is one of the important factors to be considered in any planned program of selection. In plants the body form bears a close relationship to the

FIG. 24–1 *The effect of selection on egg size. The eggs at the top are the average size laid by a flock of chickens at the Mt. Hope Experimental Farm in Williamstown, Mass. After three generations of selection for large size, the eggs were, on the average, as large as those shown at the bottom. Such a program of selection is of little commercial value, however, unless accompanied by a selection for other factors, such as high egg production and lack of broodiness.*

practical aspects of cultivation. In tobacco, plants which grow tall and straight produce a greater number of large, well formed leaves than plants which are low and branching. In fruit trees, however, the low, branching form is desirable, both from the standpoint of production and for ease in harvesting. Of course, it is possible to modify the form of a plant somewhat by pruning, but this cannot be done with animals. Form is perhaps most important in ornamental plants, and the weeping willow, for instance, is widely used in landscaping because of its form while ordinary willows are seldom used. In meat-producing animals body form is indicative of market value. A certain body build in beef cattle is required to produce the finest quality of meat and the largest percentage of dressed beef.

Productivity

In most commercial plants and animals, the yield of a saleable product is the primary concern of the breeder, and selection for productivity often takes precedence over all other factors. The number and size of apples, oranges, grapes or other fruit; the number of bushels of corn, wheat, or oats per acre; the quantity of milk, eggs, or wool per animal — these are all vital characteristics which must be considered in any program of selection which is to have commercial value.

Quality of Product

In addition to quantity, it is necessary for the breeder to consider the quality of his produce. High milk yield with a very low butterfat content is less desirable than a lower yield with a high butterfat content. Sweet, juicy oranges are more desirable than a larger number of sour, pithy ones. A high gluten content makes wheat more valuable for baking purposes. In some cases the desired quality varies with the uses to which the product is put. Corn varies in the amount of oil, starch, sugar, and protein which it contains. A variety yielding a high amount of oil is desirable for the purpose of producing salad oil from corn. Another variety with a high sugar content is desirable for human consumption in the fresh condition. A high starch content is valuable in plants from which cornstarch is made, and a high protein content is desirable in corn used as feed for livestock. As a result of the demand for a variable product, selection has been directed toward different ends, and today we have varieties which are outstanding for the production of each of these four substances.

Hardiness

This characteristic includes such things as ability to withstand extremes of temperature or moisture, resistance to disease, and general bodily vigor. Any person who has raised crops in the early spring has noticed the variations in the resistance of plants to frost. Among a group of plants equally exposed and of about the same size, some will be killed while others will survive a late frost. Some plants withstand drought better than others. These are characteristics which are influenced by heredity, so that through selection plants may be produced which are frost- or drought-resistant. Also, disease resistance in both plants and animals is dependent upon hereditary characters. The establishment of rust-resistant strains of wheat saves the wheat farmers millions of dollars yearly because of a smaller loss due to this infection.

Shipping and Storage Qualities of the Product

For most fruits and vegetables, the shipping and storage qualities are very important. No matter how high the quality of the product as it comes from the field or the orchard, it has little commercial value if it cannot be shipped without damage or deterioration and if it cannot be stored for a reasonable length of time. There is a variety of peach, the Greensboro, which ripens early, is sweet, juicy, free-stoned, and delicious. Most peach growers, however, raise the Elberta, which is relatively dry, but ships and keeps much better than the Greensboro. In fruit, quality must often be sacrificed for shipping and keeping qualities.

Early Maturity

This is another valuable quality which must often be considered in a program of selection. Early fruits and vegetables usually command premium prices. Watermelons harvested at the height of the season will not bring one-half the price of the early melons. Early maturity in animals is of value also. The sooner a hen matures and begins laying, for instance, the more valuable she will be.

Economy in Use of Food

Another important factor in the selection of animals is economy in the consumption of food. A milk cow that converts a maximum amount of her feed into milk is a valuable animal. Records need to be kept to compare food intake with milk yield in order to select for this factor. Hogs are selected for maximum amount of ham and bacon produced in relation to their intake of food. There is always a loss of food value when plant foods are converted into animal foods, and we can never hope to approach the point where the entire quantity of feed which an animal receives is converted into animal food that can be consumed by man. Nevertheless, through selection we can increase the amount which is thus converted.

Selection for Combinations

These seven points serve to emphasize the complicated number of factors which must be considered in any program of artificial selection which is to have maximum commercial value. Hays and Sanborn, at the Massachusetts Experiment Station, raised the average annual egg production from 145 eggs to 235 eggs in a flock of chickens through selection. They did not concentrate on number of eggs alone, however. Indeed, such a program would be of little commercial value, for in selecting for number of eggs other factors would be ignored which would produce defects in the flock that would outweigh any advantage gained by increasing the number of eggs produced. In this particular program the following series of factors was considered and selection was based on the over-all favorableness of all of the factors: number of eggs, size of eggs, egg hatchability, persistency in laying, early sexual maturity, high sexual intensity, lack of winter pause, and non-broodiness. Thus it becomes apparent that the principle of artificial selection for the improvement of domestic animals and cultivated plants is more complicated than would appear at first sight.

MODERN GENETIC TECHNIQUES

The Progeny Test

Animal breeders often run into a problem in selection which can only be solved by the **progeny test.** We learned in Chapter 9 that sex-limited characteristics are inherited through both parents even though only one parent may express the character. In many domestic animals only one sex is commercially valuable in certain respects. Thus if we are selecting for egg yield and egg size in chickens, it is easy to measure these factors in the hens, but the rooster in the pen has just as much to do with the yield and size of the eggs as the hen. If we ignore this male parent, we lose half the effectiveness of our selection. Yet how can we determine a rooster's genes for egg production when he never produces eggs? We can do this by the same method we would use to test for hidden hereditary traits in other types of experimental organisms — by breeding and by studying the offspring. We call this technique the progeny test.

If we wish to test a rooster for genes influencing egg yield, we mate him to a group of hens of a known annual rate of egg production. Then we study the rate of egg production of the progeny. If we found that the hens which descended from this rooster consistently had a higher annual egg yield than their female parents, then we would assume that the rooster contained genes for high egg yield. And if he were the best rooster we could find, on the basis of the progeny test, he would be selected for mating to increase the egg yield in the flock. On the other hand, if he failed to raise the egg production, we would assume that his genes were about the same as those of the hens to which he was mated, and, if he lowered the production, we would conclude that he was inferior in this respect.

This principle is well illustrated by a specific case in the breeding of dairy cattle. A large dairy paid over $100,000 for a young bull which had taken many prizes in the show ring and which had a fine pedigree that gave every indication that the bull would be a valuable breeding animal. As his female calves matured and began giving milk, however, records showed that their milk production was always below that of their mothers. The prize bull just didn't have the genetic makeup required to produce daughters with a high yield of milk. Another bull who was very inferior by all show ring standards consistently raised the milk production of his daughters. Without a progeny test the dairyman would have continued using the prize bull for many years and would have greatly lowered the value of his herd for milk production.

Of course, when we breed a bull with a prize-producing cow, we can hardly say that the bull is inferior to the average if there is a decrease in milk production in the daughters. Whenever cows are well above the average for their breed in milk production, it is necessary to evaluate the bulls on the basis of average production. If the milk production of the daughters is half way between the average and that of their mothers, the bull would be rated as average. Higher production than this would indicate a bull above the average, even though he might lower the production when bred to cows with a very high rate of production.

As previously mentioned, no program of selection can be of maximum value which considers only one factor. Hence, other important factors must be considered in the progeny test as well as in the direct selection. For instance, in considering milk production it is necessary to consider butterfat production also. To illustrate: there were two bulls who were sons of a famous bull, King Regis Pontiac. The daughters of both these bulls had a high milk yield, but tests of ten daughters of one of them showed an average annual butterfat production of 975 pounds, whereas tests of eleven daughters of the other bull showed only an average of 528 pounds. The former significantly increased the butterfat production of the offspring of all the cows to which he was mated; the latter decreased it in every case. Hence, even though they had common parents, these two bulls differed greatly in their value, due to the segregation of the genes which influence butterfat production.

Inbreeding

Close inbreeding must almost always accompany artificial selection for the improvement of commercial breeds of plants and animals. Of course, in some plants asexual propagation is possible through budding, grafting, or growth of cuttings; and once a desirable type has been obtained, it is possible to continue it indefinitely without genetic change. In plants which must come up from seed, and in animals, however, inbreeding is resorted to as a means of retaining a desirable genotype. The number of individuals which show the desirable characteristics in any sample population is usually small and, if such individuals have been produced by artificial control of the breeding stock, they are probably very closely related. Hence close inbreeding becomes a necessity. Also, when through accident or intent one outstanding individual is produced, the closest kind of inbreeding is needed in order to concentrate the desirable qualities as much as possible by keeping together the fortuitous aggregation of genes that has made the progenitor outstanding.

In the popular mind it is often falsely believed that inbreeding *creates* harmful characteristics, since harmful characteristics often do become evident in a program of inbreeding. We have already learned a genetic explanation for this. Most genes which produce harmful characteristics are recessive, and any individual is likely to be heterozygous for many of these. Inbreeding promotes homozygosity of these recessive genes so that the harmful phenotypes may be expressed. Should a race be relatively free of such harmful recessive genes, there would be no such dire effect. The Ptolemies of Egypt, in order to perpetuate the fine qualities of their line, practiced brother-sister marriages for many generations, ending with the famous siren of the Nile, Cleopatra. And the records do not indicate that any harmful qualities appeared in the family as a result of this closest form of human inbreeding.

Most animal breeders note a decrease in size, vigor, and fertility among the descendants of inbred stock. This can be counterbalanced to some extent by combining inbreeding with a program of selection among the offspring so as to eliminate the deleterious effects of homozygous harmful recessive genes. Through such a program uniformity may be obtained in a race without much sacrifice of quality. Figure 24–2 shows the effects of various systems of inbreeding on the attainment of homozygosity. This chart assumes a beginning with a random population containing about 50 per cent heterozygous genes. Of course, the closest form of inbreeding is self-fertilization. Within eight generations of self-fertilization practically all individuals in the line will be homozygous for all their genes. Mendel did not have to worry about obtaining homozygous stocks in his experiments with garden peas, for these plants normally undergo self-fertilization, and the seed from any variety is homozygous because of this fact. Since self-fertilization is impossible in the higher animals, the closest form of inbreeding possible among them is the brother-sister or parent-offspring cross. It can be seen from the chart that about sixteen generations are required to attain practically total homozygosity through such matings. As the degree of relationship becomes further removed the increase in homozygosity becomes less pronounced. When the relationship between mates is further removed than first cousins, there is a negligible increase, and an equilibrium is reached at a point which is only slightly higher than that found in random-bred stock.

For most prize, pedigreed animals, the ancestry record frequently shows that individual animals appear more than once in a given line of ancestry, and this is good evidence of the extent of the practice of inbreeding. Thus a pedigree of a famous shorthorn bull, Comet, shows many duplications among the

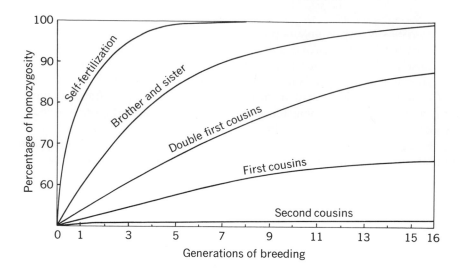

FIG. 24-2 *Effect of various degrees of inbreeding on homozygosity. The chart assumes a 50 per cent homozygosity to begin with. With self-fertilization, homozygosity is attained in about eight generations; this is not possible in most animals. As the degree of relationship diminishes, the increase in homozygosity diminishes also. When the second cousin relationship is reached, there is scarcely any distinction from non-related crosses, which would remain at about 50 per cent.* MODIFIED FROM WRIGHT.

thirty ancestors from the past four generations. The cow, Favourite, appears six times, Foljambe appears four times, and Haughton, R. Barker's bull, and Phoenix appear three times each. There are only nine individuals among the thirty ancestors!

Outbreeding

The commercial benefits which may be gained by selection and inbreeding within any particular strain of a population are necessarily limited to those genes within the strain which influence the desirable character, plus mutations which are so relatively infrequent that we will ignore them for the purposes of our illustration. This was demonstrated vividly by the pioneer work of the great Danish botanist, Johannsen. When he planted a mixture of seeds of a common garden bean, the princess bean (*Phaseolus vulgaris*), he noted that among the beans he harvested there was a considerable variation in size which, when plotted according to weight, produced the normal, bell-shaped curve of distribution. From these beans he selected the largest, the smallest, and a number of grades between, and planted them the following year. He then found variation of size among the beans from each of these plants, for while those coming from plants grown from the largest seeds were larger on the average than those from a mixed population, there was still variation within this group. Johannsen again selected for size within the groups, but found no further segregation of characteristics in the following generations. Within any one group, the beans grown from the smallest seeds were just as large on the average as those from the largest seeds. Thus selection was effective for only one generation. This is accounted for by the fact that self-fertilization occurs in the beans, and, as a result, they are normally homozygous. Johannsen had merely isolated homozygous strains for size which were present in the mixed seed with which he started his experiments. He obtained nineteen **pure lines,** as he called them, each of which bore beans of a certain average weight which could not be altered by selection. The fluctuations within the pure lines were due wholly to environmental factors.

Johannsen concluded that selection could sort out the various hereditary potentialities in the germplasm, but that it could not yield anything new, either by adding to or detracting from the species. Among those forms of life in which self-fertilization does not occur the potentialities of selection do not end so abruptly after only one generation. They may continue for many generations, but it is obviously impossible to select for genes that are not present. Under these circumstances the problem of the commercial breeder centers on a way to introduce new genes into a population in order to increase the possibilities of selection in any one direction.

One method which comes to mind as a possible means of introducing new genes is the artificial induction of mutations. Because of the preponderance of harmful mutations, however, and the expense involved in such methods, this is not practical for the average commercial breeder. A much more practical possibility lies in the introduction of new genes into the population through **outbreeding.** It is often possible to outbreed a desirable type to another type less desirable and then, through selection, to exceed in degree the desirable characters of the original. In this manner it is possible to pick up new genes which may be of great value to man even though they may come from a variety of the stock that is in the main somewhat inferior.

As an illustration, let us consider some of the work of East and Jones of the Connecticut Agricultural Experiment Station. In the Connecticut Valley a good deal of tobacco is grown which is particularly suitable for tobacco wrappers. Two of the popular varieties have been the Sumatra and the Broadleaf, but neither quite satisfies all the requirements of the grower, the manufacturer, and the consumer. For instance, the Broadleaf has a large leaf which is desirable, but a pointed tip which is not; whereas the Sumatra has a good, rounded tip, but the leaf is small. The ideal leaf would be both large and rounded. By crossing these two varieties together and applying progeny selection for several generations, a new variety was established, the Round Tip, which combines the good qualities of the two original varieties and includes some improvements over both. The large, rounded leaf was obtained, and in addition there was a larger number of leaves per plant. The Broadleaf averages about eighteen leaves, the Sumatra about twenty, and the Round Tip about twenty-six leaves per plant. The Round Tip is also superior to both parents in its resistance to disease and in the extent of its root system.

FIG. 24–3 *Hybrid vigor in sorghum. A head of Dawn Kafir sorghum is shown on the left, and one of Yellow Milo sorghum on the right. The head in the center is from a hybrid obtained by crossing the two. This plant is more vigorous and resistant to disease than either of the parent types.* FROM GEORGE M. REED IN JOURNAL OF HEREDITY.

Outbreeding for Vigor and Fertility

Outbreeding may also be used to restore the vigor and fertility which is usually lost to some extent in any program of inbreeding, no matter how careful the selection. Almost invariably, the hybrid of two inbred varieties will exhibit an increased vigor and fertility, as in the case of the hybrid Sorghum shown in Figure 24–3. Hence, quite often a commercial breeder will maintain two closely inbred stocks which are crossed together to produce a hybrid for market. The value of such hybridization is very evident in the field of corn production (see Figure 24–4). This has been known for centuries. Even the Indians and the early white settlers in this country frequently mixed corn of different colored

FIG. 24–4 *Hybrid vigor in corn. Two inbred varieties of corn may be crossed to yield plants with greater vigor and superior yield. The photograph on the left shows some hybrid plants between the two parent varieties. The photograph on the right shows a hybrid ear in the center with typical ears from the parent plants on either side.* FROM D. F. JONES, CONNECTICUT AGRICULTURAL EXPERIMENT STATION.

grains before planting, for experience had shown that a greater yield was obtained when there was a mixture of kinds than when a single type of corn was planted. Today, hybridization of corn has been standardized, and the production of hybrid seed corn is a major agricultural pursuit. The increase in yield may be as great as 250 per cent when two closely inbred varieties of corn are crossed together. However, the value of such hybridization cannot be continued generation after generation by planting seeds from the hybrids, for the hybrid vigor **(heterosis)** results from the heterozygosity of genes.

One explanation for hybrid vigor lies in the fact that most harmful genes are recessive and most beneficial genes are dominant. And hybridization tends to bring out the beneficial qualities of both varieties and to suppress the harmful qualities of each. An equally prominent theory is that there is a heterotic interaction between particular alleles when they come together in the hybrid to produce heterozygous expressions which are superior to those produced by homozygous genes. It is easy to understand why such vigor cannot be prolonged through successive generations of inbreeding of the hybrids. The inbreeding will again produce homozygosity, and this will result in a phenotypic expression of some of the harmful, recessive genes which were masked in the heterozygote. Also, homozygosity will cause a loss of the vigor resulting from possible heterotic interactions between alleles.

In corn it has been shown that there will be a gradual decline of vigor and yield for about seven generations if seeds from a hybrid and its descendants are planted successively through this period of time. At the end of this time the yield will probably be somewhere near the average for the two original parent varieties. This is shown very strikingly in Figure 24–5.

FIG. 24–5 *Loss of hybrid vigor through seven generations. This photograph shows how the hybrid vigor in corn diminishes with inbreeding as indicated by the size of the stalk. The hybrid is shown on the left. Note that the decline becomes less pronounced in the last three generations. There is no further decline after this as homozygosity has been attained.* D. F. JONES, CONNECTICUT AGRICULTURAL EXPERIMENT STATION.

FIG. 24–6 *Double-cross hybridization in corn. By using four inbred varieties it is possible to produce corn which is superior to the single hybrid. This photograph shows four ears from four inbred varieties, the two hybrids produced by crossing them, and one ear from the double-cross hybrid produced by crossing the two hybrids.* D. F. JONES, CONNECTICUT AGRICULTURAL EXPERIMENT STATION.

The principle of hybridization for increased vigor and yield in corn has been carried even a step further. If a cross between two inbred strains will yield a superior hybrid by bringing out beneficial dominant genes from each of the parents, will not the cross between two hybrids be of even greater benefit by combining the beneficial genes of four varieties in one individual? Such a line of reasoning sounds logical, and Figure 24–6 shows that it is sound. This technique, called **double-cross hybridization,** is widely practiced in the production of superior seed corn. Indeed, much of the hybrid seed corn on the market today is produced by such a method. The benefits resulting from added yield and improved quality more than compensate for the extra cost involved in the 2-year program of hybridization which is required.

Outbreeding for Specific Types

In animals, outbreeding is often practiced in order to produce some specific type which is desirable for market purposes. A cross between the white Shorthorn and the black Angus cattle yields a blue-roan hybrid which is noted for its vigor, rapid growth, economical utilization of food, and the high quality of beef which it produces. Such animals are usually bred solely for the market, and are not used for breeding purposes because of the variability of the offspring which would result from the segregation of genes in succeeding generations. Dairy cattle are sometimes crossed with beef cattle in order to produce calves superior for veal production. In domestic swine, sows of the bacon-producing breeds, such as the Yorkshire or Tamworth, are often outbred to boars of the

FIG. 24–7 *Double-cross hybridization in domestic swine to secure a superior meat-producing animal. This sow was produced from a Landrace-Duroc hybrid crossed with a Landrace-Poland-China hybrid.* BUREAU OF ANIMAL INDUSTRY, U.S. DEPARTMENT OF AGRICULTURE.

lard-producing breeds, such as the Duroc-Jersey or Poland-China. This type of cross yields a larger number of vigorous, early-maturing offspring with a superior market value when compared with the offspring of pure breeds. Figure 24–7 illustrates double-cross hybridization in domestic swine.

Those vacationers who drive through the cattle-raising regions of the central part of Florida are often amazed to see huge humpbacked Brahma bulls grazing along with cows of the better known American breeds. Hybridization between such cattle produces animals somewhat intermediate in appearance which are valuable for their tolerance of tropical heat and their resistance to many cattle diseases, particularly those carried by insects and other arthropods. The Brahmas are also used extensively for hybridization in Australia and South Texas, where pure breeds of Brahmas are maintained as a source of Brahma bulls for rodeo performances.

The same principle of outbreeding may be applied within the breed by establishing two strains of stock by selection and inbreeding, which then may be crossed together to produce market animals or animals which produce for the market. The advantages of such a method are well illustrated by a practice in poultry raising. Two inbred strains of Wyandottes have been developed for the express purpose of crossbreeding for egg production. One strain is identified by a dominant, sex-linked gene which produces silver plumage; the other strain is homozygous for the recessive allele for gold plumage. The silver hens are mated with the gold cocks to produce egg-laying pullets. In this type of cross it is possible to distinguish the sexes immediately after hatching, for the female chicks will always be gold and the male chicks will always be silver. This makes it easy to remove the males from the group. These hybrid pullets have the vigor which is characteristic of hybrids, are resistant to disease, are early maturing, and are heavy layers.

To Create New Breeds

The foregoing illustrations are instances of the use of outbreeding to take advantage of hybrid vigor, but outbreeding may also be used for the purpose of creating new breeds, as illustrated in Figure 24–8. When hybrids are bred to

FIG. 24–8 *A new variety of marigold produced by hybridization. The large yellow African marigold on the right was crossed with the dwarf French marigold on the left to produce the beautiful red and gold hybrid in the center. The parent plants are different species and the hybrids are almost totally sterile.* W. ATLEE BURPEE COMPANY.

produce a second generation, there is a segregation of their genes, and a great variety of offspring will appear. It is possible to establish a breed with the desirable qualities which will appear in the hybrid by selection from succeeding generations, and thus eventually to obtain a new breed incorporating from the two original breeds the characteristics which are desired. This is a long, tedious, and expensive process, however, and in spite of the most careful selection there is practically always some loss of vigor and fertility as a result of the necessary inbreeding. Nevertheless, this is the manner in which many of our present-day breeds have been established — through hybridization and selection from other breeds. Most breeders of today, however, prefer to stick to improvement within the breed because of the many difficulties involved in establishing a new breed.

In some cases different species are crossed and yield a hybrid of superior vigor and commercial value. The **mule** is a good example of such an interspecific cross. The mule is produced by crossing the mare, *Equus equus,* to the jack, *Equus hemionus.* A reciprocal cross may also be made between the horse and the jenny. This results in a hybrid, usually called the **henny,** which is somewhat different from the mule, possibly as a result of the smaller size of the female parent. The cross of the jack and the mare is used almost exclusively for the commercial production of hybrids. There is considerable variation in mules, depending upon the quality of the two parents which produce them, but they are generally superior to horses in strength, physical endurance, resistance to disease, and the ability to work under unfavorable conditions such as extreme heat. There are some who maintain that the mule is also superior in intelligence to either of its parents. They cite the notable stubbornness of the mule as evidence of its greater intelligence. Mule-breeders who consistently turn out superior work animals maintain a breeding stock of jacks and mares which have been selected especially for their ability to produce high quality mules. There is no question here of the possibility of establishing a breed of hybrids, for the mule

FIG. 24-9 *Colt born to a mule. Mules are usually sterile, but in rare cases they have been known to produce offspring. The mule mare in the photograph at the right was bred to a Percheron stallion and the colt shown at the left was born. The horse-like appearance of the colt suggests that the ovum from which it came must have received nearly all horse chromosomes though the process of segregation in meiosis, hence the fertility.* FROM W. S. ANDERSON IN JOURNAL OF HEREDITY.

(like interspecific hybrids in general) is sterile. A study of the seminal fluid from the male mule shows that its sperms are non-functional. Several cases have been reported, however, of a female mule giving birth to a colt after mating with a jack. It is understandable how, in extremely rare instances, all the chromosomes of the horse or the donkey might segregate into one germ cell and produce a viable gamete. Figure 24-9 shows such a colt. Its horse-like appearance seems to indicate that the ovum from which it came must have received nearly all horse chromosomes through the process of segregation in meiosis.

While man has been selectively breeding domestic plants and animals for thousands of years, the new methods made possible by recent advances in genetics hold great promise for the future.

PROBLEMS

1. Why does it take longer to obtain homozygosity through brother-sister matings than through self-fertilization?

2. Why is the vigor of heterosis lost when hybrid plants are bred together for a number of generations?

3. Why is selection through progeny testing of great value in some animals and of little value in others?

4. A certain poultryman decides that he would like to have all his flock of a particular feather pattern. Accordingly, he instigates a program of selection to establish this pattern. To his surprise he finds that his egg production drops rather sharply after the program has been going for some time. How would you account for this?

5. Suppose you have a sow to breed and two boars are available. One is of

rather high quality, but is the litter mate of the sow. The other boar is of lower quality, but is not related to the sow. Which boar would you choose and why?

6. Cocker spaniel dogs were originally bred and selected for their ability to hunt the woodcock. In recent times they have become major show animals and family pets. They do not show much of the hunting ability today that they once possessed. Why?

7. The Dutch belted or "Lakenfeld" cattle of Holland have been bred for centuries by aristocratic families who take great pride in the particular belted pattern which these cattle exhibit. These cattle do not equal some of the other breeds in productivity. Can you suggest a possible reason?

8. Most pure breeds of dogs are more susceptible to diseases than mongrels. What genetic explanation can you offer for this?

25

Survey of Human Heredity

TO MOST STUDENTS HUMAN HEREDITY is naturally one of the most interesting phases of the study of genetics. Recognizing this fact, we have used the inheritance of human characteristics as illustrations of many genetic principles throughout this book. But a great many other human characteristics have been studied genetically besides those we have so far mentioned, and since some knowledge of these may be useful as well as interesting, we shall devote this chapter to a catalogue of some of these and to giving some information about their mode of inheritance.

At the outset it must be stressed once more that much of our information is incomplete regarding the inheritance of certain traits, and present conclusions are often tentative. This is to be expected because of the lack of evidence from experiments, but other difficulties also may prevent an easy classification of inherited characters. One of these difficulties is that the same trait may behave in different ways at different times. For instance, the inheritance of **syndactyly** (the presence of a flap of skin between the second and third toes) acts as a sex-linked recessive in many family pedigrees, but in some others it is inherited as an autosomal recessive. As a result we must assume that two of the genes involved in the growth of the skin on this part of the body may mutate to produce indistinguishable phenotypic effects. Indeed, there are even a few cases where *three* genes are known to produce similar effects — one an autosomal dominant, the second an autosomal recessive, and the third a sex-linked recessive. Since we know that many genes are involved in the production of nearly every characteristic, there is nothing surprising about this or contradictory to the principles of genetics which we have learned, though this variety of causes for the same effect does somewhat confuse the picture in those few cases where it exists.

The student of human heredity will find that many quantitative variations of human characteristics can be explained only by multiple gene inheritance. Multiple genes also participate in reducing the penetrance or altering the expressivity of many human variations which depend primarily on a single gene difference. Finally, surgery is often employed early in life to remedy inherited defects. This, of course, does not alter the genes, but it may make it very difficult to determine whether or not a person possessed a certain characteristic at birth. Clubfootedness, for instance, may be hereditary, but skilled orthopedic treatment and surgery in early childhood can correct the condition, so that an adult may not even know that he was born with clubfeet. Hence, these tabulations of the method of inheritance of human characteristics are presented with the warning that they must be taken with reservation as to their absolute accuracy. They may be of value, nevertheless, as a preliminary guide in the survey of family histories which may extend our knowledge of human heredity.

HAIR

Color

The great range of continuous variations in hair color suggests that this human character is influenced by multiple genes. Children resulting from marriages between persons of a dark-haired race and persons with very blonde hair are typically dark-haired. This indicates an over-all dominance of the genes for dark hair over the genes for light hair. In a mixed population, such as our own, the children of parents with opposite extremes of hair color are often intermediate because the dark-haired parents are heterozygous for a number of genes involved in this character. Marriages between two light blondes in our population almost invariably yield light blonde children, a fact which again points to the recessive nature of blonde genes as a whole.

Red hair, however, often seems to result from a single pair of recessive genes. The children of two red-headed parents practically always have red hair. Also, red-haired children may be born to parents who are not red-headed and, in such families, approximately one-fourth of the children show this trait, as would be expected on the basis of simple Mendelian inheritance. The few cases where red-haired parents have borne children with other hair colors could be explained by the fact that there may be recessive genes for red hair at two different loci. Cockayne studied a number of families in which one parent had red hair and at least one child had red hair. This combination would indicate that the other parent was heterozygous. He found forty-six red-haired children and forty-eight children who did not have red hair in such families, a result which is very close to the 1 : 1 ratio which would be expected between homozygous recessive and heterozygous dominant persons. Incidentally, hair color may have some relation to the endocrine secretions. A baby born of a dark-haired mother may have dark hair at birth because of the influence of the hormones of the mother even though its genotype indicates blonde or red hair. When this first hair falls out and is replaced by a new growth the color will be lighter.

There seem to be two primary pigments in the hair, each of which is subject to quantitative variation through multiple genes and multiple alleles. One of these is a black melanin which may result in black, brown, sepia, or light blonde hair with varying shades between. Upon this is superimposed another melanin pigment which may range from sandy-red to yellow. Various genes influence the intensity and the quality of these pigments in such a way as to produce the wide variation in hair color which is characteristic of man.

We learned in Chapter 17 that the recessive gene for albinism interferes with the production of any melanin in the hair as well as in the skin and the eyes. This gene, therefore, is epistatic to all of the genes involved in the production of this pigment in the hair. Albinos often have some color in their hair as a result of other pigments. Negro albinos, especially, tend to have sandy-red hair, and white persons often have some reddish or golden-yellow pigment.

Under the general topic of hair color we may also mention the graying of hair. There is not complete agreement as to the exact cause of this phenomenon, for one theory holds that it results from the production of a substance known as leucokeratin, while another maintains that it is due to the formation of air bubbles in the hair shaft. We do know that the age at which the hair begins turning gray is influenced by heredity. It is difficult to establish the exact method of inheritance, however, because of the environmental effects of diet and other factors. In certain family pedigrees, premature grayness is inherited as an autosomal dominant.

Some persons are born with a white forelock of hair which is given the genetic designation of **blaze.** This isolated island of white hair grows out from the central portion of the scalp where it joins the forehead and is usually combed back over the head to produce a white streak in a head of hair otherwise normal in pigmentation. This condition results from an autosomal dominant gene. A similar condition, called **white spotting,** produces the white forelock and also produces colorless spots on the skin in other parts of the body. This also is inherited as a simple dominant.

Form of Hair

The form of hair is dependent primarily upon its shape in cross-section. Straight hair is rounded, while wavy, curly, and kinky hair show progressive degrees of flattening. No doubt a number of genes are involved in hair form, but in the white race the evidence indicates that there is one pair of alleles which can produce the difference between curly and straight hair, with the heterozygote showing wavy hair. In the children from a large number of marriages between straight-haired and wavy-haired persons, Cockayne found sixty-one straight to fifty-two wavy, a close approximation to the expected 1 : 1 ratio. When both parents had wavy hair he found straight, wavy, and curly in the following ratio of 27 : 51 : 22, a close approximation to the expected 1 : 2 : 1 ratio. The kinky hair of the Negro race generally dominates over the hair form of the white race, although there may be some degree of intermediate expression.

Presence or Absence of Hair

Next to color and form, the most striking feature of hair concerns its presence or absence. **Baldness** is a characteristic which, without doubt, can be induced by environmental agents, such as disease, but the majority of people who are bald become so because of their genes. The condition is inherited as a result of a sex-influenced gene which is dominant in men and recessive in women. The pattern of baldness which develops and the age at which it begins is also influenced by heredity. There is another condition, hairlessness **(hypotrichosis),** which is characterized by the complete absence of hair on the head from birth or shortly thereafter and a scanty growth of hair on other parts of the body. Such a condition is inherited as an autosomal recessive which seems similar in its effects to genes for hairlessness in domestic mammals, such as cats. A dominant form of hairlessness has also been reported from some human races including Australian natives. The hairlessness of the Mexican hairless dog is also due to a dominant gene.

The opposite condition, excessive amounts of hair on the body **(hypertrichosis),** may be produced by a dominant gene in man. Also, other genes may produce excessive amounts of hair only on certain body parts. Heavy, bushy eyebrows result from a dominant autosomal gene. Hypertrichosis of the ears is a trait induced by a gene on the non-homologous portion of the Y-chromosome. Hence it appears in men only and passes directly from father to son. The amount of hair on the face and body is variable among the different races. The Ainus of northern Japan are the hairiest people on earth, the white race comes next, followed by the Negroes, while the American Indians, Mongoloids, and Eskimos have little hair on the face and body. The intermediate condition of race hybrids suggest multiple gene inheritance for these differences.

The growth of hair on the middle joint of the fingers is an interesting human characteristic which may easily be studied by amateur geneticists. **Mid-digital hair** may occur on fingers in various combinations. It is most common on the ring finger; next comes the ring finger plus the middle finger; next comes the ring finger, plus the middle finger, plus the little finger; and least common is hair on all four fingers. A multiple allele hypothesis has been proposed to account for the inheritance of these combinations. This hypothesis assumes the presence of hair on all four fingers as dominant to its presence on three; on three, in turn, dominant to its presence on two; two dominant to one; and one dominant to none. In making such a study, a hand lens should be used to bring out small hairs that might not be easily visible with the naked eye.

Arrangement of Hair on the Head

Near the top part of the back of the head there is a crown or **whorl of hair** which rotates in a clockwise direction in most people. In a few cases, however, the whorl may be counterclockwise as a result of a recessive gene. Occasionally,

FIG. 25-1 *Widow's peak (left) is characterized by a downward point of hair in the center of the forehead. The gene for this seems to be dominant over the gene for a continuous hair line (right).*

a person will have two whorls rather than one. This seems to be inherited as a recessive character also. At the center of the forehead the hairline may dip down to form a point which is called a **"widow's peak"** (see Figure 25-1). This characteristic is inherited as a dominant character in some families.

EYES

Color

The color of the iris of the eye is a clearly defined human character which draws much interest. Indeed, it was one of the first human characters which was suggested as an example of Mendelian inheritance in man. Davenport, in 1907, studied a number of family pedigrees and suggested that brown eyes are dominant over blue eyes in the white race. There are many variations of eye color, however, which may be influenced by other genes. The blue color of the iris is due to a pigment in the back (or retinal layer) of the iris as seen through a semiopaque, colorless layer in the front of the iris. This blue layer is present in most persons, but in those with darker eyes, it will be masked by the development of melanin in the front part of the iris. Nearly all babies of the Caucasian race are born with blue eyes, which may darken later as the melanin pigment develops in the front layer of the iris. Gray eyes seem to be a variation of the blue. Green and hazel eyes appear when the melanin partially masks the reflection from the rear of the iris. Brown and "black" (dark brown) represent an almost complete masking. There is also variation in the distribution of the melanin — it may be smoothly distributed, it may be in spots, or it may occur in a ring around the outer edge of the iris.

In most families where both parents have pure blue or gray eyes, the children all have blue or gray eyes, thus indicating the recessive nature of this character. The darker colors generally are dominant over the blue or gray, but there is considerable variation in the expression of the darker colors because of modifying genes. Of course, when the recessive gene for albinism is homozygous, there is no formation of melanin. In some albinos the iris is pink because the blood in the retinal layer becomes visible. In other albinos there are some reflecting bodies in the retinal layer and the eyes will be a pale blue.

Coloboma iridis is the name given to a condition in which there is a fissure in the iris extending from the pupil out to the "white" of the eye. This makes the pupil appear as a slit extending from the center of the iris out to its outer border, while in the normal eye it appears as a black circle in the center of the iris. It results from a sex-linked recessive gene.

Vision

Nearsightedness, **myopia,** is a very prevalent hereditary defect of the eyes. It may be brought about by either one of two independent factors. When the eyeball is too long, a normal lens will bring distant objects to a focus in front of the retina, thus causing a fuzziness of the image upon the retina. Such a condition seems to be inherited as an autosomal recessive. An excessive curvature of the cornea is a less common cause of nearsightedness, and seems to be caused by a dominant gene. There is also a sex-linked form of myopia which has a reduced penetrance.

Farsightedness, **hyperopia,** results when the eyeball is too short for the curvature of the lens. In a normal eye, when the muscles within the eye are relaxed, the image from distant objects is in focus on the retina. When the eyeball is too short, however, the point of focus is behind the retina. Hence, there must be a constant strain on the eye muscles to accommodate for normal vision, and it is sometimes impossible to bring nearby objects into sharp focus at all. This condition seems to be a simple dominant character.

Astigmatism is a defect of vision caused by unequal curvature of the cornea, which causes objects in one plane to be in sharper focus than objects in another plane. It also seems to be inherited as a dominant.

Blindness is, of course, the most serious of eye defects. It may be induced by many environmental causes and it may be inherited in many different forms. Keeler lists over twenty different hereditary causes of blindness. One of the most common of these is **glaucoma,** which is responsible for more hereditary blindness than any other cause. In glaucoma the fluid within the chambers of the eye developed a great internal pressure which results in destruction of the optic nerve. This is a case in which preventive surgery may be used to inhibit the development of an inherited defect. An eye operation may provide a channel for a better circulation of the humors within the eye and thus relieve the undue pressure on the optic nerve. The character is dominant, but usually does not make its appearance until middle age, although in some families it may appear as early as 15 years of age. **Cataract** is another eye defect which leads to blindness. The condition is characterized by the development of an opaque condition of the lens or the cornea. We know that there are environmental agents which may lead to the production of cataract, but it is undoubtedly influenced by heredity also. The tendency to develop cataract is inherited as a dominant, and the age at which it is likely to develop is also inherited. In some families it is even present at birth. Another inherited cause of blindness is **retinitis pigmentosa,** which is characterized by the development of a pigment alongside the blood vessels of the retina. This condition is found to result from a dominant gene in some families and from a recessive gene in others. **Optic atrophy,** or degeneration of the optic nerve, which also results in blindness, is due to a recessive sex-linked gene. **Microphthalmia,** blindness due to small nonfunctional eyes, is inherited as a recessive, sex-linked trait according to the pedigrees which have been studied.

Red-green color blindness is inherited as a sex-linked recessive, as we learned in Chapter 10. Detailed studies of this condition, however, show that there are different variations of this type of color blindness, with at least two, closely linked, gene loci being involved, and with several multiple alleles at each locus. There appear to be three kinds of color-sensitive bodies (cones) in the human retina. These are sensitive to red, green, and blue, respectively. Yellow seems to be formed in our sensations when there is a simultaneous stimulation of cones sensitive to red and blue. In the most common type of color blindness, known as the **deuteran type,** the green-sensitive cones appear to be defective and an afflicted person is actually "green-blind." A variant of the deuteran type, however, seems to have all three functioning cones, but the red and green cones are somehow linked so that they convey a single sensory stimulation and this causes confusion of these two colors. The **protan type** of color blindness is less frequent, about one-fifth as frequent, as the deuteran type. The protan type is due to an impairment of the functioning of the red-sensitive cones and causes confusion between red and green as is true of the deuteran type. These two major types of color blindness can be distinguished by the use of special color charts and other tests of color vision. The normal allele, in each case, is dominant over the other alleles, and the gene for a less extreme form of color blindness is dominant over its alleles for a more extreme form of color blindness. It appears that the dominance is not complete, however, for it has been found that some women carrying one of the color-blind genes at one of her two loci, but with a normal allele at the other locus, have poor color aptitude. In a study by Waaler in Oslo, Norway, it was found that, of about 9,000 boys, about 6 per cent showed one of the forms of red-green color blindness.

There is also a sex-linked *total color blindness* which involves still another locus on the sex chromosome. This **tritan type** of color blindness is very rare, however, and is found in only about 0.002 per cent of a large group of men who have been studied. These have all three types of cones, but there seems to be a sensory linkage which makes distinction of colors of all shades difficult.

Another sex-linked recessive gene results in an impairment of vision known as **congenital night blindness.** Persons showing this character see very well in good light, but very poorly in dim light. Night blindness may also result from a deficiency in vitamin A in the diet, but feeding vitamin A in large amounts to a person who is congenitally night-blind brings no relief.

In some persons vision is impaired by an uncontrolled movement of the eyes, known as **nystagmus.** The eyes may undergo up and down, lateral, or rolling movements. In some families this character is inherited as a sex-linked recessive, in others as an autosomal dominant.

Eyelids

The eyes of the specialized Mongoloids (Chinese, Japanese, and so on) are commonly referred to as "slant-eyes." Actually the position of the eyes is the same as that in other races, but there is a fold of skin from the upper eyelid which extends down over the inner corner of the eye to produce this distinctive appearance. Interracial marriages indicate that this character is inherited as a simple dominant. Another inherited defect of the upper eyelid is called **ptosis** or drooping eyelids. Persons with this condition are unable to use the muscles which raise the upper eyelid, which therefore remains drooped down over the eye and leaves only a small slit between the upper and lower lid for vision. There are various degrees of expressivity of this gene. In some persons the drooping is sufficient only to give a sleepy appearance, but in others it is so pronounced that the head must be thrown back for them to see straight ahead. This condition appears generally to be a dominant.

In the early human embryo the eyelids are fused, but normally they separate at about the seventh month of development. There is an inherited condition called **cryptophthalmos** in which the lids fail to separate and the baby is born unable to open its eyes. This appears to be due to a recessive autosomal gene.

EARS

Form

Many continuous variations in the size and form of the ears as well as their position on the head indicate multiple gene inheritance of ear characteristics. A few of these seem to correspond primarily to variations in a single gene. **Free ear lobes** seem to dominate over **attached ear lobes,** but there is variation in the degree of freedom of those which are not attached. The outer, rolled rim varies continuously in its size, and sometimes is almost lacking. In some persons, moreover, there is a very distinct point which projects inward from this rim. As noted earlier this is called **Darwin's point** and is inherited as a dominant trait with somewhat variable expressivity, since it shows in only one ear in many persons. There is also some reduced penetrance. Nearly all persons have a small enlargement of the cartilage at this point, but it usually does not stand out enough to show distinctly. Keeler reports an interesting pedigree showing inheritance of natural **"earring holes."** In the center of the lobe of the ear of an affected person there is a distinct depression in exactly the same spot where the young ladies pierce their ears for earrings. This condition is inherited as an autosomal dominant with somewhat incomplete penetrance and variable expressivity. About half the people showing the character express it on only one side.

Hearing

The outer portions of the ear are the more noticeable, but the important functional parts are embedded in the bone within the skull. **Deafness** is, of course, the most serious defect of the ears. Since hearing involves the correct coordination of many parts of the ear, deafness may result from a number of causes. Various types of ear infections and other environmental agents may induce deafness. One inherited form of deafness, however, is present at birth. Since children so afflicted cannot hear the spoken word, they normally do not learn to speak and become deaf-mutes. At least two genes are known which produce such deafness, and both of them are recessive; hence a person will be a deaf-mute if he is homozygous for either of the two genes. Some special training institutions have been developed to teach such persons to speak and to read lips. The author once held a rather lengthy conversation with a person who had been so trained, only to learn later that the person had heard not a word that was said. Usually, however, the speech of persons born deaf is noticeably different from that of persons with normal hearing. Another inherited form of deafness makes its appearance after maturity, beginning with a slightly defective hearing and progressing to total or near total deafness in old age. This is known as **otosclerosis** and is inherited as a dominant. It is usually accompanied by a ringing, or buzzing, in the ears. It is induced by an abnormal growth of bone around the bones of the middle ear which transmit sound vibrations and make hearing possible. This type of deafness is usually first noticed at about 30 years of age. Still another type of deafness results from atrophy of the auditory nerve, which transmits the impulses from the hearing organs of the ears to the brain. This usually begins to manifest itself at about 40 years of age and progresses for several years, finally resulting in complete deafness. It seems to be inherited as an autosomal dominant.

NOSE

We often hear the expression, "As plain as the nose on your face," which emphasizes the fact that the nose is one of the most conspicuous of facial features. As with so many other human characteristics, we find continuous variations of many sorts in the length, width, and shape of the tip of the nose which result from multiple genes. A number of the characteristics of the nose which show simpler inheritance may be mentioned. A high, convex bridge of the nose seems to be dominant over a straight bridge or a concave bridge. The root of the nose, the point at which it joins the forehead, in some persons is quite high, forming a suitable attachment for eyeglasses; while in others it is low. There is variability in this character, but high root seems to dominate low root. A straight tip of the nose seems to be dominant over an upturned tip. The wings which spread out from the point of the nose are also quite variable, and may be wide and spreading or narrow. The widespread wings seem to show dominance over the more narrow wings. Also, some wings may be higher than the septum of the nose, so that the openings of the nostrils are clearly visible from a side view. In other cases the wings are level with or lower than the septum which divides the nostrils. Some geneticists think that the high wings are recessive to the lower types.

MOUTH

Lips

The lips form the gateway to the mouth and are a very prominent part of a person's facial expression. Numerous multiple genes determine the general shape and size of the lips. Full lips seem to be dominant over thin lips in some family pedigrees. **Harelip** is one of the most common abnormalities of the lips. When the upper lip is first formed in early embryogenesis, it consists of three portions, two lateral and one median. During the second month of embryonic life these processes normally fuse to form the upper lip. On some occasions, however, they fail to fuse, most frequently on the left side, and a baby is born with a distinct cleft extending up toward the nose. The deformity takes its name from the mouth of the hare, but the cleft in the animal's upper lip is median, while that in a human lip lies to one side or the other of the median line. Harelip is often accompanied by a cleft palate in the roof of the mouth and connecting the mouth cavity with the nasal cavity, and it is more common in men than in women. The evidence indicates that the conditions can be induced by certain environmental agents, such as low oxygen concentration or vitamin deficiency, at a critical stage of development — at the stage of fusion of the three parts of the upper lip. On the other hand family pedigrees and twin studies leave no doubt but that it can be influenced by heredity in some instances. When a couple has one child with harelip, the chance of the next child having this affliction is about one in six. In some cases at least a recessive gene is involved which has greater penetration in the male. Thus, environmental agents at times can produce a phenocopy of a trait which is influenced by heredity at other times.

Teeth

There are many genes which influence the development and conformation of the teeth. Variations in single genes may produce pronounced effects. One dominant gene causes the absence of some of the upper incisors and molars. Another dominant gene causes the absence or reduction of the two lateral incisors in the upper jaw. A sex-linked recessive gene causes the absence of the canines and perhaps some other surrounding teeth. A dominant sex-linked gene causes the absence of enamel on the teeth. This causes them to wear down rapidly so that they often barely protrude through the gums. Still another dominant gene causes defective dentine, so that the dentine is soft and opalescent, the enamel has a bluish color, and the teeth wear down easily. Brown enamel of the teeth results from a recessive sex-linked gene. Tooth decay is without doubt influenced by environmental factors such as diet, but the susceptibility of the teeth to decay seems to be due to a dominant gene.

Tongue

As with most other body characteristics the shape and size of the tongue respond to so many different genes that it is not possible to isolate many effects of individual genes. Some people have the distinctive ability to roll their tongue into a U-shape when they extend it from the mouth. This ability seems to depend upon a dominant gene. A much rarer gene gives a few individuals the ability to fold the tongue from front to back. This is also a dominant trait. The sense of **taste** is also associated with the tongue. It has been discovered that people vary in their ability to taste a certain chemical, phenylthiocarbamide. To some this chemical has a distinctly bitter taste, to others it is quite tasteless. This is an easily measured human characteristic which has been extensively investigated. Ability to taste this substance is dominant; inability, recessive. About 70 per cent of Americans are tasters.

CHIN AND CHEEKS

The chin is often thought to denote some aspects of a person's character — the pugnacious, aggressive, protruding chin and the weak, timid, receding chin — but the genes which affect this portion of the human anatomy seem to have no direct relationship to inherited traits of personality. The receding chin seems to be inherited as a recessive trait. The chin also varies in its length, from the lower lip to its lowest point, with multiple genes making the variation continuous. Some persons have a distinct depression, or dimple, in the lower part of the chin. This seems to be inherited as a dominant. One pedigree showed this character in an extreme form; there was actually a hole in the chin bone at this point. It too was found to be a dominant trait.

From a front view, the chin may be square or rounded, wide or narrow, with variations between. A study of the Hapsburg dynasty of Europe shows that for 6 centuries there was a rather consistent transmission in the royal family of a narrow, protruding chin, together with an overhanging lower lip. Dominance of this trait certainly seems to be indicated. A double chin, however, clearly reveals the mutual influence of heredity and environment — inheritance of the necessary genes plus excessive intake of fattening foods.

The cheek bones may be high, low, or intermediate. Many races, such as the American Indians, are characterized by high cheek bones. The gene complex for high cheek bones seems to be dominant. There is quantitative variation in the extent of the fat pads of the cheeks, both as to thickness and placement. In some persons the pads are so low as to result in pendulant cheeks which hang downward. Dimples in the cheeks are inherited also, apparently as a dominant, but with some variation in expressivity. They may occur on one cheek or both, and in rather rare cases there may be two on one cheek.

HANDS AND FEET

Length of Fingers and Toes

There are genes which produce over-all variations in the length of the fingers and toes relative to the hand or foot, and other genes which influence the relative lengths of the individual fingers and toes. Most of the genes which influence the fingers also influence the toes, but there are a few which affect only one

FIG. 25–2 *Dimpled cheeks seem to be inherited as a dominant, although there is variable expressivity.*

of the two types of digits. There is one dominant gene for a trait, called **arachnodactyly,** which produces extremely long fingers and toes. Persons with this condition are said to be spider-fingered or spider-toed. Another dominant gene, for **brachyphalangy,** causes very short fingers, because of an extreme shortening of the middle phalanx of these digits. On the basis of a study of the children of one couple, both of whom had the condition, this gene is believed to be lethal when homozygous. Another dominant gene causes a shortening of the fingers and toes through the apparent absence of the middle joint. This is known as **brachydactyly.** X-ray photographs show that this trait involves a fusion of a rudimentary middle joint with one of the other joints of the digits. Still another dominant gene, **symphalangy,** causes a fusion of the joints without any shortening. This produces stiff fingers. An interesting sex-influenced character concerns the relative length of the index finger in relation to the fourth finger. A long index finger seems to be inherited as a dominant in men and as a recessive in women (see Chapter 10). A **second** toe longer than the big toe appears to be inherited as a dominant.

Position of Fingers and Toes

Some persons are unable to straighten out their fingers because of an abnormal shortening of the flexor tendons. Possession of such flexed fingers, known as **camptodactyly,** is inherited as a dominant with reduced penetrance. The possession of crooked little fingers is a recessive trait. Hypermobility of thumb joints is another characteristic seemingly inherited as a dominant.

Odd Numbers of Fingers and Toes

Polydactyly, or the presence of extra fingers and toes, was one of the first inherited traits to be traced in a human pedigree, and we have a very complete pedigree for this trait, collected by Maupertuis in the first half of the eighteenth century. The extra digit may be appended to the little finger or toe, or it may be attached to the thumb or the big toe. The condition is inherited as a dominant with variable expressivity and penetrance. Some persons show it in their hands, but not on their feet, and vice versa. Another dominant gene results in a reduction of the total number of fingers. This seems to be due to a fusion of digits and also varies in its expressivity. The most extreme expression results in **"lobster claw,"** a condition in which there are only two digits. Other persons carrying this gene may have three or four digits, and in a few cases it may be expressed only in the deformity or the absence of a nail of a single finger.

Nails

Extremely short nails, **brachymegalodactyly,** seem to be inherited as a dominant. The complete absence of a thumb nail also seems to be inherited as a dominant. And in some families there may be an excessive growth of the nail-forming tissue to give very thick nails, **hyperkeratosis subungualis,** another trait which is inherited as a dominant. Again, when viewed from the side, nails may show a convex curvature or may be straight. The curved condition seems to be dominant over the straight.

Other Features of Hands and Feet

In some families there is a fusion of the skin, and sometimes of the bone also, between some of the toes, usually the second and third toe. This condition,

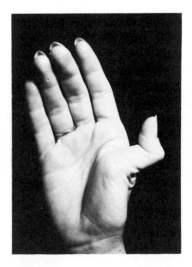

FIG. 25–3 *Hypermobility of the joints of the thumb. This extreme mobility seems to be inherited as a dominant.*

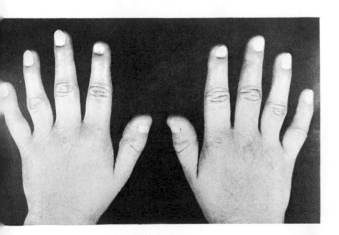

FIG. 25–4 *Absence of thumb nails. Although there is a depression in the region of the thumb nail, there is no nail of any kind. The trait is probably inherited as an autosomal dominant.* FROM H. H. STRANDSKOV IN JOURNAL OF HEREDITY.

syndactyly, does not extend to the tip of the toe, however. In some families this trait shows sex-linked inheritance. Another gene, which may cause this same condition in either sex, seems to be an autosomal recessive. In hammertoe, another inherited condition of the feet, one of the toes, usually the big toe, overlaps the others. The condition appears to be inherited as a dominant.

Swollen tips of the fingers and toes (clubbed fingers and toes) is another inherited condition. X-ray photographs show that there is a broadening of the bones of the terminal phalanges in this condition. Similar conditions arise as a result of pulmonary or heart trouble. Clubbed thumbs, where the thumbs are affected, but not the other fingers, is due to the influence of still another dominant gene.

Clubfoot results in serious deformity if corrective measures are not taken early in life. The most common type causes an inward turning of the foot so that the person must walk on the outer side of the foot. This is due to a shortening of the muscles or the tendons attached to the muscles on the inner side of the leg. Other types of clubfoot causes the afflicted person to walk on the front part of the foot, on the heel, or on the inner edge of the foot. Abnormal pressure on the embryonic foot will cause clubfootedness, but some family pedigrees show that it may result from heredity also. In some families it is inherited as a recessive, in others as a dominant with reduced penetrance. About one baby out of

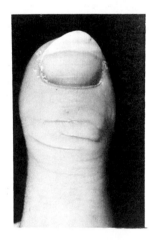

FIG. 25–5 *Clubbed thumb. In some persons only the thumb is swollen as shown here. The trait seems to be inherited as a dominant.*

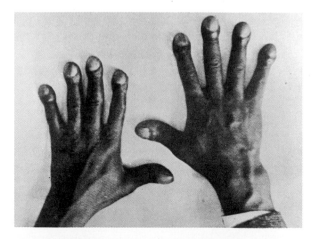

FIG. 25–6 *Clubbed fingers. This condition may arise as a result of certain cardiac diseases, but it may also arise from inheritance without any cardiac disease, as in the case illustrated above.* FROM J. T. WITHERSPOON IN JOURNAL OF HEREDITY.

every 1300 is born clubfooted. Some authorities estimate that over half of these cases are due to heredity.

Flat feet may result from environmental conditions, but in some families babies are born with flat feet, so that obviously the defect cannot be due to undue pressure during walking or standing. Since man is the only animal that has a definite arch to the foot, the flat foot represents a condition which is closer to that of other animals. It is apparently inherited as a recessive trait.

Handedness

The use of one hand in preference to the other is determined not by the greater development of the hand itself, but by the functional dominance of one side of the brain over the other. Since the nerve pathways cross in the brain-stem, a person who is right-handed would be left-brained, in the sense that the left side of his brain would dominate the right. Heredity definitely plays a part in this, and many family pedigrees indicate recessive inheritance of left-handed-ness. The fact that many monozygotic twins are opposite in their handedness indicates that something else can be involved — an effect called "mirror-imag-ing" appears to cause the right side of one twin to resemble the left side of his partner.

THE SKIN

Color

Skin color, like so many other human characteristics, is inherited through multiple genes working in conjunction with environmental agents such as sun-light. The differences in pigmentation which distinguish the white race and the Negro, however, seem to rest upon four to seven pairs of genes which are inter-mediate in their expression. The typical mulatto condition, with an inter-mediate shade of skin, results in the children of a marriage between members of the two races, though of course there is some degree of variation in the mulatto coloration due to various modifying genes which may be present in either of the two parent races. This topic is discussed in greater detail in Chapter 13. In Negroes a dominant gene has been observed which produces large spots of white on the skin, a **piebald** condition that prevails regardless of the genes for Negro pigmentation which are present. Also the recessive gene that produces **albinism** causes a very fair skin, no matter in what race it appears, because of the epistatic nature of this gene over the other genes for skin pigmentation. Keeler reports an interesting characteristic among the Cuna Indians on the isthmus of Panama, and calls it the **"moon-child"** character, since it causes a skin color which is paler than the normal and influences the pigmentation of the eyes as well as numerous other body characteristics. This trait seems to be inherited as a simple recessive and the gene for it is possibly an allele of the gene for al-binism, but there are no records of crosses between an albino and a moon-child to enable us to check on this point. There is another dominant gene which pro-duces unpigmented spots on the skin, a condition called **viteligo.** These spots are smaller than those produced by the piebald gene and tend to vary during the life of the individual, becoming smaller, then larger, and appearing on different parts of the body. Piebald spots, on the contrary, tend to remain constant. **Freckles** are the most common form of skin spotting. In freckles, pigment tends to accumulate in isolated little islands which become very prominent when darkened by exposure to sunlight. The non-pigmented areas between the freckles burn, but do not tan to any great extent. Freckles seem to be inherited as a dominant in some families.

The iris of the eye, the hair, and the skin all arise from the same embryonic layer, and genes which affect one of these body parts often affect all three. Blue eyes, blonde hair, and fair skin typically go together and characterize certain human races, whereas dark hair, eyes, and skin characterize other races. There are some genes, however, which may affect one of these body parts without influencing the others. Among the mixed descendants of mingled races blue eyes may sometimes be seen with dark skin and hair. This indicates that there are some genes which affect all three of these ectodermal structures, though other genes may influence each separately.

Skin Abnormalities

Xeroderma pigmentosum is an inherited condition characterized by an extreme sensitivity of the eyes and skin to light. A baby that inherits the gene complex for this condition will be born with normal skin, but exposure even to such light as would normally come into a room through the windows causes the development of a severe rash on the skin. As the child grows older certain regions of the rash frequently become malignant, and death nearly always results before maturity is reached. The gene for this condition is recessive, although freckles may show in the heterozygote.

Epiloia is a condition inherited as a dominant autosomal character. Externally the gene is manifested in a typical "butterfly" rash which appears on the skin of the face. The rash takes a form which roughly resembles the shape of a butterfly, with the nose as the body and the rash on the cheeks forming the wings. This condition is often accompanied by severe mental deficiency and epilepsy, and a tendency to develop tumors in the heart, kidneys, or other body parts. Most people with this affliction die before maturity, but in some the symptoms are sufficiently mild to permit the person to attain adulthood and to reproduce the character.

Ichthyosis congenita is inherited as an autosomal recessive lethal. A baby homozygous for this gene is born with a smooth, reddish brown skin, which is so thick that the child cannot nurse normally because of the firmness of the skin around its lips. Soon the skin develops deep, bleeding fissures which easily become infected, and the baby usually dies within a few weeks after birth.

Ichthyosis vulgaris is a much less severe condition which is characterized by the development of scales on the skin. The porcupine men of the circus usually have inherited this condition. In the northern European races the condition is inherited as a dominant autosomal trait with some degree of reduced

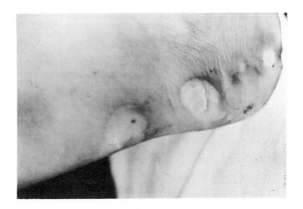

FIG. 25-7 *Epidermolysis bullosa. This disease is characterized by the formation of large watery blisters on the skin. This photograph shows some of the blisters on the side of the foot. This is the simplex form of the disease, which is inherited as a dominant.* FROM STURE M. JOHNSON.

penetrance, but in the Latin races a sex-linked recessive form of the condition is more common.

Psoriasis is a rather common skin disease in which red, scaly patches are found on the extensor surfaces of the arms and legs. The condition usually makes its appearance at adolescence. Pedigrees show definitely that it can be inherited, and some pedigrees indicate dominant autosomal inheritance, though the majority of them indicate recessive autosomal inheritance.

Epidermolysis bullosa is a rather rare skin disease characterized by the formation of large watery blisters on the skin. In the simplex form of the disease the blisters arise on the surface of the skin only and heal without scarring. In the **dystrophic** form the blisters are deeper in the skin and usually result in scars when they burst and heal. The simplex form is inherited as a dominant with some reduced penetrance. The dystrophic form may occur as an incompletely sex-linked recessive or in a dominant or recessive autosomal form.

Hyperelasticity of the skin is another skin abnormality which appears occasionally. It may be so pronounced that the skin can be pulled out from the body like a rubber sheet for 8 inches or more without pain. Upon release the skin returns to its normal position. This seems to be due to a deficiency of some of the connective tissues in the subcutaneous regions. One form of this character is associated with various other skin abnormalities collectively known as the **Ehlers-Danlos syndrome.** It seems to be inherited as an autosomal dominant.

The absence of sweat glands in the skin is a serious affliction in warm weather, for a person with this condition will be quickly overcome with heat prostration unless he keeps his clothing wet or spends his time in a bathtub or a swimming pool. Hogs are somewhat deficient in sweat glands, and for this same reason must have water in which to immerse themselves in warm weather. One form of this disease, **anidrotic ectodermal dysplasia,** is accompanied by a sparse growth of hair on the head and body and a deficiency of the teeth. In some families this is dominant; in others it is a sex-linked recessive. The mammary glands are modified sweat glands, and in some persons with this disease the nipples on the mammary glands are absent. In one family, this trait seems to be a sex-linked dominant.

THE SKELETON

Stature

There are about two hundred bones in the adult human body, and their size, shape, and arrangement to a large extent determine the body build. Body stature as a whole is due to the combined action of multiple genes and the environment. In the light of observed family pedigrees certain genes for large stature seem to be recessive. For while small parents sometimes have quite tall children, tall parents almost always have tall children. There are certain genes, however, which are epistatic to the entire group of multiple genes which influence stature in most persons. For instance, a person may inherit a series of genes for large stature, and yet may be a dwarf because of the effect of the genes at a single locus. **Chondrodystrophic dwarfism** seems to result from the influence of a dominant gene. The head and trunk of the person with this type of dwarfism are normal in size, but there is a great reduction in the length of the limbs, and this results in short stature. It is accompanied by a deformity of the long bones because of the inhibition of normal growth. The legs are usually set wide apart at the hips and curve inward as they extend down toward the feet, thus giving a somewhat bowed appearance. A similar condition has been achieved by selection in one breed of dogs. The daschund has short bowed legs and was bred for this character so that it could get into the burrows of foxes and badgers easily.

The **ateliotic dwarf,** on the other hand, is well proportioned but small in all parts of the skeleton. This characteristic is due to a deficiency of the growth hormone of the pituitary, which influences the growth of the skeleton. Such persons are commonly called midgets or "Tom Thumb" dwarfs. A similar condition exists in some mice, as a result of the homozygous condition of a recessive gene. The extensive evidence of Hanhart indicates that a recessive Mendelian factor is also responsible for this condition in man. There are some records (though of uncertain authenticity) that normal-sized children have been born of parents both of whom were midgets. If this is true, it indicates that the condition may arise as a result of recessive genes of at least two independent loci. In other words, in different pedigrees different genes might produce the same phenotypic effect.

Another type of dwarfism, **osteo-chondrodystrophy,** results from an overall irregularity in bone development which produces deformities of the trunk and limbs. Thus sometimes the limbs are of normal length, but the trunk is abnormally shortened. In some family pedigrees, this condition is inherited as a sex-linked recessive.

Bone Abnormalities

Some persons have such fragile bones that they suffer from repeated fractures no matter how careful they may be. A baby may fracture its leg when it is caught in the sheet as it is being picked up; a school girl may fracture her arm as she bumps against her desk; and a growing boy may have leg bones so fragile that he cannot stand, since his own body weight will cause them to break. Persons with this condition, known as **osteopsathyrosis,** usually suffer so many fractures that their bodies become somewhat deformed because of the imperfect knitting of the bones. This is regularly associated with a blue color of the sclera (white) of the eye, and with **otosclerosis** (a defect in the ear bones) which results in deafness. This is a good example of the multiple effects of a single gene. The condition is inherited as a dominant trait. It should be pointed out that as a result of a different gene otosclerosis may occur without the other symptoms, and not all cases of brittle bones are accompanied by otosclerosis and the blue sclera.

A few persons have the ability to fold their shoulders inward until they touch or almost touch under the chin, as shown in Figure 25–8. This is due to the absence of the clavicles or collar bones. This condition, **dysostosis cleido-cranialis,** is inherited as a dominant.

An abnormal curvature of the spine may result from various environmental influences, but certain types of curvature may be inherited. Lateral curvature seems to be inherited as a dominant. A looseness of the ligaments which hold the bones in place may result in various defects of posture, and also account for the ability of affected persons to perform unusual contortions. This also seems to be related to the condition of **orthostatic albuminuria,** which is characterized by the presence of albumin in the urine only when the person has been sitting or standing for some time. A study of persons with this type of albuminuria made by Beck and Glass showed that about 93 per cent of the cases also had hyperextensile joints and spinal deformity. This study indicated simple dominant inheritance with high penetrance, although more pedigrees are needed to establish the inheritance conclusively. The pleiotropic (multiple) nature of this gene is further indicated by the fact that 96 per cent of those with orthostatic albuminuria also had infected tonsils, 94 per cent had vasomotor disturbances, and 75 per cent had low blood pressure.

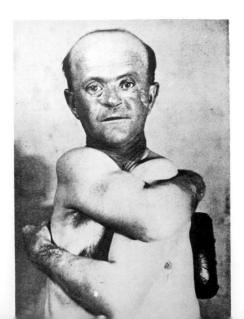

FIG. 25–8 *A folding man. This man has no collar bones (dysostosis cleidocranialis) and can fold his shoulders almost together. There is also a soft spot on his head where the bone is not completely ossified. These are partial expressions of a dominant autosomal gene.* FROM ARTHUR KELLEY IN JOURNAL OF HEREDITY.

An extremely hollow chest, a trait commonly known as "funnel chest," results from a deep depression in the breast bone. This also is a dominant.

Cartilaginous and bony outgrowths (**exostoses**) may appear on the bones during the growing period of life. They usually cease to form when maturity is reached, but those which have already been formed remain throughout life. The tendency to develop such outgrowths is inherited as a dominant characteristic.

Rickets is a well known and fairly common condition which results in varying degrees of bone deformity because of a deficiency of vitamin D in the diet. Studies of families living in similar conditions, however, indicate that the tendency to develop rickets is inherited. Studies on this disease, as it occurs in twins, made by Verschuer show a concordance of 88 per cent in monozygotic twins, but only 22 per cent in dizygotic twins. This indicates not only that heredity plays an important part in the susceptibility to rickets, but that the penetrance is very high. It appears that the susceptibility is transmitted as a dominant. Persons inheriting this tendency must have more vitamin D than other persons in order to prevent the development of rickets.

Spina bifida, a deformity in which the vertebrae fail to grow around the spinal cord, is a fairly common occurrence. It is caused by an incomplete fusion of the lateral halves of the neural groove in embryonic life. This may leave an opening in the back with an exposed spinal cord. Such a condition is usually fatal in infancy, but milder forms are more common. There is some evidence that this deformity is due to an intermediate gene, and the milder forms of the condition represent the heterozygous expression while complete exposure of the spinal cord is the homozygous expression.

There are many forms of **arthritis,** a condition which results in soreness and stiffness of the joints between the bones. In extreme forms there may be a complete fusion of the bones at the joints. Environmental factors, such as foci of infection, play a part in the development of this common disease which cripples so many people, but studies of family histories show that it is influenced by heredity. The susceptibility to some types of arthritis seems to be inherited as a dominant.

THE MUSCLES

Muscles and Body Build

While the skeleton determines body stature and general posture, the muscles play an important part in the general shape of the body. The genes for muscular development as a whole are sex-limited in their expression. The muscles of a woman usually contain heavier deposits of fat than those of a man, and the body of a woman has a heavier deposit of subcutaneous fat, with the result that the muscles do not usually appear hard and knotty even though she may exercise vigorously. The development of muscles is largely dependent upon physical exercise, but the placement of the muscles on the body, and the extent to which they can be developed through exercise, are inherited. In many families the placement of the muscles on the hips and shoulders gives the female members a characteristic shape. A different type of body shape may show in the males. The shape may be influenced by diet, of course, as well as by glandular disturbances which cause abnormally large or abnormally small deposits of fat among the muscles and in the skin. The tendency toward obesity and its reverse, extreme thinness, are inherited.

Muscle Abnormalities

Pseudohypertrophic muscular dystrophy is a serious affliction which is characterized by the gradual wasting away of the muscles, beginning in childhood and usually resulting in death in the early teens. Since the trait is dependent upon the influence of a sex-linked recessive gene, it has never been known in girls, for a girl would have to receive the gene from both parents, and boys with this trait do not normally reach maturity and so rarely have the opportunity to produce offspring.

Peroneal atrophy is characterized by a progressive wasting away of the calf muscles, usually beginning sometime between the ages of 10 and 30. In some family pedigrees the trait is definitely inherited as a sex-linked recessive, in others it is an autosomal dominant, and in a very few cases it seems to be an autosomal recessive. In the members of some families the muscles of the forearm and hands are also affected. The recessive autosomal form of peroneal atrophy is more severe than the dominant form, and the recessive sex-linked form is more severe than either of the other two.

In some persons there is a complete absence of the long **palmar muscle** of the forearm. This deformity can easily be detected, since the muscle is connected with the hand by a prominent tendon that runs down the central region of the wrist. When the hand is tightly flexed, this tendon is clearly visible on the wrist if the muscle is present at all. The muscle is missing in about 20 per cent of the white population of the United States. Its absence seems to be inherited as a dominant with a slight reduction of penetrance in some cases. The absence of this muscle causes no handicap in the use of the hands and arms.

Heredity also plays a part in the occurrence of **inguinal hernia.** About one man in each twenty-five has such a hernia and about one woman in each one hundred and fifty. Rupture is often thought to result from undue strain on the muscles of the abdomen, and such may be the immediate cause, but family histories show the importance of heredity in this matter. The testes of a man are within the body before birth and must descend through an opening in the abdominal muscle shortly before or shortly after birth. These muscles then normally grow together to seal most of this opening. In some cases, however, the opening remains rather wide and, upon undue strain, a portion of the intestine may descend through the opening to produce inguinal hernia. Such a weakness of the muscles of the abdominal wall seems to be inherited as a dominant trait with some sex-limitation. A woman does not need a large canal for the descent of testes, and her abdominal muscles grow together in this region with much less likelihood of weakness.

THE NERVOUS SYSTEM

We have already mentioned many of the inherited variations in the nervous system under other headings. Vision, hearing, and taste are senses which are intimately connected with the nervous system, and some of the abnormalities of the muscles which we have studied have a nervous origin. On the other hand, many of the defects of the nervous system which we will describe in the following pages, may actually have their origin in variations in other systems, such as the endocrine system. Our method of grouping, therefore, is somewhat arbitrary and is arranged more nearly according to the affected body parts than the ultimate cause.

Mental Ability

Of all human factors influenced by heredity, one of the most important for the future of mankind is intelligence, and yet intelligence is so complex and variable, so much a composite of many talents and aptitudes, that we cannot even define it properly. Nevertheless, there is a variety of ways in which special aptitudes and general intelligence can be tested, and some of these have proved extremely useful in genetic studies. The most broadly useful of these tests is based on what is called the **intelligence quotient,** or I.Q., which is little more than an approximate measure of general mental ability, but a useful one nevertheless. A person with a rating of 100 on the I.Q. scale is considered to be of approximately average intelligence, a rating of 130 or higher is clearly supernormal, a person with a rating of 70 down to 50 is classified as feeble-minded, from 50 to 20 as imbecile, and persons with scores below that point are rated as idiots. Numerous other tests have been devised for such special talents as musical, mechanical, or mathematical aptitude, verbal facility, and others, but the I.Q. tests have provided us with more data than any of these. It has been found by tabulations of the I.Q.'s of large numbers of persons that general intelligence in our population tends to follow the bell-shaped Gaussian curve, with the largest numbers concentrated near the point of the average, and smaller numbers ranging toward the two extremes.

General Intelligence

Moreover, enough evidence has been accumulated to show that the extreme variations in mental capacity among human beings are partly hereditary and partly environmental. Training, of course, plays an important part in bringing out inherent potentialities, but even among persons with similar training there are great variations in general intelligence, special aptitudes, emotional characteristics, and so on. Since these variations are in general continuous rather than clear-cut or discrete, we may conclude that multiple genes are involved, even though, as with body stature, there are many cases in which variations in a single gene may have a pronounced effect on mental ability. Thus a person may be born with a whole gene complex for a brilliant mind, yet also inherit a single pair of recessive genes—for amaurotic idiocy, for instance—which would be epistatic to the entire complex and thus make the person an idiot.

While the whole problem of the inheritance of mental characteristics is extremely complicated, a good deal of light has been thrown in recent years on the relative significance of hereditary and environmental factors in determining mental capacity. In general, bright parents tend to have bright children, and dull parents, dull ones, though this statement should be hedged with qualifications. Two studies by Freeman, Holzinger, and Mitchell of American children reared in foster homes and institutions showed that most of the children with higher I.Q.'s had fathers of presumably higher I.Q.'s as far as this could be judged by their occupations. On the other hand, children reared in foster homes rated as good ranked higher than those reared in homes rated average or poor. Another study, made by Lawrence in England, was based on two groups of children, one from an orphanage where the environment was uniform, the other from an elementary school where the children's home environments were varied. In both groups, the children were rated according to the occupations of the fathers, which ranged from highly skilled and professional work down through unskilled. The spread of intelligence in both groups was about the same, though slightly greater in the elementary school, suggesting the possible effect of varied environments. In both groups most of the children of highly skilled workers and professional men ranked above the 100 I.Q. level, though some were much lower; and the fact that more children of this type from the elementary school were in this lower group than from the institution indicates the probable depressing effect of poor home environments. In sum, a good environment seems to elevate the I.Q. of children and a bad one to depress it, but there is a definite correlation between the occupation of the fathers and the I.Q.'s of the children.

On the other hand, the children of parents at either extreme of the intelligence-rating scale will show a tendency to regression toward the mean of the population as a whole. This is a general tendency which is found in all forms of life where characteristics are due to multiple genes. The children of high I.Q. parents will generally show a distinctly higher I.Q., however, than the children of low I.Q. parents. To illustrate, Penrose found that a group of men in highly skilled clerical occupations had a mean I.Q. of 117.1 and the children of these men had a mean I.Q. of 109.1. On the other hand, a group of unskilled laborers had a mean I.Q. of 86.8 and their children had a mean of 92.0. Of course, the degree of regression in both these groups may be greater than would be expected if the I.Q. of both parents was considered, for it is quite possible that the mothers of these children were nearer the mean for the population as a whole than their husbands.

A number of studies on the problem of intelligence have been conducted by collecting data on numerous sets of twins, both monozygotic and dizygotic, where one or both members of the pair show the same mental disease or other characteristic, and then by working out the concordance (that is, the corrected expectancy rates) for both types of twins. By this method it has often been possible to conclude that a given mental trait, such as feeblemindedness or manic-depressive psychosis, is strongly influenced by hereditary factors, and to discover also whether environmental factors are extensively involved. More will be said about studies of this kind in the following pages.

Other studies, dealing with the general intelligence of twins, indicate a high correlation in intelligence between identical twins, though the correlation is higher for twins reared together than for those reared apart. Newman found a correlation of .881 for a group of the former, and of .670 for a group of the latter. The inference of this is not surprising — that heredity is a major factor, but that a difference in environment also has its effect. Newman also found that while identical twins differ from each other by an average of only 5.9 points on the I.Q. scale — a difference no greater than might appear in the scores of the same individual on two different tests — dizygotic twins differ by nearly twice that much, or 9.9 points, and sibs of like sex by 9.8 points. Surely this very close correlation of twins with exactly the same inheritance, and fairly close correlation of others with half the same inheritance, are clear evidences of the importance of heredity in mental capacity. Interestingly, too, the correlation between one-egg twins reared apart is greater than that between two-egg twins brought up together — which would suggest that sharing all instead of half the parents' genes has slightly more influence on the intelligence factor than environment has.

Special Aptitudes

Special aptitudes, as distinct from over-all mental capacity, are independently inherited and may have no close correlation to general intelligence. This phenomenon is well-illustrated by the interesting case of a woman moron with an I.Q. of about 60 who could never learn to tell time by a clock, but could play the piano very nicely by ear. Although low in over-all intelligence, this woman possessed a musical ability which exceeded that found in most persons of normal mentality. Some family histories show the definite transmission of a gene complex for musical ability, although too many factors are involved to allow the identification of any single gene which is responsible. Some parents give their children musical training year after year in the determination to make musicians out of them, but if the inherited potentiality is lacking, the training cannot do much. Other children develop into outstanding musicians in spite of the fact that they may get little formal training in music. Similarly, abilities in mechanics, mathematics, drawing, painting, sculpture, and other special fields seem to be inherited individually and with no close correlation to general intelligence.

Mental Retardation

As we have said, the intellectual capacity of human beings shows a continuous type of variation, as measured by the I.Q. tests, and mental retardation or feeblemindedness is not a discrete category but an arbitrarily determined range within the I.Q. scale — from 50 to 70, as we have said. Again studies of twins shed valuable light on the relative parts played by heredity and environment in the production of this defect. Among 202 pairs of monozygotic twins studied, in which one or both members were mentally retarded, it was found that the concordance, the corrected expectancy rate, was 93.6 per cent. And among 237 pairs of dizygotic twins, the concordance was 50 per cent. Thus once more we see the importance of heredity in the intellectual make-up. Interestingly enough, however, the concordance for sibs is only about 16 per cent. All these figures taken together would indicate that in addition to the hereditary factor, there may be something in the circumstances of a twin pregnancy which has some influence on the production of mental retardation in the children. As an example, there could be a lowered oxygen concentration due to the crowding of the two embryos at a critical stage of brain development.

The influence played by heredity in mental retardation would appear to be due to a number of genes in most cases. Therefore, there are some marriages between persons of inherited subnormal mentality that result in children of normal intelligence because of the gene segregation. Likewise, normal or even superior parents may occasionally produce children who inherit mental retardation because of this gene segregation.

MENTAL DISORDERS

In addition to general intelligence, which of course is a broad function of the nervous system as a whole, a number of mental disorders and other nervous defects are in varying degrees hereditary. Many defects of the nervous system may actually have their origin in variations in other systems, such as the endocrine. A few mental disorders can even be traced to a single gene. Let us briefly consider some of those about which most is known.

Amaurotic Idiocy

This is a type of mental defect which exists in two forms. In the infantile form, the child is normal at birth, but symptoms of the disease begin to appear within several months. There ensues a gradual decline in mental ability, impairment of vision leading to blindness, convulsions, progressive muscular weakness, and emaciation. Death usually comes before the second birthday. There is also a juvenile form of the disease which does not begin to have its effects until a child is around 6 or 7 years of age. At this time there begins a progressive loss of vision and this is followed by mental deterioration. Muscular incoordination then develops, and the muscles gradually waste away. Finally there is almost complete mental obliteration, and death usually occurs before the afflicted individual reaches 21 years of age. Each of these forms of amaurotic idiocy seems to result from the action of autosomal recessive genes. There are two distinct genes, one for each form of the disease, and apparently the two are not alleles.

Since amaurotic idiocy is relatively rare, is recessive, and is easily recognized, it serves very nicely to illustrate the effect of consanguineous marriages on the expression of harmful human characteristics. Studies in Sweden have shown that the incidence of cousin marriages among the parents of affected children is about 15 per cent, while the incidence of cousin marriages among the general population is less than 1 per cent. It is also interesting to note that the gene is more frequent in some races than in others. Among the European Jews, for instance, it has a far higher incidence than among European non-Jews. This does not mean, of course, that the Jews carry any more harmful recessive genes than other groups; in fact, there is evidence that other Europeans carry more genes for susceptibility to tuberculosis than do the Jews. Rather, the incidence of this trait serves to illustrate the fact that there may be variation in the different population groups in the frequency of any particular gene, be it harmful or beneficial.

Huntington's Chorea

This disease, which is one form of Saint Vitus's dance, is characterized by an uncontrolled twitching of the voluntary muscles of the body accompanied by mental deterioration. Unfortunately, from the standpoint of posterity, this condition usually does not develop until a person is in his thirties and may already have had children. It is inherited as a simple autosomal dominant, and hence about one-half the children of an afflicted parent will develop the disease in the course of time, as illustrated in the pedigree shown in Figure 25–9. Because of the serious nature of this nervous affliction, no person with a parent, brother, or sister who has developed it should have children.

FIG. 25–9 *Pedigree of the inheritance of Huntington's chorea. This severe nervous affliction usually does not appear until maturity, after the person has had the opportunity to have children.*

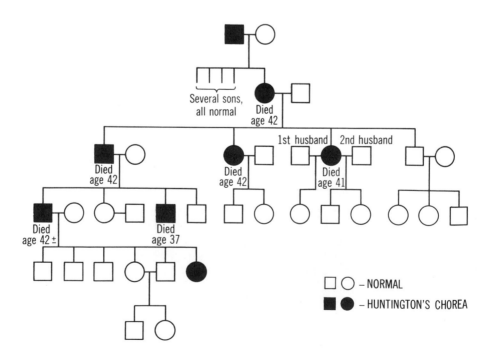

Schizophrenia

This is one of the most common mental diseases and may to some extent affect as much as 1 per cent of the population. Probably more than half the persons who must be kept in mental hospitals suffer from this disease. It manifests itself in varying degrees and ways, but it is always characterized by a tendency to retire from the world of reality, and in some forms there is an almost complete insensibility to surroundings. The onset of this disease is often accompanied by some mental stress. For this reason many think that it is environmentally induced, but repeated studies of family pedigrees indicate that there must be an hereditary predisposition before a person can develop the disease. If one member of a pair of identical twins develops the disease, it is very likely that the other member will develop it at about the same age, even though the two are living apart and in somewhat different environments. According to recent tabulations, the concordance in monozygotic twins is 86 per cent, while in dizygotic twins it is only 15 per cent. The question might be raised whether the greater concordance among monozygotic twins might be due to prenatal factors. It has been shown that monozygotic twins have a somewhat smaller chance of surviving the early embryonic stages than the dizygotic twins. There is no evidence, however, to indicate that there is sufficient difference in prenatal environment to alter the incidence of this condition among the two types of twins.

One plausible theory to explain the role of heredity in schizophrenia is concerned with the blood chemical, serotonin, which is known to be required for normal brain functioning. Those persons who are homozygous for a certain gene may produce a chemical antagonistic to serotonin when they are under great emotional stress. The reduced serotonin level of the blood brings about hallucinations and other symptoms of schizophrenia. A transfusion of blood from a schizophrenic will bring on temporary schizophrenic symptoms in a normal person. Also, injections of a chemical antagonistic to serotonin in monkeys brings on schizophrenic-like symptoms. Thus, a person inherits the potentiality of becoming a schizophrenic; without undue stress he may never show the symptoms, but under stress the chemical is produced and the symptoms appear. Persons heterozygous for the gene may develop a mild form of schizophrenia known as a schizoid personality, thus indicating that the gene has some degree of intermediate expression.

Phenylketonuria

This is a form of idiocy or imbecility resulting from a metabolic block in phenylalanine metabolism. A recessive autosomal gene is responsible (see Chapter 17 for a detailed discussion of this condition).

Manic-depressive Psychosis

Almost all persons are likely to have alternate moods of depression and elation, but the peaks of the two extremes are much more pronounced in some than in others. In its most extreme form this becomes manic-depressive psychosis and persons suffering from extreme depression often commit suicide to terminate their mental sufferings. This disease, which is more common in women than in men, is probably due to endocrine disturbance rather than to any actual defect of the nervous system, since no evident pathological condition of the brain is discoverable by autopsy. There is no question about the influence of heredity on this mental condition. Pedigrees show the tendency for recurrence of the illness in families and one study shows a concordance of 84 per cent among pairs of monozygotic twins, but of only 14.3 per cent among dizygotic twins. Another study showed even higher figures: 96.7 and 26.3 per cent, respectively. The slight discordance among the monozygotic twins indicates that an environmental agent must be involved. A dominant gene seems to be involved, with its penetrance dependent upon this environmental agent.

Senile Dementia

Many persons have normal or brilliant minds during the greater part of their lives, but in old age show a progressive deterioration of mental faculties which may lead to complete loss of normal mentality. This condition is probably due to the degeneration of brain tissue which is a part of the general degeneration of the body organs which occurs in old age. There is a certain degree of such degeneration in most old persons, but it is much more pronounced in some than in others, and the tendency to develop pronounced dementia is influenced by heredity. Kallmann made a study of a large number of twins in New York City and found a concordance of 42.8 per cent among 33 pairs of monozygotic twins over 59 years of age. Of 75 pairs of dizygotic twins over 59 which were included in his study, there was only 8.0 per cent concordance. This indicates the influence of heredity, but as with feeble-mindedness, it is not possible to trace the condition to any one gene.

OTHER DEFECTS INVOLVING THE NERVOUS SYSTEM

Epilepsy

This disease of epilepsy is characterized by sudden seizures known as epileptic fits, which in the most extreme form run to unconsciousness and muscular spasms. Some persons have the disease in a milder form in which the fits are minor. Brain injury is known to be an environmental agent which can induce the onset of epilepsy, but the majority of cases arise without such injury and have an hereditary basis. Many pedigrees have been collected which establish beyond question the importance of heredity in this condition. A great advance in our understanding of the disease has been made possible through the invention and use of the electroencephalograph, a machine which records the electrical brain waves. The pattern of waves in most persons shows a fairly regular rhythm, but all epileptics show great irregularity in the waves, though some persons who are not epileptics show the same kind of irregularity. Sufficient studies have been made to indicate that the irregular waves are inherited as a dominant trait. Perhaps all persons who show the irregularity are potential epileptics and those who do not show the condition lack some other factor, either a modifying gene or some environmental factor required for the expression of epilepsy. The condition is frequently found in pedigrees showing various forms of feeblemindedness, and epileptics are frequently subnormal mentally, but epilepsy is definitely not incompatible with normal mentality. Some highly gifted men, such as the novelist Dostoevski and the painter van Gogh, were epileptics.

Spinal Ataxia

The hereditary form of this disease results from a degeneration of the sensory neurons in the spinal cord, and with loss of sensation from the muscles comes defective control of the muscles, so that the afflicted person has difficulty in maintaining his equilibrium. He sways when he stands and staggers when he walks. As the affliction progresses, the person may lose all power of purposeful action through inability to control his voluntary muscles, and may become a helpless invalid. In most cases the disease is inherited as an autosomal recessive, but there are a few pedigrees which show autosomal dominant inheritance. In the dominant form of the disease the degeneration seems to be in the brain rather than in the spinal cord.

Spastic Paraplegia

This disease also is characterized by loss of control over the muscles of the body, so that simple tasks are performed with great difficulty and if the affected person is able to walk at all, it is with a shuffling gait. Mental retardation may accompany the condition. There is a recessive form of the disease which has its onset at about 11 years of age and a dominant form which begins at about 19.

Hypertrophic Neuritis

This is another disease which develops between birth and maturity. It involves an enlargement of the spinal nerves followed by ataxia of the arms and legs and eventual muscular atrophy. Recessive genes seem to be involved in its inheritance.

Shaking Palsy

This disease, *paralysis agitans,* usually develops between about 50 and 60 years of age. There is a gradual loss of control of the muscles and tremulous motions replace purposeful movements. The hands held in the lap may begin to tremble and then to shake violently, going completely out of the control of the will of the afflicted person. Some family pedigrees show dominant inheritance and others indicate a recessive form of the disease.

THE BLOOD

The inheritance of the blood types and other variations due to the antigen-antibody constitution of the blood cells and plasma has already been discussed in Chapter 12. In the present chapter we shall merely list some of the better known inherited characteristics of the blood.

Anemia

Anemia is a blood characteristic which may have many different causes and many different expressions. In any form of anemia, however, there is an insufficient amount of hemoglobin in the blood to transport oxygen in the quantities needed for the best performance of the cells of the body. In **pernicious anemia,** there is an insufficient number of red blood cells to accomplish this purpose because these cells are not produced by the red bone marrow rapidly enough to supply the body needs. The trouble lies in the deficiency of an antianemic factor formed in the stomach and involves the utilization of vitamin B_{12} and other food factors. The factor is stored in the liver and liver extract can be used to supply it. A recessive gene with variable expressivity can cause a low uptake of vitamin B_{12} and lead to the anemia.

Hemolytic icterus, another form of anemia, is much less common than the pernicious type. It seems to result from a hyperfragility of the red cells, which causes their excessive destruction by the spleen. This causes the spleen to hypertrophy because of the excessive labor it is called on to perform which, in turn, leads to increased destruction of red cells. Thus a vicious cycle is established. The condition is dominant but shows variation in expressivity and may be corrected by removal of the spleen. Death frequently comes when this disease reaches an acute stage, and removal of the spleen before the condition is critical is a necessary precaution when early symptoms of the disease are detected.

Thalassemia is a special type of anemia which appears in infancy or childhood. It is accompanied by an enlarged spleen and liver, abnormally shaped red blood cells, and an excess of leucocytes in the blood. The severe form of this disease results from the homozygosity of a gene for the condition, but a milder form of the disease results when a person is heterozygous. Thus we would class this gene as intermediate in dominance. The disease seems to be confined to those who are of Syrian, Greek, or Italian ancestry, and provides another example of the irregular distribution of genes in population groups. About 4 per cent of the Italians in New York City who have ancestry from southern Italy have the minor form of this disease. The high frequency in certain areas is related to a degree of protection from malaria conferred on the heterozygote. This protection is similar to that found in the sickle type of hemoglobin which was discussed in Chapter 18.

During the phase of World War II which took place in the South Pacific malaria was a major problem. Denied our primary source of quinine by the advancing Japanese, our search for new drugs to prevent and treat malaria resulted in the synthesis of a drug known as primaquine which proved to be very effective. Some people who took this drug, however, developed a severe anemia. Chemical studies of the blood of these people have revealed a deficiency of a certain enzyme in their red blood cells. This enzyme, glucose-6-phosphate dehydrogenase, is concerned with the breakdown of glutathione, an intermediate product in cellular respiration. The deficiency of this enzyme is rare among whites, but almost 10 per cent of the American Negroes have it. This fact indicates that the deficiency must have some selective advantage in the environment of the Negro in Africa, possibly a reduced susceptibility to malaria. The G-6-PD deficiency is caused by a sex-linked gene which shows some degree of expression in heterozygous females. Ordinarily the deficiency causes no detectable effect on the individual, but, when certain substances are taken into the body, there is a sudden destruction of many red blood cells and the severe anemia results. Sulfanilamide, primaquine, and naphthalene (used in moth balls and certain moth sprays) can cause this red cell destruction. Also, the intake of the common broad bean (lima bean) in a raw condition, or even inhaling the pollen from plants producing this type of bean will bring on the anemia. Gradual recovery comes about when the offending substance is removed. Many more males are affected than females, but some heterozygous females show some degree of anemia when exposed to the substance involved.

Other Blood Abnormalities

Normally red blood cells are circular in outline, but in some families there may be a high percentage of oval-shaped cells. This condition, known as **ovalocytosis,** seems not to interfere with the oxygen-carrying capacity of the cells, for such persons do not show any symptoms of anemia from this cause. The condition seems to be inherited as a dominant with variations in expressivity. In some persons the capillary walls of the kidneys rupture easily and allow blood to filter into the urine. The disease, **hemorrhagic nephritis,** is often accompanied by albuminuria. A dominant gene seems to be responsible. **Hemophilia,** a disease in which the blood fails to clot properly, has been studied extensively in Chapter 10. **Telangiectasia** is a disease characterized by an extreme fragility of the capillaries of the skin of the face and mucous membranes of the mouth and nasal cavity. Nose bleeding is usually the most severe symptom, but this is often accompanied by red spots on the face, tongue, and lips due to small hemorrhages under the skin in these regions. Sometimes the skin may break and serious bleeding may result. Pedigrees indicate autosomal dominant inheritance of the condition.

In many elderly persons the veins of the legs bulge out under the skin because of the decline in the elasticity of the walls of the veins. Environment plays a part in the development of these **varicose veins,** and heavy lifting and long hours of standing tend to accentuate the condition. Many persons never develop it, however, even though they may be in occupations which place a strain on the veins of the legs. Family histories indicate a dominant inheritance of the condition with incomplete penetrance. Some who inherit the predisposition for varicose veins do not develop them because they avoid the conditions which would cause them to develop.

Hemorrhoids are caused by varicose veins in the walls of the rectum and may occur independently of the condition in the veins of the legs. Again this is a condition which is influenced greatly by environment. Childbirth brings on hemorrhoids in most women, but these usually disappear in a short time. Occupations requiring long hours of standing or sitting tend to cause congestion of the blood in the rectal region, and will bring out hemorrhoids if there is an inherited predisposition for them. The predisposition seems to be inherited as a dominant character.

High blood pressure is a rather common affliction among people beyond middle age, but may be present in youth. Serious abnormalities are likely to accompany it — kidney trouble, heart trouble, and apoplectic stroke, to name a few. Like many other human characteristics, hypertension or high blood pressure is greatly influenced by environment, but its tendency to recur in families strongly suggests an heredity background. Lenz made an extensive study in Germany and concluded that the predisposition for the development of high blood pressure is inherited as a dominant. There are perhaps a number of genes which contribute to the nature and severity of the symptoms.

DISEASE

The word "disease" is a very broad term which may embrace almost any type of body abnormality. We have already listed many human diseases and discussed their relation to heredity under other headings in this chapter. These all come under the heading of non-infectious diseases, for disease germs are not involved in their incidence. The following list of inherited human traits will mention some of the non-infectious diseases not already touched upon and will discuss the relation of heredity to the infectious diseases.

Disease Due to Endocrine and Metabolic Disturbances

Diabetes mellitus is one of the common diseases which result from an endocrine disturbance. Persons with this disease produce an insufficient amount of insulin in the pancreas and sugar metabolism is defective as a consequence. Excess sugar accumulates in the system and is excreted in the urine. Diabetic coma and death may result if proper treatment is not administered. Large numbers of pedigrees of families which show this disease have been collected, and the influence of heredity upon it is not to be questioned. A predisposition to become diabetic seems to be inherited as a recessive, but the development of the disease may be avoided if the intake of carbohydrate foods is moderate. No doubt, other genes influence the degree of predisposition.

Diabetes insipidus is characterized by the production of large quantities of a very dilute urine accompanied by an extreme thirst because of the loss of water through the kidneys. This condition is due to an insufficient secretion of a hormone from the posterior lobe of the pituitary gland. This hormone (pitressin) regulates the reabsorption of water in the renal tubules. When it is deficient, an inadequate amount of water is reabsorbed and in consequence an excessive amount of water is eliminated in the urine. Most family pedigrees indicate dominant inheritance of this disease, but with some reduction in penetrance.

Gout results from a perversion of purine metabolism resulting in excessive production of uric acid. Persons with this disease have an abnormally high uric acid content in the blood, there are attacks of acute arthritis, and sometimes chalky deposits form in the cartilages of the joints. However, many persons have **hyperuricemia** (high uric acid content of the blood) who do not have gout. A certain threshold must be reached before the uric acid is sufficient to cause the disease. Hyperuricemia is inherited as a simple autosomal dominant, but only about 10 per cent of those with hyperuricemia reach the threshold necessary for the production of gout. More men develop gout than women because men normally have a higher uric acid content of the blood than women and with defective purine metabolism it is more likely that they will reach or exceed the threshold and develop gout.

Allergic Diseases

There are a great number of diseases which come under this heading. Hay fever, asthma, cyclic vomiting, migraine headache, hives, eczema, and colitis are some of the diseases which may have an allergic basis. All allergic conditions result from a response of the body to the presence of a foreign antigen. Antibodies are produced which react rather violently when a sensitized person is exposed to the antigen. People vary considerably, however, in the ease with which they become sensitized to foreign antigens. All of the people living in Indianapolis inhale about the same amount of ragweed pollen during the latter part of August and September, but only a certain proportion of them become sensitized. Those who do may react to the presence of the pollen in the air by severe hay fever symptoms or bronchial asthma. Heredity seems to be a primary factor in determining which people become sensitized and which do not. Thus, no one inherits hay fever or antibodies to any particular foreign antigen (with the exception of the **AB** blood type antigens), but rather inherits a capacity to become sensitized easily. A parent may have hay fever, one child may have allergic asthma, and another may develop an extensive skin rash every time he eats eggs. All of these could result from the same inherited tendency to become easily sensitized. One interesting theory advanced by Wiener, Zieve, and Fries holds that persons who develop allergies early in life are homozygous for a gene that makes them easily sensitized. Those who develop allergies after puberty, on the other hand, are heterozygous for this gene and those who do not become easily sensitized are homozygous for the normal allele. There is a form of migraine which is inherited as a dominant and it does not have an allergic basis.

Cancer

In spite of the great amount of scientific research which has been done on this disease there is still much about it which is not known. There seems to be little doubt that heredity plays a part in the predisposition to cancer. Not only is the tendency to develop cancer inherited, but twin studies show that the age at which cancer develops, the type of cancer which develops, and even the particular body organ which is affected are influenced by heredity. Monozygotic twins showed a concordance of 61.3 per cent for the development of cancer while dizygotic twins showed a concordance of 44.4 per cent. When analyzed for the particular type of cancer which developed, it was found that 58 per cent of the monozygotic twins had the same type of cancer, but only 24.2 per cent of the dizygotic twins had the same type of cancer. Multiple gene inheritance seems to be indicated.

Infectious Diseases

These are the diseases which are caused by germ invasion of the body. Since the germs are the actual agents which cause these diseases, it is not possible to inherit an infectious disease as such. Still, we find certain infectious diseases appearing repeatedly in certain families, even though the members of the family may not have contact with one another. This indicates an inherited susceptibility to specific infectious diseases. Tuberculosis is an excellent disease to illustrate this point. The germs of this disease are very prevalent, and few among our population escape some degree of infection during their lives. In some persons, however, the disease develops rapidly and quickly causes incapacitation and sometimes death. Many environmental agents may influence susceptibility to infection, but studies of families living under similar conditions show variations which are almost certainly due to heredity. The studies of twins which were discussed in Chapter 21 show the importance of heredity on the susceptibility to infection of tuberculosis. Extreme susceptibility is believed by some to be inherited as a recessive.

Poliomyelitis is a disease which can strike at certain families in a region, although large numbers of persons are unaffected. It is believed that an inherited susceptibility is a major factor in the incidence of this disease. A study made by Addair and Snyder in McDowell County in West Virginia indicates that the susceptibility to this disease is inherited as a recessive with a penetrance of about 70 per cent. This study showed that only twenty-nine persons contracted the disease among a group of families which should have included about forty, who were homozygous for the recessive gene for susceptibility.

Studies of pedigrees showing susceptibility to **diphtheria** also indicate a recessive gene for the susceptibility to this serious disease of childhood. The same is true of **scarlet fever.**

26

Eugenics

EUGENICS DEALS WITH THE APPLICATION of the laws of genetics to the improvement of the human race. We have accomplished remarkable improvements in domestic animals and cultivated plants (mainly improvements which benefit mankind) through the application of these laws. Can we not apply some of these principles to man and achieve a betterment of mankind? Theoretically, it should be possible. But when such a plan is considered we run into many difficulties. Imagine the consternation which would be evinced if we asked a rancher to maintain and improve a herd of cattle, but with the following restrictions: First, that he must allow all animals to reproduce freely according to their own inclinations. Second, that whenever any weak or deformed calves are born they must receive special care to keep them alive and must be allowed to reproduce if they are able. Third, that each bull must be allowed complete freedom in the choice of a mate, but, once he has chosen, he must not be allowed to mate with others. Fourth, each mated pair should have complete freedom either to limit reproduction or to produce as many offspring as they choose. Ridiculous as such a proposal might sound with respect to animal breeding, this is essentially the problem which faces eugenists who would improve the human species through the application of genetic laws. Of course, the third restriction is not in actual fact as binding as the above statement of it implies; nevertheless, those changes in mates which do occur are made without regard for the eugenic value to the race. In spite of all these restrictions, however, it is possible that we may do something with the factors which are under our control and which, even though they may not bring about any improvement of the human species, may at least prevent its deterioration through dysgenic forces.

The concept of eugenics was founded by Sir Francis Galton, who defined it as "the study of agencies under social control that may improve or impair the racial qualities of future generations, either physically or mentally." Most of the agencies concerned with the improvement of mankind are dedicated to social welfare which is designed to improve the condition of the living generation. Our country abounds in educational, medical, social welfare, and religious institutions which function on government funds and through the gifts of philanthropic individuals who are interested in the betterment of man. Very few exist whose primary concern is with the future welfare of mankind. Many persons, who do not understand the stability of the genes and the principles of selection, believe that an improvement of the environment of the present generation will be reflected in an improvement of future generations. This is a carry-over of the views of Lamarck regarding the possibility of inheritance of acquired characteristics. Unfortunately, we know that this theory is not in accord with the facts; the physical and mental benefits which are acquired by a man during his lifetime are interred with his bones — each generation must start afresh — and a good thing too, for the burden of acquired disabilities and bad habits would far outweigh the value of acquired abilities of a desirable type.

THE STATUS OF NATURAL SELECTION IN MAN

In our study of genetics we have emphasized the importance of natural selection as a purging agent in the species, weeding out the less fit and perpetuating those best adapted. So far as we know, there are no exceptions to this rule in all the forms of life which have been studied. We might well be concerned, therefore, with the status of natural selection in the human populations of today. Among the many human inhabitants of the earth, natural selection operates in the same fashion as with other forms of life — it carries on the same cruel yet efficient elimination of the great majority of the children which are born. In much of Asia and the nearby islands, where the great majority of the world's population is concentrated, there is practically unlimited reproduction coupled with a high rate of elimination. The standard of living of the majority of these peoples is pathetically low according to ours but, on the other hand, there is no grave eugenic problem. Among the people of North America and Europe, however, with their highly complex industrialized societies, conditions are somewhat different. Here factors are operating which tend to reduce the effective action of natural selection. Let us consider some of these.

The Development of Medical Science

There have been many phenomenal advances in medical science during the past century, and every year new discoveries are made which increase man's chances of survival. Many of the dread germ diseases which took such a heavy toll of life in the past are now controlled through the use of antibiotics. Indeed, many who read these words would not now be alive but for these drugs. Yet, we know that the susceptibility to many diseases is inherited, so that by saving susceptible individuals we allow genes that would otherwise be eliminated by natural selection to continue propagating.

Preventive medicine also plays its part. Through vaccination, quarantine, sanitation, prophylaxis, and other public health measures the dangers of infection from the serious germ diseases are greatly reduced. Many in our population may inherit a high susceptibility to bubonic plague, yet they are able to live and reproduce because the vigilance of our public health authorities prevents the entrance and spread of the germs of this dread disease in our population. Other persons may be susceptible to smallpox, but vaccination and quarantine have practically eliminated the virus of this disease within our borders.

In a similar fashion, advances in surgery permit many inherited body deformities to survive and be propagated that otherwise would be eliminated. For instance, a child may be born with a defective heart which permits unoxygenated blood to be mixed with oxygenated before it is pumped through the body. Without surgery such a child would very likely die sometime in the first few years of its life, yet a delicate operation can correct the defect and permit the child to lead a normal life and later to reproduce. Any genes which may be involved in producing the original defect are thus propagated. Again, a woman may inherit a misshapen pelvis which prevents normal delivery in childbirth. In a primitive society death would terminate her first pregnancy, but modern surgery permits the child to be delivered easily, with little more inconvenience than that which accompanies a normal childbirth, and the genes involved are then transmitted to future generations.

All such medical advances have done incalculable good for the existing people. Infant and child mortality have dropped greatly, our life span has been extended, and the physical suffering and disabilities resulting from disease have been greatly diminished. But all this has at the same time *reduced the effects of natural selection,* for many of those who would be eliminated through natural selection in a more primitive environment are now enabled to live through a normal life span and to reproduce. Certainly one would not argue against the utmost use of the skills of medical science. We must be concerned with the welfare of the present as well as with that of the future. But we should also realize the eugenic problems which will result. The continued benefits of medical science are bought at a price; through their effects we can expect a gradual increase in the number of those who could not exist without their aid. It is conceivable that eventually a population could develop in which every member would have one or more inherited defects which, except for the availability of medical science, would cause his death. By its very practice, medical science is increasing the load which it will be called on to bear in the future.

Social Welfare Agencies

In all the highly civilized countries of the world there have developed various types of social welfare agencies to help those who are unable to care for themselves. In the more primitive societies, those who are either mentally or physically unable to support themselves are very likely to die from starvation or neglect. In the United States, however, a sizeable portion of our national income goes to the support of such individuals, either through governmental aid or through private charitable contributions. Few in our country actually suffer from lack of food, clothing, shelter, or medical attention, regardless of their financial circumstances. Many of those who receive such public assistance have become destitute because of accident or other unfortunate circumstances, but it is quite likely that many of them have some defect of hereditary background which contributes to their inability to take care of themselves. The system of public support encourages reproduction in this group. The size of relief checks is generally based on the number of children in the family — a thoroughly commendable system — yet there are some indolent families who bear children as rapidly as possible in order to get the extra allotment of public funds. This practice even occurs among unmarried women at the lowest level of the social ladder who have one illegitimate child after another for the frank purpose of obtaining more public money. Even if such a degenerate attitude be solely the result of environmental factors, surely we cannot expect normal citizens to come from the kind of homes that would be shaped by thinking of this sort.

A Differential Birth Rate

During the first century of our national life there was much room for a rapidly expanding population, and large families were the rule in all social classes. Most of our population lived in rural areas, where children were a definite asset as potential helpers on the farm or in the kitchen. Girls married early in life, usually at about 16, and bore children at a rate of about one every 2 years for the entire fertile period of their lives. This, coupled with extensive immigration, caused our population to increase at a tremendous rate. Then began the industrialization of our nation. Great cities sprang up as the factories demanded a greater and greater concentration of the population into restricted areas. Modern machinery made it possible for fewer people to produce more farm products, and opportunities for the youth lay more and more in the growing cities rather than on the farms. Modern transportation facilities were developed to move the supplies from farm to city to take care of the needs of the high urban concentration of population. Modern schools were built and the educational period was extended.

As a result of these changes in our national way of life, the birth rate began to fall. No longer were children an asset; indeed, in the cities they became a very definite financial liability. Apartments and city homes were not built to accommodate large families — in fact, it became difficult to get or to keep decent living accommodations for a large family. The cost of bearing and rearing children soared. No longer were children born in the home, but in hospitals under antiseptic conditions. Mothers received special medical attention during pregnancy and there were regular visits to the doctor after the children were born to check on the welfare of both the mother and the child. Immunization shots were given the child at the proper time and special diets were provided. As Mary and Junior grew older, there was little they could contribute to the welfare of the home — no wood to chop, no cows to milk, no chickens or hogs to feed and care for. Perhaps they helped to dry the dishes or sweep the floors, but simple chores like these were hardly enough to keep them out of mischief. Special instruction in music, dancing, or art became customary to occupy their extra time (at extra expense, of course). Then, as they matured, it was thought necessary to give them the advantages of a college education. Can we wonder why the birth rate declined?

A declining birth rate is not a tragedy, however. Rather, it would be most tragic if the rate did not decline. There is a limited amount of room for people in the United States, and only so much food can be produced on the land which can be cultivated. As Malthus first brought out in 1798, there cannot be an unlimited geometrical increase of a population within a limited boundary. It is better that the population level off before it reaches a saturation point, as it has done in some countries of the world, where the population level is held down by famine, disease, and others forms of catastrophe.

Of far more serious concern than total numbers, to those interested in the future welfare of our nation, is the fact that the decline in the birth rate has been differential among the social and economic classes. Families in the lower socio-economic levels of our population tend to be larger than those in the upper socio-economic levels. While of course no one can say with assurance that any person's socio-economic level in society is an indication of the desirability or undesirability of his genotype, yet there are many in the lower groups who are unable to compete in modern society because of mental deficiencies due at least in part to heredity. Many such persons may be classified as high-grade morons according to intelligence ratings — not deficient enough to require confinement, yet not able to care for themselves adequately.

War

Any large-scale war is likely to be dysgenic in its effects on mankind. In the selection of members of the armed services, those persons are rejected who do not come up to certain physical or mental standards. Large numbers of casualties, therefore, tend to obliterate some of the best of the germplasm and to leave the weak, the deformed, and the mentally incompetent in full strength in the population. To some extent, this selection is compensated for by the fact that wars tend to bring about earlier marriages and a boom in the baby market. In our own country this trend toward earlier marriages and more babies has continued and has resulted in a serious overtaxing of our educational system. Hitler fully realized the dysgenic effects of war, and urged German girls to have babies through soldiers going off to war in order to preserve the germplasm which otherwise might die with the soldiers.

There is a great likelihood that this outcome might be changed in any possible future wars. With the weapons now at the disposal of the world powers the deaths of the civilian population would probably far outnumber military losses should any country be so foolish as to unleash an atomic war. Death would be randomized.

THE EUGENIC PROBLEM

Thus we find that our modern, highly civilized society has created conditions which tend to run counter to the ruthless force of natural selection which is nature's way of keeping the species fit. Some feel concerned lest these conditions bring about a gradual deterioration of the genetic qualities of our future generations. In ancient Sparta, according to stories, defective babies were destroyed by throwing them over cliffs into the sea in order to keep the race strong. Certainly no one in his right mind would recommend such a thing today, but some do feel that the right to live does not necessarily imply the right to reproduce without limitation, when the children are most likely to be defective also. They feel that under such circumstances the rights of the individual infringe upon the rights of society and that some means of limitation is justified. Is it possible to substitute an intelligent program of human selection for the more rigorous program of natural selection which we have thwarted to some extent? Much thought and much discussion have been given to the feasibility of such a program. In general, the proposals which have been made as partial solutions to this problem can be roughly classified into negative and positive measures.

NEGATIVE EUGENICS

Negative eugenics has as its goal the prevention of the deterioration of the human race through a reduction of the birth rate among the defectives so that they do not produce more children than are produced by the more normal members of society.

Segregation

Segregation is one method of negative eugenics which has been practiced for many years. We do not allow the inmates of mental institutions to mingle, marry, and bear children. This prevents the propagation of genes for defects from those whose condition is due to heredity. It is a fact, however, that the majority of persons suffering from various forms of mental defects are not confined to institutions. Of course, many of these defects are due to environmental agents, such as brain injury, and hence would not be transmitted to future generations anyway. The majority, however, have some hereditary background for their defects, and certainly none of them would be desirable as prospective parents. Segregation would not be feasible for such persons who are not dangerous to themselves or to society when living unconfined. Aside from the great financial burden it would entail, it would not be in conformity with our principles of humane consideration.

Sterilization

Sterilization has been suggested as an alternative proposal. Twenty-eight states now have laws which permit the sterilizing of defectives who are not violent enough to require confinement. Such persons can live free lives, with public assistance, and they can marry if they wish and have a normal marital life except for the fact that they can have no children. The sterilization operation is a simple one for a male. Using only a local anesthetic, a doctor slits the skin of the scrotum and ties off the spermatic duct through which the sperm cells travel to the ejaculatory ducts within the body. Such an operation in no way unsexes the individual; in fact, if it were performed without a person's knowledge he would probably never know that he had been sterilized, although he might wonder why no children resulted from his marriage. The testes continue producing the male hormone, and the seminal vesicles and prostate gland continue to secrete the ejaculatory fluid. In a woman the operation is more complicated, since it involves opening the body cavity to reach and tie off the oviducts. With modern surgical skill, however, this involves a minimum of inconvenience. Again, the operation does not disturb the endocrine balance, and except for childbearing the woman can carry on a perfectly normal marital life.

Of course, sterilization laws must be formulated with great care, for they could be abused if·not properly safeguarded. The sterilization program of Nazi Germany was reportedly abused in some cases when persons were declared unfit to bear children simply because they would not agree with the Nazi ideology. It was even suggested that extreme ugliness might be sufficient cause for sterilization. Our laws are formulated in an attempt to prevent such possible abuses. The number of sterilizations performed under these laws is only a very small proportion of the total number of mentally defective persons in the United States because the safeguards make such sterilizations rather difficult.

The proponents of eugenic sterilization do not hold that it will eliminate or even greatly reduce the incidence of mental defectiveness in one generation. Many recessive genes carried by normal individuals will come out in the future and would do so even if all mental defectives were sterilized today. Multiple genes are involved, however, in many cases, and those who carry a number of these genes are likely to show some mental abnormality. A continuing program of sterilization would achieve some reduction. Haldane has estimated that we could achieve a maximum reduction of about 15 per cent by such a continuing program of effective sterilization, together with segregation and other measures. This is far from elimination, but does represent a reduction; and any reduction would be desirable. Also, some defectives could be released from confinement if sterilized and could perhaps perform simple manual tasks which would enable them to become contributing members of society. Studies made in California show that many released patients have married and established a stable home-life without the burden of a large number of children whom they are not qualified to raise.

Opponents of eugenic sterilization point out the possibility of political exploitation of the power and possible misplaced judgment if we attempted to extend the measures to borderline cases. Class and race prejudice might easily become involved if decisions were not controlled by objective criteria. These opponents also point out the fact that we cannot always predict from what hereditary antecedents great men will arise. Thus Lincoln, Franklin, Keats,

Faraday, Schubert, and Beethoven, to name only a few, all came from families that were not outstanding in their attainments and that lived in poor or average environments. Hence, any widespread sterilization of those on the lower economic levels might prevent the birth of some great genius of the future. Also, those who oppose genetic sterilization point to the slowness by which the unfit would be reduced through any such program. To illustrate the slowness with which a recessive characteristic is eliminated, let us assume that a certain characteristic of this nature is expressed in about one person in each hundred thousand in the United States. Suppose we should decide that this characteristic was so deleterious to the human race that it should be reduced through a program of sterilization. If every person showing the trait were sterilized in every generation, it would require about 2000 years to reduce the number of such persons to one-half their present number. The same figures would hold for recessive mental characteristics of the same frequency in the population. If multiple recessive genes are involved in the production of a characteristic, as is surely true of many mental deficiencies, the rate of reduction would be even slower. Moreover, in spite of all our efforts at reduction, there is no way to prevent the continued recurrence of mutations in the very genes which we are trying to eliminate. Many human characteristics bring about sterility in all persons who express them, yet the characteristics continue to appear without abatement because of the equilibrium frequency which has been established between the occurrence of mutation and the elimination of the characteristics through sterility. On the whole, those persons afflicted with the most serious hereditary abnormalities would not propagate at a rate sufficient to keep their number constant, were it not for the fact that new mutation continually reintroduces the trait into the race. Finally, some persons oppose sterilization on religious or moral grounds, pointing out that, since man does not have the power to give life, he has no moral right to take away the power of procreation.

Birth Control

The freer dissemination of information about the techniques of birth control is advocated by many eugenists as a means of reducing the birth rate in the lower economic groups which have the greatest number of children and can least afford to support them. There are state laws which specifically prohibit the dissemination of information regarding means of contraception, except by a physician to a patient whose life or health would be seriously endangered by pregnancy. In recent years, however, little attempt has been made to enforce such laws. It is permissible to disseminate information which will aid in the prevention of venereal disease; however, the means of prevention of disease and of conception are so similar that contraceptives are even advertised under the thinly-veiled guise of disease-preventives. The present state of affairs makes such devices readily available to the more enlightened members of our population, but the law places a stumbling block in the way of the groups who do not have the opportunities to understand or learn about the available contraceptives. Many eugenists contend that if the laws were repealed it would be possible to convey the information to those groups which seem to need it most, and that there would then be a voluntary restriction in the size of families, so that there would not be such a disproportion in the number of children in such families as compared to those of more informed parents. North Carolina has developed an extensive program whereby birth control clinics are located strategically at places where they are most needed, and social workers visit the poorer families which are already overburdened with children and offer advice on the prevention of future conceptions. Similar programs are developing in other states.

There are opponents to such a plan, however, who present rather convincing arguments. They hold that freer dissemination of information on means of preventing conception would lead to an unprecedented wave of immorality among our youth. Freed of the restraining fear of pregnancy, girls would more easily yield to the temptation of premarital intercourse, and the moral fiber of our entire nation would be threatened. Also, such easily available information would cause many married couples to remain childless in their selfish self-gratification. Statistics show that divorce too is much more frequent among childless homes than among those with children. Childless women tend to be neurotic. Denied the natural satisfaction of one of their basic instincts they develop real and imagined ills which may be more expensive in the long run than the cost of bearing and rearing children. Also, these opponents quote the opinions of some physicians to the effect that many contraceptive devices are physically harmful to a woman and with continued use may render her incapable of bearing children at a time when she may want them. Finally, many people believe that contraceptives are morally wrong; that they defeat the basic purpose of sex, which is reproduction; that intercourse under such conditions is no more than mutual masturbation, with self-gratification as its primary aim.

In opposition to such views, those who favor the freer dissemination of contraceptive information advance arguments which seem just as convincing. First, they say that the majority of people are upright and moral creatures and that the knowledge of how to prevent conception will not lure our unmarried young women into a life of promiscuity. Girls so inclined, they argue, seem not to be deterred at present and turn to the dangers and evils of abortion as a means of escape if their information has been inadequate. There lives within the human heart a deep desire for parenthood which will not be thwarted by knowledge of a means to escape parenthood. Most childless couples are childless because they are unable to have children rather than through any deliberate intent. The great and unsatisfied demand for adoptions coming from such couples offers evidence for this view. Moreover, birth control does sometimes permit young people to marry earlier than would otherwise be possible and to space their children so as not to endanger the health of the mother or the ability of the family budget to care adequately for them. Also, there are physicians who feel that no harm can result from the proper use of contraceptives and regularly prescribe them for their patients. Finally, there are some religious leaders who hold that there is nothing morally wrong with the use of such devices for the prevention of conception.

Thus we see two opposing points of view, each championed by outstanding leaders in public life who have no other interest in the question than the welfare of mankind in the present and in the future. Rather than emphasize their differences, however, let us find the points of agreement between them, and perhaps we can find a common ground of eugenic value. Most members of both groups agree that there is a need for some means of limiting the size of families. The concern over the health of the mother and over economic circumstances of the parents, as well as the explosive growth of the world population, makes such limitation desirable. One group would hold that contraceptives offer a convenient means to this end. The other group counters that there are other ways to accomplish it without the objections which we have already listed. The latest research indicates that a woman is fertile for no more than 48 hours in any one monthly reproductive cycle, and probably for a shorter time than this. During the balance of the month she is sterile. By avoiding intercourse during this fertile period, conception can be prevented. There is individual variation and sometimes there is irregularity in the occurrence of this fertile period, but by avoiding intercourse for several days before and after the expected time of ovulation, conception can sometimes be prevented. The discovery of the relation of basal body temperature to the time of ovulation offers a means of pinpointing the time of the month when conception is possible and thus of increasing the chance for the success of this method. Still the method leaves much to be desired because of the ever-present possibility of missing the time of ovulation.

Contraception by Pill

Present contraceptive devices are comparatively expensive and require intelligent application for effective prevention of conception. These two factors often make them impractical for certain segments of the population who desire and need a means of limiting family size. A cheap, effective, and easily used method would greatly simplify this problem. A solution might lie in the birth control pill which has received some acceptance. The pill acts in a manner similar to one of the female hormones in that it prevents ovulation.

Consanguineous Marriages

The prohibition of consanguineous marriages is a negative eugenic measure which would have no effect on the actual frequency of undesirable genes, but would diminish the proportion of persons who show certain undesirable traits. Most societies have some sort of taboos or restrictions on marriage between close relatives. These may have arisen as a result of the observation that such marriages often produced defective children. As we have repeatedly emphasized in the study of genetics, most harmful traits are recessive and are, therefore, most likely to appear in the children of parents who are closely related. About the only practical extension of the present laws would be to extend the prohibition of cousin marriages to those states which do not now have such laws. If this were done, according to Haldane, it still would stop only about 56 per cent of the consanguineous matings, since many of these are incestuous or take the form of other relationships which are already prohibited by law. In the course of time, however, a law of this kind would probably cause a significant reduction in the proportion of persons showing rather rare hereditary defects. Haldane has estimated that such a law would bring an eventual reduction of 2.5 per cent of mental defects in the population as a whole — a small yet worthwhile reduction.

Genetic Counseling

A voluntary restriction of child bearing by couples who carry serious hereditary defects can be brought about through proper counseling by well qualified persons in this field. For instance, if a man has seen his father go through the agonies of Huntington's chorea, he is certainly not going to want the same fate to befall any of his children. Yet, if he does not know the hereditary basis of this condition, he may think that it was just an unfortunate affliction that came on his father. On the other hand, if he is told by a reliable counselor that the chance of his children having the affliction is one in four, he may wish to forego fatherhood rather than subject them to this rather high probability. Such voluntary restriction would reduce the transmission of many hereditary abnormalities when the information is available. There are a number of counseling clinics in the United States, but far too few to take care of the needs for this service and to have the maximum eugenic value. To be effective such clinics must have the cooperation of the medical profession as well as geneticists. There must be both a correct diagnosis of the trait being studied and a knowledge of the complicated reactions of genes which may be involved. Since many of the tests for heterozygous carriers of harmful recessive genes are difficult to perform and interpret, they must be done by experts. Also, it seems that such clinics must be supported by grants from the government or foundations, since people are generally unwilling to pay for such counseling although they will pay for the services of a physician diagnosing and treating a disease.

POSITIVE EUGENICS

Positive eugenics attempts to enlarge the proportion of the children borne by those in the population who have the most desirable hereditary characteristics. Some individuals, like the Nobel prize-winning geneticist, H. J. Muller, have suggested not only increasing the number of children in the family of such persons, but that, through artificial insemination, outstanding men could serve as fathers to many more children than would be otherwise possible. Artificial insemination is already widely practiced to permit women whose husbands are sterile or have some serious hereditary affliction to bear children. In most of these cases, however, little consideration is given to the eugenic values involved. The donor of the semen is usually chosen because he has the same general characteristics of the husband and no gross hereditary deformities. It is now possible to quick-freeze human semen and the sperms can thus remain alive. Thus, semen can be stored in a deep freeze and used at later times. Some children are living today who were conceived through the use of semen which had been stored for over a year. In all probability, the semen will continue to be functional for 100 years or longer, as long as it is kept at a very low temperature. This makes it possible to establish sperm banks with careful registries, and women can bear children from men who had died many years before and whose progeny had proved their genetic desirability. It is now also possible to preserve eggs from females and transplant a young embryo, formed by the union of sperm and egg in a test tube, into the uterus of a woman where it will continue its development. Thus, an outstanding woman could have many more children than would be possible if she bore them all herself and sterile women would be able to bear children.

One of the strongest arguments for the establishment of sperm and egg banks lies in the protection of these precious germ cells from radiation. They could be collected during early adulthood and stored in lead-lined containers in a deep freeze. In this state the germ cells would not be subjected to radiation exposure which might affect the donors. Such radiation can come from fallout, diagnostic methods, occupational exposures, and even nuclear wars. To astronauts, workers in atomic energy plants, and others who are likely to be exposed to unusually high quantities of radiation, this plan might be appealing. The big objection, of course, would be that there would be a separation of sexual life from procreation. Conception brought about by the aid of instruments in the antiseptic environment of a hospital bed could hardly have the emotional satisfaction of a more natural method. It is possible that serious social and psychological consequences would result. Also, many people raise moral objections to the use of any sort of artificial means to achieve conception, feeling that this is contrary to natural moral laws. In addition, there are certain genetic objections which must be considered. Even a genius might easily carry some very harmful recessive genes which could be widely spread if there was extensive use of his (or her) germ cells. This could result in many abnormalities in future generations when intermarriage brings these recessive genes together. There could even be the possibility that marriages between half-siblings would take place. A man and a woman planning marriage could both have the same father through artificial insemination; yet, since the donor of the semen is usually kept secret, not even the parents of the couple would be able to inform the latter of their common progenitor. The chance of abnormal children from such marriages is, of course, very great, and it is conceivable that serious legal snarls could result. Laws passed before the days of modern science did not anticipate the difficulties which might arise as a result of donor germ cells. For instance, a man might leave his wife and refuse to support children which were not from his own germ cells. Or some children might be denied a man's inheritance because they were conceived through the use of semen from another man. Thus, we can see that positive eugenics has many discouraging problems to solve. Perhaps its greatest hope lies in an enlightened and educated public.

There are many international aspects of eugenics which need to be understood by all who have anything to do with the formulation of our foreign policy, and indeed by all of us who wish to understand some of the basic reasons for much of the unrest in the world. Unfortunately, these aspects have not been understood in the past, and we have had some bitter experiences as a result.

Rats and People

We may begin to understand some of these implications by considering some rats living in Baltimore — a far cry, perhaps, from problems of international eugenics, yet it is possible to view a population of rats with an objectivity which we cannot achieve when viewing human populations, and we can at the same time learn lessons which have a bearing on some pressing human problems of the size and growth of populations and on possible measures of control. Let us consider a square block in one of the older sections of town. There are sparkling, clean white doorsteps at the front, but many of the back yards are littered with rubbish, and the alley is lined with garbage cans, some of them without covers or with poorly fitting covers. A public health survey reveals the fact that eighty-seven rats live in this square block. This is not such a large number — many blocks have more — but it is a stable number. Each year a certain number are killed by dogs, others by traps, still others in battles among themselves for the available food and shelter, and some die of starvation and other causes. Here is natural selection in a stabilized population in a stable environment. If a new family moves in with an extra garbage can, this may become a ninety-six-rat block. If a new dog is brought in it may drop to an eighty-three-rat block. Slight variations which increase or decrease the chance for survival will be reflected in the total number of rats in the block. All of the rats have a great struggle to get enough food to keep alive, and a change which provides more food in the block will be reflected temporarily in larger and healthier rats, though soon the numbers will increase to a point where they all live on the same semi-starvation plane — there are just more of them. Periodic poisoning campaigns like those conducted in many cities have no permanent effect on the number of rats per block. It is almost impossible to poison all the rats in a block — some just don't have a taste for the poisoned food put out, find other sources of food, and pass the poison by. Let us suppose we put out poison in our eighty-seven-rat block of Baltimore and it proves so effective that we kill 90 per cent of the rats — only nine survive. For a time, the rats in this block will be very large, sleek, and healthy. There is food for eighty-seven and only nine to consume it. Soon nature takes its course, however. A healthy female may bear twelve to fourteen young in a litter every 6 weeks, and a few migrants may come in from other blocks. Inside 6 months there will once more be eighty-seven rats in this block! This number can be reduced permanently only

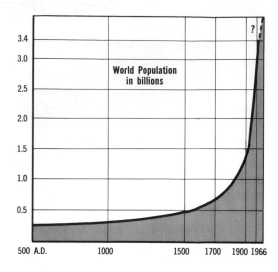

FIG. 26-1 *Our exploding population. This diagram gives a graphic representation of the explosive nature of the growth of the world's population during the past 300 years. Continuing growth at this rate will bring about many problems.*

by cleaning up the neighborhood, destroying the nesting places of the rats, and placing tight-fitting covers on the garbage cans. Rats seldom cross a large city street, so little trouble will be experienced with migrants.

The reader may, at this point, feel that this is all a very good public health lesson, but what does it have to do with eugenics? The answer is that human populations are subject to the same laws which regulate the sizes of populations of rats and other animals. A group of primitive people living in a limited range, such as an island, will establish a population equilibrium which will change only when some factors in the environment become altered for better or for worse. Man differs from these lower animals, however, in having a brain which permits him to substitute intelligence for the forces of nature which limit population. Now appears to be a good time to use that intelligence to keep the world's population under control.

Our Exploding Population

When you sat down to breakfast this morning there were about 270,000 new mouths to share the world's food supply with you — mouths which were not seeking food yesterday morning. This is the number of babies that were born, about 187 every minute. During this same period, about 142,000 people died, leaving a net gain in the world's population of about 128,000. This is enough people to populate a sizeable city. Within a year the increase will total about 47 million; in only 4 years an increase at this rate would equal the population of the United States. These figures serve to emphasize the explosive rate of the population in the world today, an explosion which has more potential danger to the people of the world than the explosions of nuclear weapons. Today about two-thirds of the world's population is chronically undernourished, and, with the prospect of rapidly expanding population, there is little hope that expanding the productivity of the land can keep pace with the increasing demand for food.

This explosive increase in population has begun in comparatively recent times — for many centuries the world's population was relatively stable at somewhat less than half a billion people. There was unrestricted reproduction, but this was balanced with a high death rate that cut short the human life span. Famine, plagues, floods, and other ravages of nature kept the population under control. By 1700 the population stood at about 600 million. Then civilized man began improving his environment. New continents were opened for expansion; the industrial revolution brought about mechanized farming with greater food production; improvements in medicine, sanitation, and public health services reduced the deaths from disease; engineering achievements tamed flood waters and opened vast new areas to agriculture through irrigation; transportation facilities were fantastically improved, making food distribution rapid and easy. All of these things brought about a gradual increase in man's life span, but the birth rate continued unabated. It required all of the time from man's origin on the earth, estimated at about 200 million years, until about 1830 for the human population to reach 1 billion. It required only an additional 100 years, to 1930, for the second billion to be added. Then, only 30 years, to 1960, were required to add the third billion. At the present time the world population is about 3.3 billion. How long can such a steep rate of multiplication continue? Although the growth rate is a world-wide problem, many persons in the United States are indifferent to it especially since a surplus of food exists at this time. It is true that our yearly growth rate, about 1 per cent, is less than the world rate of about 1.7 per cent. Still, if our rate continues, our present population of about 200 million will jump to 400 million in about 50 years and to about 1 billion after another 50 years. This means that, in less than a century, we could have a population density greater than that found in China and India today. We are already having problems with water supply and other resources, and our standard of living would surely sink to a very low level should so many people populate the country.

It is quite obvious that there must be a halt to this explosive growth of the world population. The question is, "What shall bring about a halt to this explosive growth?" Shall we leave it to the forces of nature which so efficiently keep the population number of wild animals under control by the cruel forces of natural selection? Shall we wait for plagues, wars, or famine to bring the human population in line with the capability of the earth to maintain it or for governments to tell couples how many children they can have? Will a time come when there will be laws restricting births? It is hoped that, through a world-wide program of education as to the dangers of an exploding population, there will be voluntary controls which will make unnecessary any governmental restrictions or controls by the forces of natural selection. There are encouraging signs that such an educational program, even though rather limited in scope at present, is having an effect. Leaders of oriental countries are beginning to take cognizance of the problem. Having legalized abortion, Japan has changed from a nation with one of the world's highest birth rates to a country with one of the lowest. India's leaders are seeking ways of reducing family size that will be in accord with the economic status and religious beliefs of the people. Even the

Chinese communists are taking steps to reduce the population growth which has thwarted their desperate efforts to produce enough food to feed their people well. In the United States the birth trend is turning downward. In 1957 the birth rate reached a peak of 25.3 per thousand. Each year since, it has shown a slight decline; a figure of 19.7 was reported for 1965, which is still explosively high. Thus, although the population problem is still very grave, there is an ever-increasing realization that it exists. Let us hope that eventually there will be voluntary control.

SELECTED REFERENCES

For the student who would like to pursue his study of genetics somewhat further, the books listed below are recommended. Also, the *Scientific American* magazine publishes many worthwhile articles on the subject of genetics. These articles can be located by reference to the index of bound copies of this magazine.

Allen, J. M., *The Molecular Control of Cellular Activity,* New York: McGraw-Hill Book Company, Inc., 1962.

Baker, W. K., *Genetic Analysis,* Boston: Houghton Mifflin Company, 1965.

Clarke, C. A., *Genetics for the Clinician* (Blackwell Scientific Publications), Philadelphia: F. A. Davis Company, 1962.

Dobzhansky, T., *Evolution, Genetics, and Man,* New York: John Wiley & Sons, 1955.

McElroy, W. D., and Glass, H. B., *Human Heredity,* Chicago: University of Chicago Press, 1954.

Neel, J. V., and Associates, "Genetics and Epidemiology of Chronic Diseases," *U. S. Department of Health, Education and Welfare,* 1965.

Neel, J. V., and Shull, W. J., *Human Heredity,* Chicago: University of Chicago Press, 1954.

Neel, J. V., *Changing Perspectives on the Genetic Effects of Radiation,* Springfield, Illinois: Charles C. Thomas, Publisher, 1963.

Reed, S. C., *Counseling in Medical Genetics,* 2nd ed., Philadelphia: W. B. Saunders Company, 1963.

Roberts, J. A. F., *Introduction to Medical Genetics,* 3rd ed., London: Oxford University Press, 1963.

Stern, C., *Principles of Human Genetics,* 2nd ed., San Francisco: W. H. Freeman & Company, 1960.

Wagner, R. P., and Mitchell, H. K., *Genetics and Metabolism,* 2nd ed., New York: John Wiley & Sons, 1964.

Whittinghill, M., *Human Genetics and its Foundations,* New York: Reinhold Publishing Corporation, 1965.

Winchester, A. M., *Heredity and Your Life,* New York: Dover Publications, Inc., 1963.

Winchester, A. M., *Heredity, an Introduction to Genetics,* College Outline Series, New York: Barnes and Noble, Inc., 1960.

INDEX